Hadamard Matrices and Their Applications

Hadamard Matrices and Their Applications

K. J. Horadam

PRINCETON UNIVERSITY PRESS

PRINCETON AND OXFORD

Published by Princeton University Press, 41 William Street, Princeton, New Jersey 08540

In the United Kingdom: Princeton University Press, 3 Market Place, Woodstock, Oxfordshire OX20 1SY

Library of Congress Cataloging-in-Publication Data

Horadam, K.J., 1951–
  Hadamard matrices and their applications / K.J.Horadam.
     p. cm.
  Includes bibliographical references and index.
  ISBN-13: 978-0-691-11921-2 (hardcover : alk. paper)
  ISBN-10: 0-691-11921-X (hardcover : alk. paper)
  1. Hadamard matrices. I. Title.

QA116.4.H67 2006
512.9′434—dc22   2006049331

British Library Cataloging-in-Publication Data is available

This book has been composed in Times-Roman in LaTeX

The publisher would like to acknowledge the author of this volume for providing the camera-ready copy from which this book was printed.

Printed on acid-free paper. ∞

pup.princeton.edu

Printed in the United States of America

10 9 8 7 6 5 4 3 2 1

DEDICATION

To my parents
Eleanor Mollie Horadam (in memoriam) and Alwyn Francis Horadam
mathematicians both
for their love and inspiration.

Contents

Preface

A Hadamard matrix is a square matrix with entries from $\{1, -1\}$, for which the inner product of any pair of distinct rows is 0.

Hadamard matrices have exerted a fascination over us for the past one-and-a-half centuries. Transparently easy to describe, ubiquitous and utilitarian, they nonetheless continue to elude the most basic identification: do they exist in all possible orders? Though the answer is widely believed to be "yes", no proof has yet been found, and the Hadamard Conjecture, that for every natural number n there exists a Hadamard matrix of order $4n$, remains one of the great unsolved problems of mathematics.

In daily life, the practical use of Hadamard matrices is constant and largely invisible. The Walsh-Hadamard Transform is in common use as a fast discrete transform. Error-correcting codes (Reed-Muller codes) used in early satellite transmissions — for example, in the 1972 Mariner mission to Mars and recent flybys of the outer planets in the solar system — are based on Hadamard matrices. Modern CDMA cellphones use Hadamard matrices (Walsh covers) to modulate transmission on the uplink and minimise interference with other transmissions to the base station. New applications are everywhere about us, in pattern recognition, neuroscience, optical communication and information hiding, for example. Despite this, there is still no uniform technique for constructing all the known Hadamard matrices.

Our curiosity and ingenuity does not stop at square matrices with entries from $\{\pm 1\}$. Hadamard matrices have been extended and generalised, to nonbinary alphabets and higher dimensional arrays, and their desirable properties adapted for multilevel and multiphase applications in signal processing, coding and cryptography.

A novel perspective has been brought to the whole field over the past fifteen years. The steady infiltration of cohomological techniques throughout mathematics has spread from algebra into combinatorial design theory. It now informs and illuminates existence and construction questions for Hadamard matrices and their extensions and applications.

The cohomological approach has matured into a theory within which many, perhaps most, Hadamard and generalised Hadamard matrices may be defined by a *factor pair* of functions, or equally by a *group extension*, satisfying an additional correlation property called *orthogonality*. Such Hadamard matrices, whose entries are values taken by the factor pair, are now called *coupled cocyclic* (or, briefly, *cocyclic*) Hadamard matrices.

Conversely, if we start with a factor pair, the internal structure of its cocyclic matrix gives us a constructive, powerful and flexible technique for testing whether the matrix is Hadamard. This is the most successful general method yet known for constructing Hadamard matrices.

This book provides the first unified account of our current knowledge of Hadamard matrices and their generalisations and applications from this cocyclic point of view. The cohesion provided by this body of theory allows us to transfer knowledge from the cohomology of finite groups to uncover fundamental ideas and important new constructions for generalised Hadamard matrices.

Very recently we have seen the first traffic in the other direction — a natural equivalence relation within combinatorial design theory translates to a previously unknown but quite universal finite group action, the *shift action*, within cohomology classes. We have discovered the atomic structure of group cohomology — in dimension 2, at least!

Many of the questions we ask about generalised Hadamard matrices are driven by problems in spectral analysis, error correction, separation, correlation and encryption of digital signals and data sequences. The novel ideas and new families of generalised Hadamard matrices discovered here are applied to such current problems.

This book is for graduate students and researchers in mathematics, computer science and communications engineering. Open research questions appear regularly: there are 90 in total. Proofs of many results already appearing in the literature are only sketched, or left wholly to the reader to pursue. A reader wishing to undertake further research will be advantaged by a mathematical background including an undergraduate course in abstract algebra. Otherwise, this book can be treated as a handbook, supplemented if necessary by an abstract algebra textbook which covers groups, rings, fields, vector spaces and modules.

ACKNOWLEDGEMENTS

Foremost I would like to thank my colleague and collaborator Warwick de Launey, with whom the idea of cocyclic Hadamard matrices was born 15 years ago and with whom many of the original results were obtained. Since then my colleagues Yu Qing Chen, Dane Flannery, Udaya Parampalli and Asha Rao and my graduate students Athula Perera, Garry Hughes, Kenneth Ma, Wei-Hung Liu, John Galati and Alain LeBel have added so much to the richness and depth of our knowledge of the subject.

Thanks are also due to Serdar Boztaş, Dane Flannery, John Galati and Asha Rao for their valuable comments on earlier drafts. Advice from Princeton University Press's readers, including Neil Sloane and Charles Colbourn, has improved both organisation and currency, and is much appreciated.

Support and advice from the editorial team at Princeton University Press: Acquisitions Editor Vickie Kearn, her assistant Adithi Kasturirangan, Production Editor Lucy Day Hobor and Copyeditor Alison Anderson, has been invaluable. I am most grateful to Clea Price for early help with Figure 2.1 and the Bibliography and to Duncan Bayly for the Figures and Index.

Finally, my heartfelt thanks go to my husband, Garth Price, and my daughters, Anna and Clea Price, for their good-humoured forbearance and unfailing support during the three years it has taken to write this book.

Kathy Horadam   September 2006

Chapter One

Introduction

The purpose of this book is three-fold: to report the current status of existence and construction problems for Hadamard matrices and their generalisations; to give an accessible account of the new unifying approach to these problems using group cohomology; and to support an understanding of how these ideas are applied in digital communications. I have tried to present results and open problems with sufficient rigour, and direction to the literature, to enable readers to begin their own research, but with enough perspective for them to gain an overview without needing in-depth knowledge of the algebraic background.

The book has two Parts. In Part 1, consisting of four Chapters, our present understanding of Hadamard matrices, generalised Hadamard matrices and higher dimensional Hadamard matrices is summarised. One Chapter is devoted to introduction and explanation of the main applications of Hadamard matrices in digital signal and data sequence processing, principally for spectral analysis and signal error protection, separation or encryption.

Generalised Hadamard matrices and higher dimensional Hadamard matrices are each natural enlargements of the class of Hadamard matrices, in the direction of entries not restricted to $\{\pm 1\}$ and not restricted to 2-dimensional (2-D) arrays, respectively. Part 1 contains the basic definitions and properties of these three types of Hadamard matrices and, for each of them, a status report on recent results using classical techniques. The two ideas from which Warwick de Launey and I developed the group extensions approach to Hadamard matrices: group development of Hadamard matrices and construction of higher dimensional Hadamard matrices from relative difference sets are highlighted.

Part 2, also consisting of four Chapters, develops in detail the unifying group extensions approach to existence and construction of the three types of Hadamard matrices covered in Part 1. Some necessary algebraic background is included. This Part covers the major theoretical advances made over the past 15 years, culminating in the Five-fold Constellation, which identifies cocyclic generalised Hadamard matrices with particular 'stars' in four other areas of mathematics and engineering: group cohomology (factor pairs), incidence structures (divisible designs), combinatorics (relative difference sets) and signal correlation (perfect arrays). The work in this Part has not been collected before, or is accessible only in journal articles. Some is not yet published.

The latter half of Part 2 introduces less mature, but very exciting, theoretical results on the atomic structure of cohomology classes. These *shift orbits* have remained invisible for nearly a century, but carry the statistical information about distributions of the entries of cocyclic matrices that determines whether or not they will produce Hadamard matrices, high-distance error-correcting codes and low-correlation sequences. Finally, the first applications of the theory of cocyclic Hadamard matrices to multiphase signal and data sequence processing are presented. We construct novel and optimal families of such cocyclic generalised Hadamard matrices and their corresponding Generalised Hadamard Transforms, codes and sequences.

Half the open research problems arise in this last quarter of the book.

A summary of each Chapter follows.

Chapter 2 covers basic definitions and properties of Hadamard matrices, in abbreviated form. There are many excellent texts [288, 1, 123, 315], reviews [68, 69] and databases [212, 287, 297], describing Hadamard matrices and their numerous constructions in more detail; the intention here is to provide a succinct summary and update of research over the past decade or so. Direct constructions of Hadamard matrices by Sylvester, Paley and Williamson and from Hadamard designs are described and illustrated.

More modern techniques of constructing Hadamard matrices, by patterning entries according to the multiplication table of a group, are treated next. This is our first link to cocycles and cocyclic Hadamard matrices. In the final section of Chapter 2, advances towards direct confirmation of the celebrated Hadamard Conjecture, and improved asymptotic support for it, are outlined, as is progress on the circulant Hadamard conjecture.

The purely intellectual excitement and challenge of finding new Hadamard matrices and homing in on confirmation of the Hadamard Conjecture is heightened by the knowledge that they are marvellously useful. *Chapter 3* is devoted to two of their three principal applications: Hadamard transform spectroscopy and object recognition, and coding of digital signals. Applications in design of experiments are not included. Most emphasis is placed on coding of digital signals or data sequences for error correction, separation, correlation or encryption.

Each application area is introduced briefly to explain how the Hadamard matrix is applied, but in enough detail, and in the language of the application, to explain current trends. My aim is to bridge the two worlds: to translate the physical application into terms a pure mathematician will appreciate and the theoretical structure into terms an applied mathematician, computer scientist or communications engineer can adapt and use.

Chapter 4 moves us from Hadamard matrices to generalisations where matrix entries are not restricted to $\{\pm 1\}$. More than one direction for enlargement of the class of Hadamard matrices has flourished, but generalisations to maximal determinant matrices, weighing matrices, orthogonal designs and nonsquare matrices will not be covered. The two main formulations we treat are *complex Hadamard matrices* (invertible, with entries on the complex unit circle) — especially those with entries which are roots of unity, called *Butson matrices* here — and *generalised Hadamard matrices* (with entries from a finite group N, for which the inner quotient of any distinct pair of rows in the integral group ring $\mathbb{Z}N$ equals $\lambda \left(\sum_{u \in N} u \right)$, for some fixed integer λ). To complicate matters, in the literature the term complex Hadamard matrix often refers only to a Butson matrix with entries in $\{\pm 1, \pm \sqrt{-1}\}$, of which those with uniformly distributed rows are also called *quaternary* generalised Hadamard matrices. Although complex Hadamard matrices will be revisited on occasion, the principal subject of this book is generalised Hadamard matrices. Jungnickel's seminal 1982 result, relating generalised Hadamard matrices, class regular divisible designs and relative difference sets, underscores the richness of the interconnections between these areas and the group extensions approach described in the second part.

This Chapter follows the structure of Chapter 2, for each of Butson, complex Hadamard and generalised Hadamard matrices in turn, illustrated with numerous examples. One section covers their applications to multiphase signals and sequences. The final section is new work, unifying the two formulations in the invertible *Generalised Butson Hadamard* matrices, which include all complex Hadamard matrices and all invertible generalised Hadamard matrices, and their *Generalised Hadamard Transforms*.

Chapter 5 enlarges the class of Hadamard matrices from 2-D to n-dimensional arrays with entries from $\{\pm 1\}$. It deals with *n-dimensional proper Hadamard matrices*, introduced by Shlichta in 1971, which have the property that all 2-D subarrays obtained by fixing any $n-2$ coordinates are Hadamard matrices.

Despite a strong presumption of their utility — based on that of Hadamard matrices — and their formative role in development of the group extensions approach to Hadamard matrices, remarkably little is known about higher dimensional proper Hadamard matrices. The first monograph on the subject is Yang [334]. A summary of construction techniques, relationships between these techniques, equivalence classes and applications to Boolean functions useful for cryptography and to error-correcting array codes is presented.

Higher dimensional proper Hadamard matrices were central to the discovery of cocyclic Hadamard matrices by Warwick de Launey and myself. His effort to characterise those Hadamard matrices which would generate higher dimensional proper Hadamard matrices led him to isolate functions which must satisfy specific relations between their values and which I subsequently identified as cocycles.

A 2-dimensional *cocycle* between finite groups G and N, with trivial action, is a function $\psi : G \times G \to N$ satisfying the equation

$$\psi(g, h)\psi(gh, k) = \psi(h, k)\psi(g, hk), \quad \forall\, g, h, k \in G.$$

We then rederived this equation by asking when an abstract combinatorial design could be functionally generated from a single row. This *cocyclic development* of matrices includes group development of matrices, which was described in Chapter 2. The *cocyclic matrix* developed from $\psi : G \times G \to N$ is

$$[\psi(g, h)]_{g, h \in G}.$$

A cocycle whose matrix is Hadamard is called *orthogonal*.

The first Chapter of Part 2, *Chapter 6*, concerns cocycles, which arise naturally in many areas: surface topology, algebra and quantum mechanics, for instance. The usual unit studied in group cohomology is a cohomology (equivalence) class of cocycles, not the individual cocycles comprising it, so the examples, properties and constructions collected here do not appear in cohomology texts and are listed for the first time.

Some time is spent on the practicalities of computing cocycles. One of the advantages of the group extensions approach to Hadamard matrices is that the internal structure of a cocyclic matrix promises efficiency in computer searches for generalised Hadamard matrices, cutting down the search space over exhaustion dramatically. But first we need to find and list the cocycles. Three algorithms are presented: one, the Flannery-O'Brien algorithm, was developed to exploit the ideas presented in this book and is distributed as a module in the computer algebra package MAGMA.

The Chapter continues by showing that most of the direct constructions of Hadamard matrices listed in Chapter 2 are cocyclic, for some group G and $N = \{\pm 1\}$. To date, cocyclic construction is the most successful general method known, both theoretical and computational, for finding Hadamard matrices. In particular, the most productive single construction of Hadamard matrices, due to Ito, is cocyclic over the dihedral groups. The Cocyclic Hadamard Conjecture follows: that for each odd t there is a group G of order $4t$ such that a G-cocyclic Hadamard matrix exists. The Chapter concludes with a status report on 12 research questions posed by the author in earlier papers on cocyclic Hadamard matrices.

Cocycles are special cases of *factor pairs* of functions. *Chapter 7* contains the full description of the theory of orthogonal factor pairs and the generalised Hadamard matrices they determine. The theory has been complete for only a few years. Sufficient background information on group extensions, factor pairs and cohomology of finite groups is included to make the book self-contained.

The limiting class of generalised Hadamard matrices obtained using the group extensions approach is the class of *coupled cocyclic* generalised Hadamard matrices. We can do no better than this. Whilst not every generalised Hadamard matrix is a coupled cocyclic matrix, I know of only one counterexample, a matrix of order 6 with entries from the group $\mathbb{Z}_3$ of integers modulo 3. I know of no Hadamard matrix which is not cocyclic — but the sheer number of inequivalent Hadamard matrices even for small orders makes it unlikely all will be cocyclic.

The Chapter's central purpose is to convey the pervasive influence of cocyclic generalised Hadamard matrices, by locating them (in four different guises) within combinatorics, group cohomology, incidence structures and digital sequence design. This is done by proving mutual equivalences — the *Five-fold Constellation* — between coupled cocyclic generalised Hadamard matrices, semiregular relative difference sets, orthogonal factor pairs, semiregular class regular divisible designs with regular action and well-correlated arrays. These equivalences have been established in increasing generality over the past decade by de Launey, Flannery, Perera, Hughes and the author, with the fullest expression due to Galati. The general form of the fifth equivalence — with well-correlated arrays — is given here for the first time. Such universality helps to explain the tremendous variety of uses to which we can put these matrices.

Chapter 8 deals with the way in which different definitions of equivalence class interrelate within the Five-fold Constellation. There are preexisting concepts of equivalence for generalised Hadamard matrices, for transversals of subgroups in groups, and for factor pairs and group extensions arising naturally from theoretical considerations in each area, and they do not coincide. The equivalence relation for transversals is revealed to be the strongest relation. It becomes a very productive and novel way of investigating each of the 'stars' of the Constellation.

When equivalence of transversals is transcribed to an action on factor pairs, it forms orbits termed *bundles*. These bundles are copied around the Five-fold Constellation. For splitting factor pairs, bundles define equivalence classes of functions $G \to N$, which form the basis of a new theory of nonlinearity. For semiregular relative difference sets, the resulting taxonomy allows us to establish a classification program for their equivalence classes and begin to populate it. This problem is at the heart of research in relative difference sets.

Two components of bundle action can be isolated, one an action by automorphism groups of G and N and the other a differential G-action called *shift action* which arises from translation and renormalisation of transversals. Thus a bundle is an automorphism orbit of shift orbits, and vice versa. These components, though not wholly independent, can be extracted and investigated in more general situations.

Shift action is a remarkably universal action and should be identifiable in more contexts than in fact appears to be the case. Shift action operates wholly within the natural equivalence classes of factor pairs, partitioning each one into shift orbits — its atomic structure. So, it is invisible from the point of view of cohomology theory, but it is critical to our study. Shift orbits (and the bundles they generate) carry the statistical information about distributions of the entries of cocyclic matrices that determines whether or not they will produce Hadamard matrices, high-distance error-correcting codes and low-correlation sequences.

Some external sightings of shift action in disguise have been made: in differential cryptanalysis and in the Loewy series for p-groups. LeBel's thesis [217] identifies shift action within the trivial cohomology class with a natural action in a quotient algebra of the standard module of a group ring.

In the final, and longest, Chapter, we begin to reap the rewards of all the preceding hard work. *Chapter 9* contains a multitude of new constructions and applications of cocyclic complex and generalised Hadamard matrices, and a tantalising set of new problems, too.

Initially we look at several recent applications of cocycles, not necessarily orthogonal, to computation in Galois rings, to elliptic curve cryptography and to the developing field of *cocyclic codes* over nonbinary alphabets.

Then splitting orthogonal factor pairs are applied to establish a general theory of nonlinear functions suitable for use as cryptographic primitives. These include planar, bent and maximally nonlinear functions, and surprising and beautiful connections with finite presemifields and projective planes are uncovered.

In turn, these help identify large classes of new cocyclic generalised Hadamard matrices. We are next led to the discovery of families of optimal codes, such as the q-ary codes meeting the Plotkin bound found by Udaya and myself and the extremal self-dual binary codes found by Rao.

Finally, differential uniformity, an important measure of the resistance of a block encryption cipher to differential attack, is extended to array encryption ciphers, and a class of orthogonal cocycles proposed as array S-box functions.

I hope the reader will find this field as rich and exciting as I do. Good luck and good hunting!

PART 1
Hadamard Matrices, Their Applications and Generalisations

Chapter Two

Hadamard Matrices

A *Hadamard matrix of order n* is an $n \times n$ matrix H with entries from $\{\pm 1\}$ such that

$$HH^{\top} = nI_n, \tag{2.1}$$

that is, for which the real inner product of any pair of distinct rows is 0. Examples for the smallest orders $n = 1, 2$ and 4 are

$$[1], \quad \begin{bmatrix} 1 & 1 \\ 1 & -1 \end{bmatrix}, \quad \begin{bmatrix} 1 & 1 & 1 & 1 \\ 1 & -1 & 1 & -1 \\ 1 & 1 & -1 & -1 \\ 1 & -1 & -1 & 1 \end{bmatrix}.$$

Hadamard matrices have excited interest for almost 150 years, since the first examples were published by Sylvester in 1867 [303]. Sylvester also noted that if H is a Hadamard matrix, so is

$$\begin{bmatrix} H & H \\ H & -H \end{bmatrix}; \tag{2.2}$$

the examples above illustrate two iterations of this construction. Then, in 1893, Hadamard [133] published examples in orders 12 and 20, showing that the matrices which have come to bear his name could exist in orders other than the powers 2^t previously demonstrated by Sylvester. Hadamard was interested in finding the maximal determinant of square matrices with entries from the unit disc, and he showed in [133] that this maximal determinant $n^{n/2}$ was achieved by matrices with entries ± 1 if and only if they satisfied (2.1). So Hadamard matrices are extremal solutions of a problem in real analysis.

Moreover, Hadamard proved that such matrices could exist only if n was 1, 2 or a multiple of 4. This observation has formed the basis of one of the great unsolved problems in mathematics, for we simply do not know when Hadamard matrices exist. All information presently available supports the proposal that the converse of Hadamard's observation is true: that if n is 1, 2 or any multiple of 4, there exists a Hadamard matrix of this order. Evidence supporting this famous Conjecture is presented in Sections 2.1, 2.3 and 2.4 below.

Research Problem 1 *The Hadamard Conjecture. Show that if n is a multiple of 4, a Hadamard matrix of order n exists.*

Following the proof of Fermat's Last Theorem, the Hadamard Conjecture is also one of the longest-standing open problems in mathematics.

The aim of this Chapter is to provide a succinct summary and update of research on Hadamard matrices over the past decade or so. It covers basic definitions, properties and constructions of Hadamard matrices, and finishes with a status report on the Hadamard Conjecture.

There is a very large number of techniques for constructing Hadamard matrices, but as yet, no infinite arithmetic sequence is known in which all the terms are orders of Hadamard matrices. A good overview of the topic, up to 1992, appears in Seberry and Yamada [288], where the constructions are roughly classified into three types: multiplication (recursion) theorems, 'plug-in' methods and direct constructions. A centennial survey emphasising the historical development of these techniques appears in Craigen and Wallis [69] and a clear summary of the direct constructions is given in Hedayat, Sloane and Stufken [144, Chapter 7].

The foundation of the multiplicative and plug-in techniques is the tensor product.

DEFINITION 2.1 *If $A = [a_{ij}]$ is an $m \times m$ matrix and $B_1, B_2, \ldots, B_m$ are $n \times n$ matrices, with entries from a ring R, their* tensor product $A \otimes [B_1, B_2, \ldots, B_m]$ *(also called their* Kronecker *or* direct *product) is the square matrix of order mn defined by*

$$
A \otimes [B_1, B_2, \ldots, B_m] = \begin{bmatrix} a_{11}B_1 & a_{12}B_1 & \ldots & a_{1m}B_1 \\ a_{21}B_2 & a_{22}B_2 & \ldots & a_{2m}B_2 \\ \vdots & \vdots & \ldots & \vdots \\ a_{m1}B_m & a_{m2}B_m & \ldots & a_{mm}B_m \end{bmatrix}, \qquad (2.3)
$$

or $A \otimes B = [a_{ij}B]$, when $B = B_1 = B_2 = \ldots = B_m$.[1]

For example, Sylvester's construction (2.2) is the tensor product $\begin{bmatrix} 1 & 1 \\ 1 & -1 \end{bmatrix} \otimes H$.

We begin by presenting the original four major families of Hadamard matrices discovered during the past century or so: the Sylvester, Paley, Hadamard design and Williamson families.

2.1 CLASSICAL CONSTRUCTIONS

Some elementary constructions of Hadamard matrices follow easily from (2.1) and (2.2).

LEMMA 2.2 *Let H be a Hadamard matrix of order n, so by (2.1) it is invertible over $\mathbb{Q}$, with $H^{-1} = n^{-1}H^{\top}$. Set $\mathcal{S}_1 = \begin{bmatrix} 1 & 1 \\ 1 & -1 \end{bmatrix}$. Then*

1. *the negation $-H$ of H is a Hadamard matrix;*

2. *the transpose $H^{\top}$ of H is a Hadamard matrix;*

3. *[133] if H' is a Hadamard matrix of order n', the tensor product $H' \otimes H$ is a Hadamard matrix of order $n'n$;*

---

[1] Some authors use the alternative convention and define $A \otimes B = [Ab_{kl}]$, where $B = [b_{kl}]$.

4. *for* $t \geq 1$, $(\otimes^t \mathcal{S}_1) \otimes H$ *is a Hadamard matrix of order* $2^t n$.

2.1.1 Sylvester Hadamard matrices

The earliest known, and still by far the most significant, family of Hadamard matrices are those of order 2^t for $t \geq 1$, due to Sylvester. They are constructed by iterating the tensor product of $\mathcal{S}_1$ with itself (that is, by setting $H = [1]$ in Lemma 2.2.4 above). These matrices are all symmetric.

DEFINITION 2.3 *The* Sylvester Hadamard matrices *are the matrices in the family* $\{\mathcal{S}_t = \otimes^t \mathcal{S}_1 : t \geq 1\}$.

These matrices have numerous alternative descriptions or variants, almost as many as they have applications.

One, which will prove very useful to us, represents the entries in terms of the inner product of their index coordinates. If we index $\mathcal{S}_t$ by the integers $0 \leq i \leq 2^t - 1$, and write each i in its binary representation (see Definition 3.2) as a vector (or string) of length t over the binary field $GF(2)$ and let $\langle i, j \rangle$ be the inner (dot) product of i and j over $GF(2)$, then

$$\mathcal{S}_t = [(-1)^{\langle i,j \rangle}]_{0 \leq i,j \leq 2^t-1}, \tag{2.4}$$

so that, for instance,

$$\mathcal{S}_2 = \begin{bmatrix} (-1)^{\langle 00,00 \rangle} & (-1)^{\langle 00,01 \rangle} & (-1)^{\langle 00,10 \rangle} & (-1)^{\langle 00,11 \rangle} \\ (-1)^{\langle 01,00 \rangle} & (-1)^{\langle 01,01 \rangle} & (-1)^{\langle 01,10 \rangle} & (-1)^{\langle 01,11 \rangle} \\ (-1)^{\langle 10,00 \rangle} & (-1)^{\langle 10,01 \rangle} & (-1)^{\langle 10,10 \rangle} & (-1)^{\langle 10,11 \rangle} \\ (-1)^{\langle 11,00 \rangle} & (-1)^{\langle 11,01 \rangle} & (-1)^{\langle 11,10 \rangle} & (-1)^{\langle 11,11 \rangle} \end{bmatrix}.$$

2.1.2 Paley Hadamard matrices

These next two families of Hadamard matrices were found by Paley [256] using the *quadratic residues* (that is, the nonzero perfect squares) in a finite field $GF(q)$ of odd order.

The *quadratic character* χ on the cyclic group $GF(q)^* = GF(q) \backslash \{0\}$, defined by $\chi(g) = 1$ if g is a quadratic residue in $GF(q)$ and $\chi(g) = -1$ if g is a quadratic nonresidue, is extended to $GF(q)$ by setting $\chi(0) = 0$. When $q = p$ is a prime and $g \in \mathbb{Z}$, the Legendre symbol $\left(\frac{g}{p} \right)$ is also used to denote $\chi(g)$.

The version of Paley's constructions given here, and a more accessible proof, may be found in [144]. Other definitions of the Paley matrices appear in the literature, but they determine *equivalent* [2] Hadamard matrices.

LEMMA 2.4 *For q an odd prime power, and an ordering* $\{g_0 = 0, g_1, \ldots, g_{q-1}\}$ *of* $GF(q)$, *set* $Q = [\chi(g_i - g_j)]_{0 \leq i,j < q}$. *Let S be the* $(q+1) \times (q+1)$ *matrix* $S = \begin{bmatrix} 0 & \mathbf{1} \\ \mathbf{1}^\top & Q \end{bmatrix}$, *where $\mathbf{1}$ is the all-1s string.*

---

[2] See Section 2.2 for the definition of equivalence.

1. *(Paley Type I Hadamard matrix)* If $q \equiv 3 \bmod 4$, *then*

$$P_q = \begin{bmatrix} 1 & -\mathbf{1} \\ \mathbf{1}^\top & Q + I_q \end{bmatrix}$$

is a Hadamard matrix of order $(q+1)$.

2. *(Paley Type II Hadamard matrix)* If $q \equiv 1 \bmod 4$, *then*

$$P'_q = \begin{bmatrix} S + I_{q+1} & S - I_{q+1} \\ S - I_{q+1} & -S - I_{q+1} \end{bmatrix}$$

is a Hadamard matrix of order $2(q+1)$.

Note that Q is skew-symmetric ($Q^\top = -Q$) when $q \equiv 3 \bmod 4$ and symmetric when $q \equiv 1 \bmod 4$.

For example, the quadratic residues in $GF(11)$ are $1, 3, 4, 5, 9$, and the Paley Type I Hadamard matrix P_{11} of order 12 is

$$\begin{bmatrix} 1 & -1 & -1 & -1 & -1 & -1 & -1 & -1 & -1 & -1 & -1 & -1 \\ 1 & 1 & -1 & 1 & -1 & -1 & -1 & 1 & 1 & 1 & -1 & 1 \\ 1 & 1 & 1 & -1 & 1 & -1 & -1 & -1 & 1 & 1 & 1 & -1 \\ 1 & -1 & 1 & 1 & -1 & 1 & -1 & -1 & -1 & 1 & 1 & 1 \\ 1 & 1 & -1 & 1 & 1 & -1 & 1 & -1 & -1 & -1 & 1 & 1 \\ 1 & 1 & 1 & -1 & 1 & 1 & -1 & 1 & -1 & -1 & -1 & 1 \\ 1 & 1 & 1 & 1 & -1 & 1 & 1 & -1 & 1 & -1 & -1 & -1 \\ 1 & -1 & 1 & 1 & 1 & -1 & 1 & 1 & -1 & 1 & -1 & -1 \\ 1 & -1 & -1 & 1 & 1 & 1 & -1 & 1 & 1 & -1 & 1 & -1 \\ 1 & -1 & -1 & -1 & 1 & 1 & 1 & -1 & 1 & 1 & -1 & 1 \\ 1 & 1 & -1 & -1 & -1 & 1 & 1 & 1 & -1 & 1 & 1 & -1 \\ 1 & -1 & 1 & -1 & -1 & -1 & 1 & 1 & 1 & -1 & 1 & 1 \end{bmatrix}. \qquad (2.5)$$

Combining these constructions using tensor products (Lemma 2.2.2) gives a very large family of Hadamard matrices, but, as always, the tensor product increases the 2-power factor in the order.

DEFINITION 2.5 *Let* $\{q_i, i \in I\}$ *and* $\{q'_j, j \in J\}$ *be finite sets of prime powers congruent to* 3 *mod* 4 *and* 1 *mod* 4*, respectively. A matrix of the form*

$$(\otimes_{i \in I} P_{q_i}) \otimes (\otimes_{j \in J} P'_{q'_j}),$$

which is a Hadamard matrix of order $\prod_{i \in I}(q_i + 1) \prod_{j \in J} 2(q'_j + 1)$*, is called a* Paley Hadamard matrix *.*

2.1.3 Hadamard designs

The Paley Type I Hadamard matrices form one of three main known families of Hadamard matrices which may be constructed directly from square block designs, so a little combinatorial design theory is now introduced. For deeper coverage see, for example, the survey texts [24, 58, 96].

DEFINITION 2.6 *A (square)* (v, k, λ)-design *is a pair* $\mathcal{D} = (P, B)$ *consisting of a set* $P = \{p_1, \ldots, p_v\}$ *of* v points *and a set* $B = \{B_1, \ldots, B_v\}$ *of* v blocks *each containing* k points $(1 < k < v)$, *such that each pair of distinct points is contained in exactly* λ *blocks.*

The full automorphism group $\mathrm{Aut}(\mathcal{D})$ *of* $\mathcal{D}$ *consists of all bijections of* $P \cup B$ *which preserve the point, block and incidence structure. An* automorphism group *of* $\mathcal{D}$ *is a subgroup of* $\mathrm{Aut}(\mathcal{D})$. *If* G *is an automorphism group of* $\mathcal{D}$ *such that for each pair of points* p, p', *there is a unique* $g \in G$ *with* $p^g = p'$, *and similarly for blocks, the design* $\mathcal{D}$ *is called* regular *with respect to* G *and* G *is called a* regular *(or Singer) group for* $\mathcal{D}$.

An incidence matrix $A = [a_{ij}]$ *of* $\mathcal{D}$ *is a* $v \times v$ *matrix with entries* $0, 1$, *having* $a_{ij} = 1$ *if and only if* $p_j \in B_i$.

It follows that a $v \times v$ matrix A with entries $0, 1$ is an incidence matrix of a (v, k, λ)-design if and only if

$$AA^\top = (k - \lambda)I + \lambda J, \quad AJ = kJ, \tag{2.6}$$

where I is the $v \times v$ identity matrix and J is the $v \times v$ all 1s matrix (for proof, see, for example, [314, Theorem 2.8]). Note that other authors [314, 24] index rows by points and columns by blocks, so their incidence matrices are the transpose of ours.

To illustrate the construction, consider the Paley Type I Hadamard matrix P_7, defined from the set $\{1, 2, 4\}$ of quadratic residues mod 7. Multiplying each column except the first by -1 gives a Hadamard matrix of the form

$$\begin{bmatrix} 1 & \mathbf{1} \\ \mathbf{1}^\top & -(Q + I) \end{bmatrix} = \begin{bmatrix} 1 & 1 & 1 & 1 & 1 & 1 & 1 & 1 \\ \hline 1 & -1 & 1 & 1 & -1 & 1 & -1 & -1 \\ 1 & -1 & -1 & 1 & 1 & -1 & 1 & -1 \\ 1 & -1 & -1 & -1 & 1 & 1 & -1 & 1 \\ 1 & 1 & -1 & -1 & -1 & 1 & 1 & -1 \\ 1 & -1 & 1 & -1 & -1 & -1 & 1 & 1 \\ 1 & 1 & -1 & 1 & -1 & -1 & -1 & 1 \\ 1 & 1 & 1 & -1 & 1 & -1 & -1 & -1 \end{bmatrix}. \tag{2.7}$$

This is an example of a *normalised* matrix, that is, a matrix whose first row and first column consist entirely of 1s. The submatrix excluding the first row and column of a normalised matrix is called its *core*.

The $(0, 1)$ version of the core $-(Q + I)$ of the normalised P_7 in (2.7) is

$$A = \frac{1}{2}(J - (Q + I)) = \begin{bmatrix} 0 & 1 & 1 & 0 & 1 & 0 & 0 \\ 0 & 0 & 1 & 1 & 0 & 1 & 0 \\ 0 & 0 & 0 & 1 & 1 & 0 & 1 \\ 1 & 0 & 0 & 0 & 1 & 1 & 0 \\ 0 & 1 & 0 & 0 & 0 & 1 & 1 \\ 1 & 0 & 1 & 0 & 0 & 0 & 1 \\ 1 & 1 & 0 & 1 & 0 & 0 & 0 \end{bmatrix}. \tag{2.8}$$

Then $AA^\top = 2I + J$, $AJ = 3J$ and A is the incidence matrix of a $(7, 3, 1)$-design.

In general, the formula (2.6) allows us to equate the core of a normalised Hadamard matrix of order $4n$ and the (± 1) version of an incidence matrix of a $(4n -$

$1, 2n-1, n-1)$-design. That is, if $A' = 2A - J$ is the (± 1) matrix obtained from incidence matrix A by replacing 0 by -1, then

$$H = \begin{bmatrix} 1 & \mathbf{1} \\ \mathbf{1}^\top & A' \end{bmatrix} \qquad (2.9)$$

is the corresponding Hadamard matrix of order $4n$, and vice versa.

LEMMA 2.7 [24, Lemma I.9.3] *There exists a Hadamard matrix of order $4n$ if and only if there exists a square $(4n-1, 2n-1, n-1)$-design.*

For obvious reasons, a square $(4n-1, 2n-1, n-1)$-design is called a *Hadamard* design. Hadamard designs are doubly valuable because, by adjoining one point and suitably redefining blocks, an extended design is obtained in which every block is incident with $2n$ points, with the stronger incidence property that every 3 distinct points, not just every 2, are together incident with exactly $n-1$ blocks. Such a 3-$(4n, 2n, n-1)$-design is called a *Hadamard* 3-*design*, and conversely, every 3-$(4n, 2n, n-1)$-design is the unique extension (up to isomorphism) of a Hadamard design — see, for example, [14, 7.2] for details.

One of the most powerful tools for finding such designs is by developing them from difference sets.[3]

DEFINITION 2.8 *A (v, k, λ)-difference set in a (multiplicatively written) group G of order v is a k-element subset D of G such that the list of quotients $d_1 d_2^{-1}$ of distinct elements d_1, d_2 of D contains each nonidentity element of G exactly λ times. The* order *of the difference set is $n = k - \lambda$. A difference set is called* cyclic, abelian *etc. if the group G has the respective property.*

THEOREM 2.9 [24, Theorem VI.1.6] *Let D be a k-element proper subset of a group G of order v. Define the* development *$\mathrm{dev}(D)$ of D to be the pair $(G, \{gD, g \in G\})$. Then D is a (v, k, λ)-difference set in G if and only if $\mathrm{dev}(D)$ is a square (v, k, λ)-design with regular group G. Moreover, every square (v, k, λ)-design with regular group G may be represented this way.*

For example, the set of quadratic residues $D = \{1, 2, 4\} \bmod 7$ is a $(7, 3, 1)$-difference set in the (additive) cyclic group $\mathbb{Z}_7$, and the incidence matrix of $\mathrm{dev}(D)$ is A in (2.8) above.

A *Hadamard* (or *Paley-Hadamard*) difference set in a group G has parameters $(4n-1, 2n-1, n-1)$. Apart from the quadratic difference sets in $GF(q)$ used to construct the Paley Type I Hadamard matrices, two other parametric families are known: the Singer and twin-prime families of Hadamard difference sets.

Example 2.1.1 *The three series of (Paley-)Hadamard difference sets which cover (parametrically) all known examples are*

1. *Paley difference sets with parameters $(q, (q-1)/2, (q-3)/4)$, where $q \equiv 3 \bmod 4$, which are the sets of quadratic residues*
$$\{g^2 : g \in GF(q)^*\}$$
in the additive group $(GF(q), +)$ of the finite field $GF(q)$.

---

[3] Difference sets were originally defined in additively written abelian groups, and the usage remains, though a more accurate term would be *quotient sets*.

2. *Singer difference sets (m-sequences) with parameters* $(2^t - 1, 2^{t-1} - 1,$ $2^{t-2} - 1)$. *Here the difference set may be defined [191] as those integers* mod $2^t - 1$

$$\{i : 0 \le i < 2^t - 1, \operatorname{tr}(\alpha^i) = 0\}, \tag{2.10}$$

where α is a generator of $GF(2^t)^*$ and tr $: GF(2^t) \to GF(2)$ *is the* trace map $\operatorname{tr}(g) = \sum_{i=0}^{t-1} g^{2^i}$.

3. *Twin prime power difference sets with parameters* $(q(q + 2), [q(q + 2) - 1]/2, [q(q + 2) - 3]/4)$, *where q and $q + 2$ are both prime powers. Here the difference set may be defined [192, p. 268] as*

$$\{(g, h) \in GF(q)^* \times GF(q + 2)^* : \chi(g)\chi(h) = 1\} \cup \{(g, 0) : g \in GF(q)\} \tag{2.11}$$

in $(GF(q), +) \times (GF(q + 2), +)$, where χ is the quadratic character on the respective field.

2.1.4 Williamson Hadamard matrices

The Williamson construction is the simplest of many powerful 'plug-in' methods for finding Hadamard matrices. These techniques essentially capitalise on the success of tensoring as a generator of Hadamard matrices, by allowing judicious replacement of a matrix B_i in Definition 2.1 by several different matrices, which do not have to be Hadamard. We will vary Williamson's original template [323], by taking the overlying matrix A in Definition 2.1 to be the Hadamard matrix

$$\begin{bmatrix} 1 & 1 & 1 & 1 \\ 1 & -1 & 1 & -1 \\ 1 & -1 & -1 & 1 \\ 1 & 1 & -1 & -1 \end{bmatrix}.$$

LEMMA 2.10 (Williamson [323]) *If there exist (± 1) matrices A, B, C, D of order w which satisfy both*

$$XY^\top = YX^\top, \quad \text{for} \ \ X \ne Y \in \{A, B, C, D\} \tag{2.12}$$

and

$$AA^\top + BB^\top + CC^\top + DD^\top = 4wI_w, \tag{2.13}$$

then

$$\begin{bmatrix} A & B & C & D \\ B & -A & D & -C \\ C & -D & -A & B \\ D & C & -B & -A \end{bmatrix}. \tag{2.14}$$

is a Hadamard matrix of order $4w$.

A $v \times v$ matrix $M = [m(i,j)]_{0 \le i,j < v}$ is *circulant* if $m(i,j) = m(0, j-i \bmod v)$ for all i, j and is easily identified by the rightward circulant shift pattern of consecutive rows, or, equally, by the downward circulant shift of consecutive columns.

Williamson used 'plug-in' components A, B, C, D which were symmetric circulant matrices, but this restriction is not necessary.

Whilst the terms *Williamson* and *Williamson-type* often refer (cf. [288]) to the component matrices A, B, C, D in the symmetric circulant and unrestricted cases, respectively, we will adopt the term 'Williamson' to refer to the Hadamard matrix which results, in both cases.

DEFINITION 2.11 *The* Williamson Hadamard matrices *are the Hadamard matrices constructed in Lemma 2.10.*

The principal means by which Williamson Hadamard matrices have been found is through extensive computer searches by many authors. Djoković [98] found the first odd number $w = 35$ for which no Williamson Hadamard matrix of order $4w$ using symmetric circulant components exists. A recent distributed computer search [170] reconfirms that this is the only value missing for $1 \le w \le 45$. It is certainly possible that Williamson Hadamard matrices exist for all orders; tables listing orders (to 1992) of known Williamson Hadamard matrices appear in [288, Appendix A.2] and unknown orders $4w < 4,000$ (to 1996) in [65, Table 24.30]. The first value for which existence is unknown is $w = 47$.

Research Problem 2 *(a) The* Williamson Hadamard Conjecture. *Show that, if n is a multiple of 4, a Williamson Hadamard matrix of order n exists.*
(b) For what proportion of orders do Williamson Hadamard matrices constructed using symmetric circulant components exist?

Generalisations of Williamson's construction of Hadamard matrices have proliferated. Two of the main mechanisms used have been: first, to vary the internal structure of the component matrices and, second, to modify the template used for plugging in component matrices. In the second case, one valuable technique has been to alter the size of the template and the number of components, for example, by using an orthogonal design as a template. Another has been to replace negation of a component by some other matrix operation such as transposition of a component or reversal of its rows, as occurs in the Goethals-Seidel arrays [124]. See [288] for a comprehensive survey of these techniques.

Discussion of two more important families of Hadamard matrices, constructed by related mechanisms, will be delayed until Section 2.3.

2.2 EQUIVALENCE CLASSES

The notion of equivalent Hadamard matrices has already been mentioned several times without elaboration, but it is basic to our understanding. Most often, it is sufficient to demonstrate any proof or property for Hadamard matrices up to membership of an equivalence class, and the efficiency this allows us often goes unremarked.

DEFINITION 2.12 *Two* (± 1) $n \times n$ *matrices* M *and* M' *are* (Hadamard) *equivalent, written* $M \sim M'$, *if one can be obtained from the other by performing a finite sequence of the following operations:*

1. *permute the rows or the columns;*
2. *multiply a row or a column by* -1.

In particular, any equivalence class of $n \times n$ matrices with entries from $\{\pm 1\}$ will contain at least one normalised representative.

Any equivalence operations applied to a Hadamard matrix give a Hadamard matrix, so the Hadamard matrices of order n partition naturally into equivalence classes, each containing normalised Hadamard matrices. However, it has been a considerable challenge to gain headway in the classification of Hadamard matrices by equivalence. As computational resources increased during the thirty-five years since 1970, millions of inequivalent Hadamard matrices have been constructed, but complete lists are available only in a few small orders.

What is known is that there is only one equivalence class of Hadamard matrices for each of the orders $n = 1, 2, 4, 8$ and 12. There are 5 equivalence classes for $n = 16$, 3 for $n = 20$, 60 for $n = 24$ and 487 for $n = 28$ [144]. For representatives in each class, see, for example, Neil Sloane's website [297]. Representatives of the five equivalence classes for $n = 16$ are constructed from S_3 and three particular representatives of its equivalence class, using the tensor product (2.3) and its transpose — details appear in [14, 7.6].

These are the only complete classifications.

For $n = 32$ there are at least 3,578,006 equivalence classes; for $n = 36$, at least 4,745,357 [254]. For lower bounds in orders $8t = 40$ and above, see [215, 216, 68]. These numbers grow so rapidly that Orrick [254] proposes a weaker equivalence relation — *Q-equivalence* — to include the kinds of switching operations of equivalent Hadamard submatrices commonly used in tensor product constructions (2.3) to generate inequivalent Hadamard matrices. He shows that for each of the orders $n = 4, 8, 12, 16, 20, 24$ and 28 there are, respectively, $1, 1, 1, 1, 1, 2$ and 2 Q-equivalence classes, and for $n = 32$ and $n = 36$ the known equivalence classes occur in only 11 and 21 Q-equivalence classes, respectively. For practical reasons this weakening makes considerable sense, and may supersede Definition 2.12 as the most appropriate way of regarding two Hadamard matrices as being essentially the same. For consistency, though, we will continue to use Definition 2.12.

Published lower bounds for numbers of other equivalence classes are for $n = 44$, at least 500; for $n = 52$, at least 638; for $n = 60$, at least 256 and for $n = 68$, at least 340 [210], though Christos Koukouvinos informs me these are already massively superseded [211]. For updates, see his website [212]. The MAGMA computational algebra software system [32] contains a database of known inequivalent Hadamard matrices of orders up to at least 256 (see the MAGMA website [238]).

Research Problem 3 *How many equivalence classes of Hadamard matrices of order 32 are there? And how many of order 36? And how many Q-equivalence classes of each order?*

It is known that the number $h(n)$ of equivalence classes of Hadamard matrices of order n, while clearly not a monotone function of n, is unbounded with increasing n (see [145, Theorem 2.1]). Apparently no other analysis of the asymptotic behaviour of $h(n)$ has been undertaken. The following open questions appear in [69, Section 6].

Research Problem 4 *How does the number $h(n)$ of equivalence classes of Hadamard matrices of order n behave as $n \to \infty$? Is $h(n) \le h(2n)$ for all n? If $t \ge 1$ is fixed, is $h(2^t n)$ monotone?*

Of the elementary constructions in Lemma 2.2, $H \sim -H$ and $H \otimes H' \sim H' \otimes H$, but often $H \not\sim H^\top$; the first examples of inequivalent transposes occur at order 16.

Not all of the classical construction methods of the previous section will give inequivalent matrices. For example, Turyn proved (see [309]) that a Paley Type II Hadamard matrix is equivalent to a Williamson Hadamard matrix with symmetric circulant components. The proof given here follows [88].

LEMMA 2.13 (Turyn) *Let $q \equiv 1 \bmod 4$ be a prime power and P'_q a Paley Type II Hadamard matrix. Then P'_q is equivalent to a Williamson Hadamard matrix with symmetric circulant components.*

Proof. For $q \equiv 1 \bmod 4$ there is an ordering of $GF(q)$ in which the matrix S of Lemma 2.4 takes the form

$$S = \begin{bmatrix} X & Y \\ Y & -X \end{bmatrix},$$

where X and Y are symmetric and circulant, Y has entries ± 1 and X has zeroes down the main diagonal and entries ± 1 off the diagonal. We have

$$P'_q = \begin{bmatrix} I+X & Y & -I+X & Y \\ Y & I-X & Y & -I-X \\ -I+X & Y & -I-X & -Y \\ Y & -I-X & -Y & -I+X \end{bmatrix}.$$

Now set

$$M = \begin{bmatrix} 1 & 0 & 0 & 0 \\ 0 & 0 & -1 & 0 \\ 0 & 1 & 0 & 0 \\ 0 & 0 & 0 & 1 \end{bmatrix} P'_q \begin{bmatrix} 1 & 0 & 0 & 0 \\ 0 & 0 & 0 & 1 \\ 0 & -1 & 0 & 0 \\ 0 & 0 & 1 & 0 \end{bmatrix},$$

so that

$$M = \begin{bmatrix} I+X & I-X & Y & Y \\ I-X & -I-X & Y & -Y \\ Y & -Y & -I-X & I-X \\ Y & Y & -I+X & -I-X \end{bmatrix},$$

which is Williamson Hadamard with $A = I+X$, $B = I-X$ and $C = D = Y$. $\square$

A second example is given by the Hadamard matrix constructed from a Singer difference set, which is equivalent to a Sylvester Hadamard matrix.

LEMMA 2.14 *The Hadamard matrix constructed from the Singer difference set (2.10) is equivalent to S_t.*

Proof. Let α be a generator of $GF(2^t)^*$ and order the elements of $GF(2^t)$ as $\{0, \alpha^i, 0 \leq i \leq 2^t - 1\}$. With this ordering, define $A = [\text{tr}(gh)]_{g,h \in GF(2^t)}$ and $H = [(-1)^{\text{tr}(gh)}]_{g,h \in GF(2^t)}$. In particular, $(-1)^{\alpha^i} = 1$ if and only if $\text{tr}(\alpha^i) = 0$, if and only if i is in the Singer difference set, so H is the Hadamard matrix (2.9).

Because the trace function is a homomorphism on $(GF(2^t), +)$, A has rank t over $GF(2)$, and a linearly independent set of t rows, for example $[\alpha^j h,\ h \in GF(2^t)]$, $0 \leq j \leq t - 1$, has columns consisting of every vector in $\mathbb{Z}_2^t$. The same holds for the binary version A_{2^t} of S_t obtained by the 'log$_{(-1)}$' map $(-1)^k \mapsto k$, by (2.4). Thus there is a column permutation of the t linearly independent rows of A which gives those of A_{2^t}, which when extended to all rows of A gives the rows of A_{2^t} in some order. These column and row permutations convert H to S_t. $\square$

One consequence of the perceived intractability of the equivalence class problem (after all, it is at least as difficult as the Hadamard Conjecture, confirmation of which would demonstrate the existence of at least one equivalence class in each order) is that effort has focussed on finding combinatorial and algebraic invariants of Hadamard matrices which will distinguish between different equivalence classes. In the former category are, for example, computations of the '4-profiles' of Hadamard matrices by Lin et al. [225].

In the latter category are investigations of the automorphism groups of Hadamard matrices by many authors. From Definition 2.12 it is apparent that two Hadamard matrices H and H' are equivalent if and only if there are monomial matrices U and V such that $UHV^\top = H'$. (A *monomial* matrix is square with entries in $\{0, \pm 1\}$ and exactly one nonzero entry in each row and column.) The process is illustrated in the proof of Lemma 2.13 above. An equivalence of H with itself is called an *automorphism* of H.

The flavour of this type of analysis is given, for instance, by Tonchev [306], who shows that there are exactly 11 equivalence classes of Hadamard matrices of order 36 which have an automorphism of order 17. One of these is the equivalence class containing P'_{17}.

DEFINITION 2.15 *The automorphism group $\text{Aut}(H)$ of a Hadamard matrix H of order n is the set of ordered pairs of $n \times n$ monomial matrices (U, V) for which $UHV^\top = H$, with the group operation given by direct product: $(U, V) \cdot (U', V') = (UU', VV')$.*

If $H \sim H'$, then $\text{Aut}(H) \cong \text{Aut}(H')$. Of course, $(-I, -I)$ is always an automorphism of H, of order 2. The automorphism group $\text{Aut}(P_q)$ of the Paley Type I Hadamard matrix was determined by Kantor [196] for $q > 11$ and the automorphism group $\text{Aut}(P'_{q'})$ of the Paley Type II Hadamard matrix by de Launey and Stafford [93]. As a consequence we know the two constructions do not give equivalent Hadamard matrices unless the matrix has order 12.

LEMMA 2.16 [93, Corollary 4.1] *A Paley Type I Hadamard matrix P_q is equivalent to a Paley Type II Hadamard matrix $P'_{q'}$ if and only if $q = 11$ and $q' = 5$.*

2.3 THE FIRST LINK: GROUP DEVELOPED CONSTRUCTIONS

Attempts to construct Hadamard matrices by replacing (2.14) by more complex block matrices with similar properties have fostered deeper structural and algebraic approaches to the problem.

Their starting point has been the knowledge that if a circulant matrix has the order of its rows (or its columns) reversed, the resulting *back-circulant* matrix has the same internal pattern as the multiplication table of a cyclic group. For instance, if this reversal is applied uniformly to the rows within each template row (or columns within each template column, equally) of a Hadamard matrix of the form (2.14) with circulant components, the components all become back-circulant and the resulting equivalent matrix is Hadamard.

This raises the tantalising prospect of harnessing the massive power of finite group theory to find Hadamard matrices. We use component matrices whose internal structure follows the multiplication table of a group. Algebraic techniques from the character theory of finite groups and associated combinatorics of difference sets have been a dominant theme of this research over the past two decades.

DEFINITION 2.17 *Let G be a group of order v, written multiplicatively, with a fixed ordering $\{g_1, g_2, \ldots, g_v\}$. A matrix with entries from a set S is group developed over G (briefly: G-developed) if the rows and columns of the matrix are indexed by the elements of G and there is a map $\phi : G \to S$ such that the entry in position (i, j) is $\phi(g_i g_j)$, for $1 \leq i, j \leq v$. We denote such a matrix by*

$$[\phi(g_i g_j)]_{1 \leq i,j \leq v}.$$

A matrix is group-invariant *(or G-invariant) if the entry in position (i, j) is $\phi(g_i g_j^{-1})$, for $1 \leq i, j \leq v$.*

The term "group developed" refers to the fact that the whole matrix may be constructed from a single row (or column) by knowledge of the group. The row $[\phi(gg_i)]_{1 \leq i \leq v}$ indexed by g is obtained from the row $[\phi(g_i)]_{1 \leq i \leq v}$ indexed by the identity 1 by (left) multiplication in the group.

The most obvious question to ask is: when is a group developed matrix itself Hadamard?

For example, the back-circulant Hadamard matrix

$$\begin{bmatrix} -1 & -1 & -1 & 1 \\ -1 & -1 & 1 & -1 \\ -1 & 1 & -1 & -1 \\ 1 & -1 & -1 & -1 \end{bmatrix} \tag{2.15}$$

is group developed over $\mathbb{Z}_4 = \{0, 1, 2, 3\}$, the cyclic group of integers modulo 4, with $\phi(0) = \phi(1) = \phi(2) = -1$ and $\phi(3) = 1$. The equivalent $\mathbb{Z}_4$-invariant (circulant) matrix is obtained by exchanging columns indexed j and $-j$:

$$\begin{bmatrix} -1 & 1 & -1 & -1 \\ -1 & -1 & 1 & -1 \\ -1 & -1 & -1 & 1 \\ 1 & -1 & -1 & -1 \end{bmatrix}.$$

A group developed matrix with entries in an abelian group $(S, +)$ has constant row and column sums. A Hadamard matrix with constant row and column sums is called *regular* and must also arise as the (± 1) version of the incidence matrix of a square design.

LEMMA 2.18 *Let H be a $v \times v$ matrix with entries ± 1 and $A = \frac{1}{2}(H + J)$ be the corresponding $(0, 1)$ matrix. Assume $v > 2$. Then H is a regular Hadamard matrix if and only if both $v = 4u^2$ and A is the incidence matrix of a square $(4u^2, 2u^2 \pm u, u^2 \pm u)$-design.*

Proof. H is a regular Hadamard matrix if and only if $HH^\top = 4nI$ and $HJ = tJ$ (where $v = 4n$ and t is the constant row and column sum), if and only if $AA^\top = nI + (n+t/2)J$ and $AJ = (2n+t/2)J$. By (2.6), this holds if and only if A is the incidence matrix of a square $(4n, 2n + t/2, n + t/2)$-design. Consequently t must be even. Since $\lambda(v-1) = k(k-1)$ in any square (v, k, λ)-design, $n = (t/2)^2 = u^2$ and $t = \pm 2u$. $\square$

2.3.1 Menon Hadamard matrices

Clearly, group developed Hadamard matrices, being regular, must have a combinatorial construction from square designs. In fact, they have a combinatorial construction from difference sets.

A $(4u^2, 2u^2 \pm u, u^2 \pm u)$-difference set in G is called (following [191, p. 301]) a *Menon-Hadamard* difference set. The Menon-Hadamard difference sets are also referred to as 'Hadamard' difference sets, but since the $(4n - 1, 2n - 1, n - 1)$-difference sets of Section 2.1.3 are commonly called this, it is important to distinguish between the two.

An account of recent research into these difference sets (in abelian groups) appears in Jungnickel and Schmidt [195] (see also [194] and Chapter VI of [24]). These parameters 'provide the richest source of known examples of difference sets' [74]. A Menon-Hadamard difference set (if it exists) can be assumed, without loss of generality, to contain the identity 1 of G.

LEMMA 2.19 [152, Lemma 4.1] *Let G be a group of order $v = 4u^2$ and $\phi : G \to \{\pm 1\}$ a set map. A G-developed matrix $[\phi(g_i g_j)]_{1 \leq i,j \leq v}$ is Hadamard if and only if the set $\{g \in G : \phi(g) = -1\}$ is a Menon-Hadamard difference set in G.*

Proof. Proof is delayed until Corollary 7.33, since the result is an easy consequence of the group extensions approach of Part 2. For abelian G an indirect proof, via the equivalence of Menon-Hadamard difference sets and perfect binary arrays (see Theorem 3.23), appears as Lemma 3.25 in Chapter 3.4.2. If D is a Menon-Hadamard difference set in G, and $\phi : G \to \{\pm 1\}$ is the characteristic function of D, defined by $\phi(g) = -1$ if and only if $g \in D$, then $[\phi(g_i)]_{1 \leq i \leq v}$ is the top row of the G-developed Hadamard matrix. $\square$

For instance, in (2.15) it is easy to check that $D = \{0, 1, 2\}$ is a $(4, 3, 2)$-difference set in $\mathbb{Z}_4$.

DEFINITION 2.20 *The* Menon Hadamard matrices *are the group developed Hadamard matrices, that is, the Hadamard matrices constructed in Lemma 2.19.*

Many authors have used techniques from character theory, finite geometry and algebraic number theory together with Lemma 2.19 to look for Menon Hadamard matrices. Our natural preference for first trying the simplest approach has meant that initially the cyclic groups $G = \mathbb{Z}_{4u^2}$ were trawled for difference sets. However, apart from the smallest possible example — seen in (2.15) in back-circulant form — none have ever been found, and it is believed none exist. This remarkable observation has fanned the research flame in diverging directions: to prove that there are no circulant Hadamard matrices, and to identify those classes of groups in which Menon-Hadamard difference sets do exist. The former problem is still open, though there is near-overwhelming support for it as a result of Schmidt's recent achievements using algebraic number theory [281]. In particular, we have the following asymptotic result.

THEOREM 2.21 (Schmidt [195, Corollary 7.4]) *Let* Π *be any finite set of odd primes. Then there are only finitely many cyclic Menon-Hadamard difference sets of order* u^2, *where all prime divisors of* u *are in* Π.

Combined with results of Turyn, Schmidt's computer searches further support the nonexistence of any nontrivial circulant Hadamard matrices [195]. In particular, there are only 14 unresolved odd values for $u \leq 10,000$, of which the smallest is $u = 165$.

Research Problem 5 *The Circulant Hadamard Conjecture. Show that no circulant Hadamard matrix exists for any order greater than 4.*

Until a decade ago there was a considerable body of evidence to suggest that, even in noncyclic groups, Menon-Hadamard difference sets were rare, restricted to values $u = 2^a 3^b$. In particular, there are no abelian Menon-Hadamard difference sets with $u = p > 3$ a prime [243]. A spectacular breakthrough by Xia [327], extended by Wilson and Xiang [324] and beautifully completed by Chen [53], gives us a very large family of Menon-Hadamard difference sets in *abelian* (noncyclic) groups, and consequently a new family of orders for which Hadamard matrices exist.

THEOREM 2.22 [327, 324, 53] *For any* $a, b \geq 0$ *and any odd number* m, *there exist Menon Hadamard matrices of orders* $4 \cdot 2^{2a} \cdot 3^{2b} \cdot m^4$, *developed over abelian groups.*

What happens in the nonabelian case? It is usual to regard the dihedral groups

$$D_{2m} = \langle a, b \mid a^m = b^2 = 1, bab = a^{-1} \rangle \tag{2.16}$$

of order $2m$ as being the class of nonabelian groups which is most nearly abelian, but existence of dihedral-developed Hadamard matrices is as unlikely as existence of circulant Hadamard matrices.

THEOREM 2.23 (Dillon; Fan et al., see [74]) *If a dihedral group of order $4u^2$ contains a Menon-Hadamard difference set, then so does $\mathbb{Z}_{4u^2}$.*

Nonetheless, Menon-Hadamard difference sets in nonabelian groups have been found in orders for which no abelian examples can exist. Smith [299], using a combination of representation theory and computer search, found one for $u = p = 5$ in the group of order $v = 100$,

$$\langle x, y, z \; : \; x^5 = y^5 = z^4 = xyx^{-1}y^{-1} = zxz^{-1}x^{-2} = zyz^{-1}y^{-2} = 1\rangle,$$

which is neither abelian nor dihedral, but has Sylow 5-subgroup $\mathbb{Z}_5^2$. By tensoring, it gives rise to an infinite family of Menon Hadamard matrices over nonabelian groups.

THEOREM 2.24 [299] *For any $a, b, c \geq 0$ such that if $b > 0$ then $a > 0$, Menon Hadamard matrices of orders $4 \cdot 2^{2a} \cdot 3^{2b} \cdot 5^2 \cdot 10^{2c}$ exist.*

Our final family of Hadamard matrix constructions uses Ito's modification of the Williamson template. More template-based constructions appear in Chapter 6 (Sections 6.4.5 and 6.5.1).

2.3.2 Ito Hadamard matrices

In 1981, Ito [178, Definition 1] introduced *type Q* Hadamard matrices. These are all the Hadamard matrices equivalent to a Hadamard matrix of the form (cf. [88, 4.3])

$$\begin{bmatrix} A & B & C & D \\ B & -A & D & -C \\ C^\top & -D^\top & -A^\top & B^\top \\ D^\top & C^\top & -B^\top & -A^\top \end{bmatrix}, \tag{2.17}$$

where the order w matrices A, B, C, D are circulant. He noted [178, Proposition 3] that this matrix is Hadamard if and only if A, B, C, D satisfy (2.13) and

$$AB^\top + CD^\top = BA^\top + DC^\top. \tag{2.18}$$

Thus the Williamson Hadamard matrices with all components symmetric circulant are type Q Hadamard matrices. In [180, Example 3] Ito derived the Paley Type I Hadamard matrices as type Q Hadamard matrices of order $4w$ for all w such that $4w - 1$ is a prime power. He subsequently [181] constructed type Q Hadamard matrices of order $4w$ for all w such that $2w - 1 \equiv 1 \bmod 4$ is a prime power.

DEFINITION 2.25 *The Ito Hadamard matrices are the Hadamard matrices equivalent to a Hadamard matrix of the form (2.17), where A, B, C, D are all circulant.*

Ito cast his results in terms of special subsets in certain groups of order $8w$ that he came to call *Hadamard groups*. By 1994 he knew that his special subsets were examples of combinatorial structures called *relative $(4w, 2, 4w, 2w)$-difference sets* in the Hadamard group, with forbidden subgroup of order 2 [182]. By 1995, Flannery had isolated their equivalence with certain *cocyclic* Hadamard matrices [112].

Discovery of this interplay formed one conceptual thread which led through the labyrinth to the full theory described in Part 2 of this book.

The Hadamard group for circulant components is the group Q_{8w} of order $8w$

$$Q_{8w} = \langle a, b \mid a^{4w} = b^4 = 1, a^{2w} = b^2, b^{-1}ab = a^{-1} \rangle$$

(for odd w this is the dicyclic group). The forbidden order 2 subgroup is $\langle b^2 \rangle$ and the quotient of Q_{8w} by $\langle b^2 \rangle$ is isomorphic to the dihedral group of order $4w$ $D_{4w} = \langle a, b \mid a^{2w} = b^2 = 1, bab = a^{-1} \rangle$.

Ito's success in including the Paley Type I and symmetric circulant Williamson Hadamard matrices in a single construction led him to conjecture [184] that relative $(4w, 2, 4w, 2w)$-difference sets exist in Q_{8w} for every positive integer w. For the corresponding conjecture about Hadamard matrices, which, as we will see in Chapter 6.4.4, must be D_{4w}-cocyclic, it is sufficient to consider odd w.

Research Problem 6 *Ito's Conjecture. Show that an Ito Hadamard matrix of order $4w$ exists for every odd w.*

Complex Golay sequences are also a source of Ito Hadamard matrices. A *complex Golay sequence of length w* [64] is a pair $a_1, a_2, \ldots, a_w$ $b_1, b_2, \ldots, b_w$ of $(\pm 1, \pm i)$ sequences satisfying $\sum_{j=1}^{k} a_j \bar{a}_{w-k+j} + \sum_{j=1}^{k} b_j \bar{b}_{w-k+j} = 0$, $k = 1, 2, \ldots, w-1$, and determines two $w \times w$ circulant $(\pm 1, \pm i)$ matrices X and Y which satisfy $X\overline{X}^\top + Y\overline{Y}^\top = 2wI_w$. If $X = U + iV$ and $Y = W + iZ$, then the matrices $A = U + V$, $B = U - V$, $C = W - Z$ and $D = W + Z$ are $w \times w$ circulant (± 1) matrices which satisfy (2.13) and (2.18) [150].

Schmidt has investigated this problem from the relative difference set perspective to provide alternative proofs of Ito's results and incorporate Williamson Hadamard matrices with circulant (but not necessarily symmetric) components. Combining these with the Golay construction, he obtains the largest family of Ito Hadamard matrices known.

THEOREM 2.26 (Schmidt [280, Corollary 3.6]) *Let m be a positive integer such that $2m - 1$ or $4m - 1$ is a prime power, or m is odd and there is a Williamson Hadamard matrix with $\mathbb{Z}_m$-developed components. Then there is an Ito Hadamard matrix of order $4w$ for every w of the form*

$$w = 2^a \cdot 10^b \cdot 26^c \cdot m, \quad a, b, c \geq 0.$$

In particular, the first case $w = m = 35$ missing from the sequence of symmetric circulant Williamson Hadamard matrices is covered by the Paley Hadamard matrix P_{139}, and, because there are symmetric $\mathbb{Z}_w$-invariant Williamson Hadamard matrices for the other odd values of $w \leq 45$, there are Ito Hadamard matrices for all $w \leq 45$.

In fact, Schmidt considered Hadamard matrices of both forms (2.14) and (2.17) above in which the component matrices A, B, C, D are all group developed (more precisely, group-invariant) over an arbitrary abelian group of order w. Again, a matrix of the latter form is Hadamard if and only if A, B, C, D satisfy both (2.13) and (2.18).

2.4 TOWARDS THE HADAMARD CONJECTURE

The Hadamard Conjecture maintains its grip on our imagination both because it is delightfully simple to state and understand, and because it remains impregnable after a century of assaults. Nonetheless, there have been significant inroads made in the defences over the past two decades, both in the families of orders for which new Hadamard matrices have been found, and in what may loosely be termed the asymptotics.

Tables of orders of known Hadamard matrices, along with the construction techniques used, are published in [145, 314, 288, 144], but some errors exist, particularly (see [18] and [170]) in the examples given in earlier lists of Williamson Hadamard matrices.

Jennie Seberry (= J. S. Wallis) has an (offline) database listing the odd integers $m < 40,000$ for which a Hadamard matrix of order $2^t m$ is known to exist for some $t \geq 2$. The most comprehensive list in print is that for $m < 10,000$ in [68]. For online lists of Hadamard matrices, see the websites of Christos Koukouvinos [212], Jennie Seberry [287] and Neil Sloane [297].

Until 1977, the smallest order for which no Hadamard matrix was known was $n = 268 = 4 \cdot 67$ [145]; then until 2004 it was $n = 428 = 4 \cdot 107$ [202]. At present the smallest unknown order is $n = 668 = 4 \cdot 167$. The remaining orders $< 1,000$ for which no Hadamard matrix is known are $n = 4 \cdot 179$, $4 \cdot 191$ and $4 \cdot 223$.

Research Problem 7 *Do Hadamard matrices of orders $668, 716, 764$ and 892 exist?*

The asymptotic results give, for each odd natural number m, a lower bound t_0 for t in terms of m such that a Hadamard matrix of order $2^t m$ exists for all $t \geq t_0$. The Hadamard Conjecture would be confirmed by showing $t_0 = 2$ for all m. The first asymptotic formula was proved in 1976 by Seberry [313], who used plug-in methods in orthogonal arrays to demonstrate that for any odd m, there exists a Hadamard matrix of order $2^t m$ for all $t \geq \lceil 2\log_2(m-3)\rceil + 1$.

It took almost two decades to better this result. Craigen [66] used groups containing a distinguished central involution and sequences with zero autocorrelation to show t_0 is upper-bounded by $4\lceil \frac{1}{6}\log_2((m-1)/2)\rceil + 2$. The present bound is also due to Craigen.

THEOREM 2.27 (Craigen [66, 65]) *For any odd positive number m, there exists a Hadamard matrix*

1. *of order $2^{2b}m$, where b is the number of nonzero digits in the binary expansion of m, and*
2. *of order $2^t m$ for $t = 6\lfloor \frac{1}{16}\log_2((m-1)/2)\rfloor + 2$.*

These logarithmic bounds each start with a positive real number a ($a = 2$ in [313] and $a = 4/6$ in [66]) and demonstrate that there exists a constant c ($c = 1$ in [313] and $c = 2^{16/5}$ in [66]) such that there is a Hadamard matrix of order $2^t m$ whenever $2^t \geq cm^a$. Theorem 2.27.2 implies that we may take $a = 3/8$ and $c = 2^{26/16}$.

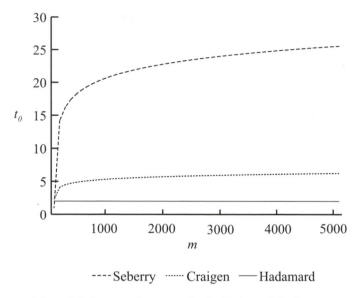

Figure 2.1 Asymptotic support for the Hadamard Conjecture

Seberry's and Craigen's asymptotic formulae for t_0 in terms of m, versus the Hadamard Conjecture, are graphed in Figure 2.1.

A separate asymptotic approach to the Hadamard conjecture has been to ask how much of a Hadamard matrix can always be constructed. More precisely, what is the largest number $\mathbf{r}(n) = r$ of rows for which there is an $r \times n$ matrix H with entries from $\{\pm 1\}$ satisfying $HH^\top = nI_r$? It is known that for sufficiently large n, about $\frac{1}{3}$ of a Hadamard matrix of order n always exists [83].

A remarkable linking of the two approaches has recently been demonstrated by de Launey and Gordon to follow from the Extended Riemann Hypothesis (that the nontrivial zeroes of the Dedekind zeta function of any algebraic number field lie on the critical line). Using the values for a and c from Theorem 2.27.2, they show that for sufficiently large n, the Extended Riemann Hypothesis implies about $\frac{1}{2}$ of a Hadamard matrix of order n always exists.

THEOREM 2.28 (de Launey and Gordon [89]) *Let $\varepsilon > 0$. If the Extended Riemann Hypothesis holds, then for every sufficiently large $n \equiv 0 \bmod 4$, $\mathbf{r}(n) \geq \frac{n}{2} - n^{17/22+\varepsilon}$.*

As the Riemann Hypothesis itself has been verified for at least the first 10^{13} nontrivial zeroes [129], and a formal proof may be achievable in our lifetimes, so too might confirmation of this result.

Chapter Three

Applications in Signal Processing, Coding and Cryptography

Practical application of Hadamard matrices goes back to 1937, when Yates developed an algorithm (a fast Hadamard transform) to determine which factors contributed the main effects in a factorial experiment.

Since then, most direct use of Hadamard matrices has fallen into one of three broad categories: for design of experiments, including factorial designs; for Hadamard transform spectroscopy and object recognition; and for coding of digital signals.

In experimental design, Hadamard matrices are building blocks for 2-level orthogonal arrays of strengths 2 or 3 (the Hadamard arrays), given, respectively, by the binary matrices A_n and C_n of Definition 3.13 below. The rows of the array represent the experiments or tests to be performed, while the columns correspond to the different variables (factors) whose effects are being analysed. Each factor takes only two values in the 2-level case. Orthogonal arrays (of different sizes, higher levels and other strengths) are one form of generalisation of Hadamard matrices, but not one to be covered here: instead, the text by Hedayat, Sloane and Stufken [144] is recommended to the interested reader. Similarly, the use of Hadamard matrices in chemical balance weighing experiments and their generalisation to weighing designs and orthogonal designs will not be discussed; see [123]. There is an enormous literature on combinatorial and experimental block designs within mathematics and statistics. For a comprehensive treatment, the standard textbook on block designs, Beth, Jungnickel and Lenz [24], is recommended, especially Chapter XIII on applications, including those for Hadamard matrices.

This Chapter is devoted to the two other main uses of Hadamard matrices, as transform matrices and masks for spectral analysis and synthesis of signals, and as codes for error protection or separation or encryption of signals. Most emphasis is on the second use. Each application area is introduced briefly to explain how the Hadamard matrix is applied, but in enough detail to support the applications of generalised Hadamard matrices described in subsequent Chapters.

Hadamard matrices are employed for spectral analysis or signal separation, especially, across a huge range of disciplines.

The most cursory search will confirm the widespread use of Hadamard transform techniques in traditional spectral analysis domains such as mass spectroscopy, polymer chemistry, signal and information processing, geophysics, acoustics, nuclear medicine and nuclear physics. Many of these applications were only hinted at 25 years ago, and it is a fascinating exercise to see how many of the predictions in the classic text by Harwit and Sloane [142, Chap. 7] have come to pass. Novel applications to digital logic design, pattern recognition, data compression, magnetic resonance imaging, neuroscience and quantum computing are emerging.

Similarly, signal spreading using Hadamard matrices, foreshadowed in [142], is well known in digital and satellite communications and is commonplace globally in CDMA mobile phones, but is an emergent technique in automated learning, ultrasonics, optical communication and information hiding.

We will describe these applications from the perspective of digital signal processing.

Digital signals and data sequences are processed for a wide variety of purposes, for example: to modify the information they carry into a more readily interpreted form; or to estimate or extract or even disguise characteristic parameters; or to remove interference or noise.

3.1 SPECTROSCOPY: WALSH-HADAMARD TRANSFORMS

3.1.1 Signal analysis and synthesis

We are interested in processing signals of finite energy which can be modelled by a periodic function $x(t)$, discrete in time t. A signal of finite duration can be treated as one period of a periodic signal. If a signal is continuous, it can be sampled at fixed time intervals to obtain a discrete-time signal, represented by a repeating sequence of values $x(0), x(1), \ldots, x(n-1)$.

Such discrete signals can be expressed as a finite linear combination of orthogonal basis functions,

$$x(t) = \sum_{j=0}^{n-1} \hat{x}(j) B_j(t), \ t = 0, \ldots, n-1, \tag{3.1}$$

where the basis functions $B_j, j = 0, \ldots, n-1$, are complex-valued functions of t and

$$\sum_{t=0}^{n-1} B_j(t) \overline{B_k(t)} = \begin{cases} n & \text{if } j = k, \\ 0 & \text{if } j \neq k. \end{cases} \tag{3.2}$$

Setting $\mathbf{x} = [x(0), x(1), \ldots, x(n-1)]$, $\hat{\mathbf{x}} = [\hat{x}(0), \hat{x}(1), \ldots, \hat{x}(n-1)]$, and $B = [b_{jk}]_{0 \leq j,k \leq n-1}$, where $b_{jk} = B_k(j), 0 \leq j, k \leq n-1$, these equations become

$$\mathbf{x}^\top = B \hat{\mathbf{x}}^\top \tag{3.3}$$

and

$$B(\overline{B})^\top = n I_n, \tag{3.4}$$

respectively, where $\overline{B}$ denotes the complex conjugate matrix of B. Consequently the *spectrum* $\hat{\mathbf{x}}$ is a representation of the signal $\mathbf{x}$ in the *transform domain* and can be recovered from the *transform matrix* $\overline{B}$ by the equation

$$\hat{\mathbf{x}} = n^{-1} \mathbf{x} \overline{B}. \tag{3.5}$$

If the basis family consists of sinusoids, recovery of one period of a continuous signal from its sampled representation depends on the fact that transmission of very high frequencies by physical systems vanishes for all practical purposes, so the signal may be assumed to be frequency bandlimited. In this case the *sampling theorem* [2, 2.6] tells us that the signal can be uniquely recovered from the sample values at a high enough sampling rate, determined by the bandwidth.

When we are trying to decompose a sampled periodic signal (3.1) into the sum of sinusoids of appropriate amplitudes and known frequencies, the usual transform is the n-point Discrete Fourier Transform (DFT) with transform matrix

$$\mathcal{F}_n = [\,(e^{-2\pi i/n})^{jk}\,]_{0 \leq j,k \leq n-1}, \tag{3.6}$$

where $i = \sqrt{-1}$. The coefficient vector $\hat{\mathbf{x}} = n^{-1}\mathbf{x}\,\mathcal{F}_n$ is the spectrum in the *frequency* domain, and it corresponds to samples equally spaced in frequency of the Fourier Transform of the signal. Until the 1960s, Fourier analysis and synthesis of signals was typically carried out using analogue equipment, but after the discovery of an efficient algorithm — the Fast Fourier Transform (FFT) — which reduced computation time by orders of magnitude, emphasis shifted to all-digital systems (see [253] for an outline of this early history).

Unfortunately, the rectangular waveforms which are most suited to digital communications are precisely the most difficult to synthesise using finitely many sinusoids. The Fourier Transform of

$$\text{rect}(t/n) = \begin{cases} 1 & \text{if } -n/2 \leq t \leq n/2, \\ 0 & \text{otherwise} \end{cases}$$

is the function $n\,\text{sinc}(tn) = \sin(\pi tn)/(\pi t)$, which has infinite bandwidth in frequency (cf. [23, Fig. 5.2 and p. 155]).

Another set of orthogonal basis functions provides a better series representation in this case. These are the Walsh functions, introduced mathematically by Walsh in 1923 [316], though they were already used by communications engineers for the transposition of conductors in open wire lines [141]. The Walsh functions form a complete orthonormal set of rectangular waveforms defined on the unit interval $[0, 1)$. It comes as no surprise that the most difficult functions to synthesise using finitely many of them are sinusoids (cf. [23, Fig. 5.2]).

Walsh functions take only the values ± 1, so their generation and implementation is simple; the corresponding fast algorithms require only addition and subtraction of input values rather than the complex addition and multiplication of input values required by the FFT.

3.1.2 The Walsh-Hadamard Transform

The Walsh functions are the basis functions for the Walsh-Hadamard Transforms (WHT). For a detailed account of their properties, see [105, Chapter 8] or the earlier [2], [23] or [141]. Whereas the basis functions for the DFT are equally spaced in frequency, the Walsh functions are ordered by their *sequency*, defined to be $\lceil z/2 \rceil$, where z is the number of zero-crossings of the function per unit interval. It follows that Walsh functions with an odd number $z = 2s - 1$ of zero-crossings and an even number $z = 2s$ of zero-crossings have the same sequency s, and are ordered

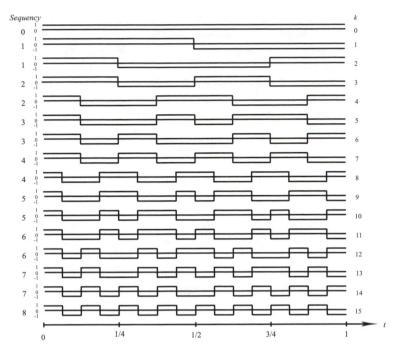

Figure 3.1 First 16 Walsh functions $\mathrm{wal}_w(k,t)$ in sequency order

with the odd function preceding the even function. The first 16 Walsh functions in sequency order are shown in Figure 3.1.

There is an analogue of the sampling theorem, due independently to Johnson and to Maqusi (see [105, p. 307]), for sequency bandlimited signals.

THEOREM 3.1 *(Sequency Sampling Theorem) A signal which is sequency bandlimited to 2^n zero-crossings per second can be uniquely recovered from its samples at every 2^{-n} seconds.*

When the first 2^n Walsh functions in sequency order are sampled uniformly, the resulting matrix with entries ± 1 is a normalised Hadamard matrix of order 2^n, the *Walsh Hadamard* matrix

$$W_n = [W_n(s,t)]_{0 \le s,t \le 2^n - 1}. \tag{3.7}$$

Any reordering of the rows determines an equivalent Hadamard matrix, and the reordering of most interest to us gives the Sylvester Hadamard matrix $\mathcal{S}_n$ of Definition 2.3.

First we need some terminology to help translate row and column indexing conventions between mathematical, computer science and engineering usage.

DEFINITION 3.2 *Any non-negative integer i, $0 \le i \le 2^n - 1$ has a unique expansion to base 2 as $i = i_{n-1}2^{n-1} + i_{n-2}2^{n-2} + \cdots + i_1 2 + i_0$, where $i_j \in \{0,1\}$.*

The binary representation *of the integer i is the string of coefficients $(i)_2 = i_{n-1}i_{n-2}\ldots i_0$ with the least significant bit at the right-hand end. Denote by $b(i)_2$ the* bit-reversed *binary representation $i_0 i_1 \ldots i_{n-1}$ of i.*

The radix-2 notation *for the integer i is the vector of coefficients $(i_{n-1}, i_{n-2}, \ldots, i_0)$. Equally, the vector or string may be treated as an element in the direct product $\mathbb{Z}_2^n = \mathbb{Z}_2 \times \mathbb{Z}_2 \times \cdots \times \mathbb{Z}_2$ of n copies of the cyclic group $\mathbb{Z}_2$ of integers modulo 2.*

Lexicographical order *of the elements in $\mathbb{Z}_2^n$ is the natural order of the integers to which they correspond under this identification.*

Denote by $i \oplus m$ the bitwise addition of $(i)_2$ and $(m)_2$, or, equivalently, addition in $\mathbb{Z}_2^n$, that is, $i \oplus m = (i_{n-1} \oplus m_{n-1}, i_{n-2} \oplus m_{n-2}, \ldots, i_0 \oplus m_0)$.

These definitions generalise to any set of bases.

DEFINITION 3.3 *Given n integers $m_0, m_1, \ldots, m_{n-1} \geq 2$, any integer i with $0 \leq i \leq \prod_{j=0}^{n-1} m_j - 1$ has a unique expansion*

$$i = i_{n-1}\prod_{j=0}^{n-2} m_j + i_{n-2}\prod_{j=0}^{n-3} m_j + \cdots + i_1 m_0 + i_0,$$

where $0 \leq i_j \leq m_j - 1, 0 \leq j \leq n - 1$.

The mixed radix notation *for the integer i is the vector of coefficients $(i_{n-1}, i_{n-2}, \ldots, i_0)$. Equally, the vector or string $i_{n-1}i_{n-2}\ldots i_0$ of coefficients may be treated as an element in the direct product $\mathbb{Z}_{m_{n-1}} \times \mathbb{Z}_{m_{n-2}} \times \cdots \times \mathbb{Z}_{m_0}$ of n cyclic groups.*

Lexicographical order *of the elements in this group is the natural order of the integers to which they correspond under this identification.*

DEFINITION 3.4 *The* Gray map *of a binary string $b_{n-1}b_{n-2}\ldots b_0$ is the binary string obtained by adding $\bmod\ 2$ to each bit, the bit immediately left of it:*

$$G(b_{n-1}b_{n-2}\ldots b_0) = b_{n-1}(b_{n-2} \oplus b_{n-1})\ldots(b_0 \oplus b_1).$$

The Gray map *of an integer i, $0 \leq i \leq 2^n - 1$, is the integer $G(i)$ corresponding to the Gray map of the binary representation $(i)_2$ of i.*

For example, the Gray map of the integers $0, 1, 2, 3$ is

$$0 \mapsto 00 = 0,\ 1 \mapsto 01 = 1,\ 2 \mapsto 11 = 3,\ 3 \mapsto 10 = 2. \tag{3.8}$$

As we will see in Chapter 4.4.3, this Gray map has also been instrumental to an exciting advance in our understanding of binary nonlinear error-correcting codes.

The Sylvester Hadamard matrix (2.4) is obtained from the Walsh Hadamard matrix by the row permutation

$$\mathcal{S}_n(i, j) = W_n(G(b(i)_2), j) = (-1)^{\sum_{k=0}^{n-1} i_k j_k},\ 0 \leq i, j \leq 2^n - 1. \tag{3.9}$$

DEFINITION 3.5 *The* Walsh-Hadamard Transform (WHT) *or* Binary Fourier Representation (BIFORE) *of a data sequence $\mathbf{x} = [x(0), x(1), \ldots, x(2^n - 1)]$ in the sequency domain is the spectrum $\hat{\mathbf{x}} = [\hat{x}(0), \hat{x}(1), \ldots, \hat{x}(2^n - 1)]$ given by*

$$\hat{\mathbf{x}}^{\top} = 2^{-n}\mathcal{S}_n\,\mathbf{x}^{\top}, \tag{3.10}$$

and its inverse is

$$\mathbf{x}^\top = \mathcal{S}_n \, \hat{\mathbf{x}}^\top.$$

Note that it is equally common to call

$$\hat{\mathbf{x}}^\top = \mathcal{S}_n \, \mathbf{x}^\top \qquad (3.11)$$

the WHT, and $\mathbf{x}^\top = 2^{-n}\mathcal{S}_n \, \hat{\mathbf{x}}^\top$ *the inverse WHT, or, for computational ease, to balance the representation in time and frequency domains by scaling:*

$$\mathbf{x}^\top = 2^{-n/2}\mathcal{S}_n \, \hat{\mathbf{x}}^\top, \quad \hat{\mathbf{x}}^\top = 2^{-n/2}\mathcal{S}_n \, \mathbf{x}^\top. \qquad (3.12)$$

Basic properties of the Walsh-Hadamard Transform and matrix are now listed (see [105, 8.2, 8.8] and [2, 6.8-9, 6.11]).

LEMMA 3.6 *Let* $\hat{\mathbf{x}}, \hat{\mathbf{y}}$ *and* $\hat{\mathbf{z}}$ *be the WHT (3.10) of sequences* $\mathbf{x}, \mathbf{y}$ *and* $\mathbf{z}$, *respectively. For fixed* m, *the* dyadic shift $\mathbf{x} \oplus m$ *of* $\mathbf{x}$ *is* $x(i) \mapsto x(i \oplus m)$, $0 \le i \le 2^n - 1$.

1. *The scaled matrix* $2^{-n/2}\mathcal{S}_n$ *is orthogonal.*

2. *The rows (and columns, similarly) of* $\mathcal{S}_n$ *are closed under pointwise multiplication:*

$$\mathcal{S}_n(i,k)\mathcal{S}_n(j,k) = \mathcal{S}_n(i \oplus j, k), \ k = 0, \ldots, 2^n - 1.$$

3. *(Parseval's Theorem) The* energy

$$\sum_{i=0}^{2^n-1} x(i)^2 = 2^n \sum_{i=0}^{2^n-1} \hat{x}(i)^2 \qquad (3.13)$$

of $\mathbf{x}$ *is preserved under Walsh-Hadamard transformation.*

4. *(Shift Invariant Power Spectrum Theorem) Let* $\mathbf{z} = \mathbf{x} \oplus m$. *Then*

$$\hat{z}(i)^2 = \hat{x}(i)^2, \ i = 0, \ldots, 2^n - 1, \qquad (3.14)$$

so the power spectrum $[\hat{x}(i)^2, i = 0, \ldots, 2^n - 1]$ *of* $\mathbf{x}$ *is invariant under dyadic shifts of* $\mathbf{x}$. *There also exists a power spectrum invariant under circular shifts of* $\mathbf{x}$.

5. *(Convolution/Correlation Theorem) The* dyadic *(or 'logical') correlation* $\mathbf{z}$ *of sequences* $\mathbf{x}$ *and* $\mathbf{y}$ *is* $z(m) = \sum_{i=0}^{2^n-1} x(i)y(i \oplus m)$, *for* $m = 0, \ldots, 2^n - 1$. *Then*

$$\hat{z}(i) = \hat{x}(i)\hat{y}(i), \ i = 0, \ldots, 2^n - 1. \qquad (3.15)$$

(The terms *energy* and *power* (energy per unit time) derive from the physical interpretation of $x(i)$ as a voltage or current signal, as a function of time, across a resistor, in which case the sum $2^{-n}\sum_{i=0}^{2^n-1} x(i)^2$ represents the average energy of the signal dissipated by the resistor and the set of values $\{\hat{x}(i)^2, \ 0 \le i \le 2^n - 1\}$ represents the spectral distribution of the power in $x(i)$ dissipated by the resistor.)

3.1.3 The Fast Hadamard Transform

Fast algorithms have been developed for the WHT and, as for the DFT, are based on factorisation of the transform matrix into sparse matrices. The factorisation of $\mathcal{S}_n$ is easily proved by induction on n (see [237, Theorem 14.4.5]).

THEOREM 3.7 *(Fast Hadamard Transform Theorem) For $i = 1, \ldots, n$, let*

$$M_{(n,i)} = I_{2^{n-i}} \otimes \mathcal{S}_1 \otimes I_{2^{i-1}}.$$

Then

$$\mathcal{S}_n = M_{(n,1)} M_{(n,2)} \cdots M_{(n,n)}. \qquad (3.16)$$

The Fast Hadamard Transform (FHT) for (3.10) is implemented in stages:

$$2^n \hat{\mathbf{x}} = \mathbf{x} \, \mathcal{S}_n = (\ldots ((\mathbf{x} M_{(n,1)}) M_{(n,2)}) \ldots M_{(n,n)}). \qquad (3.17)$$

This is an example of an 'in-place' algorithm, in which memory storage for intermediate stage calculations is not needed, because input data values can be overwritten by output data values as soon as they have been read, so it is very efficient. Since $M_{(n,i)} = I_2 \otimes M_{(n-1,i)}$ for $i = 1, \ldots, n-1$, stages can be added or subtracted to fit data sequences of varying length [321]. Hardware implementation of the FHT often goes by the name of the 'Green machine', since it originated with R. R. Green as a means of decoding the first-order Reed-Muller code used in the Mariner spacecraft on the 1969 mission to Mars. See [237, pp. 424–425] for a circuit diagram and [105, Figs. 8.8-9] for signal flowgraphs of the FHT.

Example 3.1.1   *Using (3.17) with $n = 3$ and $\mathbf{x} = [-1, 1, 1, -1, 1, -1, -1, -1]$, we have $\mathbf{x}' = \mathbf{x} M_{(3,1)} = [0, -2, 0, 2, 0, 2, -2, 0]$, $\mathbf{x}'' = \mathbf{x}' M_{(3,2)} = [0, 0, 0, -4, -2, 2, 2, 2]$ and $2^3 \hat{\mathbf{x}} = \mathbf{x}'' M_{(3,3)} = [-2, 2, 2, -2, 2, -2, -2, -6]$. So $\sum_{i=0}^{7} x(i)^2 = 8 = 2^3 \sum_{i=0}^{7} \hat{x}(i)^2$.*

Probably the simplest version of the FHT to program or perform manually is Yates' Algorithm (see [246], for example). In it, the first stage acts on $\mathbf{x}$ as follows. The first half of the intermediate output vector is obtained from the intermediate input vector by adding the entries in adjacent pairs of positions and the second half by subtracting them. Each subsequent stage operates identically, until n stages have been computed. Thus, in Example 3.1.1 using Yates' Algorithm, $\mathbf{x}$ is first transformed to $\mathbf{x}^* = [0, 0, 0, -2, -2, 2, 2, 0]$, then to $\mathbf{x}^{**} = [0, -2, 0, 2, 0, 2, -4, 2]$ and finally to $2^3 \hat{\mathbf{x}} = [-2, 2, 2, -2, 2, -2, -2, -6]$.

3.1.4 Hadamard spectroscopy

Here we deal only briefly with spectral analysis and synthesis of physical signals using Hadamard masks and the WHT. The original text [142] is recommended for its very clear explanation of the theory of Hadamard mask spectroscopy and application to optical signals. Other applications of the WHT for decoding of binary error-correcting codes and for cryptography are covered in Sections 3.2.2 and 3.5.1, respectively. Use of the WHT in a general decoding algorithm for quantum error-correcting codes is sketched in [24, Chapter XIII.5.7].

Clearly, if the signal $\mathbf{x}$ to be recovered has n components $x(0), x(1), \ldots, x(n-1)$, separated, perhaps, by time, space or frequency, it is possible to attempt a direct measurement $y(i)$ of each component $x(i)$. Each measurement may be subject to error, so is an estimate $y(i) = x(i) + \epsilon_i$ of $x(i)$. If we assume that the errors ϵ_i in each estimate are independent random variables of equal mean 0 and equal variance, a good unbiased measuring instrument (or estimator) will simultaneously minimise all the errors.

The crucial observation (due to Yates, in the context of weighing designs) is that the error variance of the measurements can be reduced by measuring the components several at a time (*multiplexing*), instead of singly. For some instruments, the mix of components in a particular measurement can be determined by whether they are present, or not, or a reversed (for example, reflected) component is present, so that a typical measurement is now of the form $y(i) = \sum_{j=0}^{n-1} a_{ij}x(j) + \epsilon_i$, where $a_{ij} \in \{0, \pm 1\}$, with corresponding matrix equation

$$\mathbf{y}^\top = A\mathbf{x}^\top + \epsilon^\top. \tag{3.18}$$

The rows of A represent n individual physical *masks* through which the signal to be recovered is passed.

The variance in ϵ is minimised (for three different measures of minimum variance) if and only if A is a Hadamard matrix H of order n [142, 3.2], so that $\mathbf{x}^\top \approx n^{-1}H^\top\mathbf{y}^\top$. In this case, the error variance in each component is reduced by a factor of n over direct measurement. That is, the root mean square *signal-to-noise ratio (SNR)* of the multiplexed measurement is increased by a factor of $\sqrt{n}$ [142, 3.2.4–5].

If, physically, A must be restricted to entries only in $\{0, 1\}$, then the core of a normalised Hadamard matrix of order $(n + 1)$, transformed by the $\log_{(-1)}$ map, is almost as good. In this case, the error variance in the components using multiplexing is reduced by a factor of $\approx n/4$ over direct measurement and the SNR increased by a factor of $\approx \sqrt{n}/2$. Further practical efficiency is gained by using a Hadamard matrix with a circulant core, since then a single mask of length $2n - 1$ may be stepped one position for each measurement rather than n separate masks of length n each being employed for one measurement [142, 3.2.6–7].

By Lemma 2.7, Hadamard matrices with $n \times n$ circulant cores are equivalent to Hadamard designs with circulant incidence matrices. The only known examples arise from cyclic Hadamard difference sets and belong to one of the three parameter families of Example 2.1.1, where $n = 4w - 1$ is either a prime p, a 'twin prime' $p(p + 2)$ or $n = 2^m - 1$. Golomb [301] has conjectured that a cyclic Hadamard difference set exists only if it has one of these forms.

Research Problem 8 *(Golomb) Show that the only cyclic Hadamard* $(4w - 1, 2w - 1, w - 1)$*-difference sets have parameters of one of the three forms:* $4w - 1 = p$, $4w - 1 = p(p + 2)$, p *prime, or* $w = 2^m$, *or else find a counterexample.*

A modern alternative for optical signals avoids moving masks in favour of a single stationary mask of length n. The mask elements are made of material whose refractive index may be switched between opaque and transparent, so the single mask may be switched through successive rows of A. This avoids the necessity of using matrices with circulant cores (though in practice they are still used) but has the disadvantage that the response time of mask elements to switching means that an 'off' element may still transmit some radiation. Hadamard transform masks of this type with $n = 127$ and $n = 255$ have been fabricated for application in laser Raman spectroscopy (Hammaker et al. [135]).

3.2 ERROR CORRECTION: HADAMARD CODES

Data sequences to be transmitted over a digital communication channel are first *modulated* — the information symbols are mapped onto signals which can be transmitted efficiently. For example, binary symbols 0 and 1 may be mapped to two waveforms which are π radians out of phase from each other. Larger data alphabets or data blocks may be keyed to multiphase signals, such as in 4-PSK or 16-PSK, where the signals are phase-shifted by multiples of $2\pi/4$ or $2\pi/16$ radians, respectively, from each other.

These signals may be affected by noise during transmission. The most common source of noise is reflection and scattering of the signal from obstacles, but ambient heat in the hardware or in the transmission medium, interference from other communications channels or atmospheric phenomena can also contribute. Once the received signal has been demodulated, some received data symbols may be incorrect.

If the channel allows information flow in both directions (a channel *with feedback*), the most common means of correcting errors is by some form of automatic repeat request (ARQ) protocol. Under this type of protocol, data are encoded for error detection before transmission, and if a transmission error is detected at the receiver, a retransmission request is generated. These protocols provide high levels of protection on channels which are error-free, apart from occasional bursts of noise of short duration. The cost of error control is reflected in some reduction of throughput.

In some situations, simple retransmission of the data is not practical or not possible. Examples of the former are when the transmitted information cannot be retransmitted economically — in either time or money — as with images from solar system or deep-space probes, and when real-time information is required, as with voice or video transmission. An example of the latter is when the transmitted data are being archived (as with transfer to a storage medium such as compact disc, hard drive or hologram) and so must be error-free for subsequent reuse. Essentially, the channel allows information flow in only one direction. Then we rely on forward error-correcting (FEC) techniques, in which pre-processing of data before modulation builds in redundancy. There should be sufficient structured redundancy in the encoded data to ensure that transmission errors can be located and, ideally, corrected. Here the cost of error control is a reduction of the ratio of data symbols to transmitted symbols. Figure 3.2 shows this model of transmission.

A very readable coverage of the engineering aspects of error control coding appears in Wicker [321].

In 1948, Shannon [290] demonstrated that ideal error-correcting codes exist for

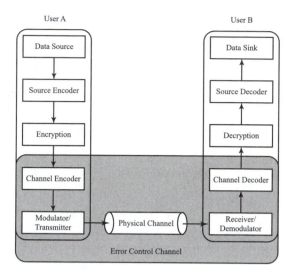

Figure 3.2 Model of coded and secured transmission over noisy channel

any discrete memoryless channel. His famous *Noisy Channel Coding Theorem* states that, under suitable conditions and at rates less than the channel capacity, there exist error-correcting codes which will transmit with arbitrarily low bit error rate. This nonconstructive proof sparked the energetic, continuing hunt for optimal error-correcting codes.

3.2.1 Error-correcting codes

What makes a code 'good' of course depends on the channel for which it is intended and the type of error patterns expected. Historically, the first-order Reed-Muller code developed from a Hadamard matrix of order 32 was used to error-protect images sent to Earth by unmanned probes of our solar system, and the FHT was used to decode them. Those codes were designed to locate and correct a high proportion of errors relative to the number of binary symbols transmitted, assuming a random pattern of errors.

Reed-Muller codes have fallen from favour over the past thirty years. Other error-correcting codes, such as convolutional codes, are preferred for most error protection applications, including deep space transmissions. However, the codes based on Hadamard matrices still dominate the tables of 'best binary codes' [226] when high distance relative to length is required. Wicker [321] suggests that the need for ever higher data rates on optical channels will again make them attractive because they have very fast decoding algorithms. Meanwhile, novel applications areas are opening up, for example, to error correction in digital watermarking systems [335] and to construction of quantum error-correcting codes [24, pp. 918–919].

A basic account of the theory of block codes for correcting random errors follows. The channels of interest are typically used to transmit q-ary symbol alphabets, where q is a prime power (usually 2^n), and are modelled as symmetric and

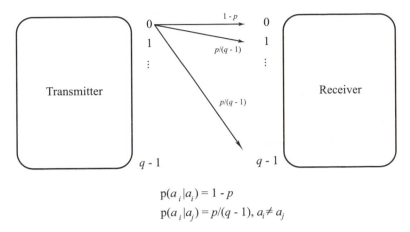

$$p(a_i|a_i) = 1 - p$$
$$p(a_i|a_j) = p/(q - 1), a_i \neq a_j$$

Figure 3.3 The q-ary symmetric channel

memoryless. Figure 3.3 is a representation of this channel model.

DEFINITION 3.8 *A q-ary channel is* symmetric *if the probability that a transmitted symbol will be received incorrectly (the* crossover *probability p) is the same for each error; ideally $p \ll 0.5$. A channel is* memoryless *if the noise process affecting a symbol during transmission is independent of that affecting preceding or succeeding symbols.*

Initially, we will model our data and code alphabet as elements of the finite field $GF(q)$, most commonly using the binary alphabet $GF(2) = \{0, 1\}$. More general alphabets will be introduced later, using $\mathbb{Z}_4$ in Chapter 4.4.3, an arbitrary finite group N in Chapter 4.4.4 and a commutative ring R with unity in Chapter 9.1.3.

DEFINITION 3.9 *Let $V(n, q)$ be the vector space consisting of all strings (or vectors) of length n over the finite field $GF(q)$, under positionwise addition and scalar multiplication. Let $k \leq n$, and let the* message set W *be a subset of size M of $V(k, q)$.*

1. *An* encoding *of W is an injective mapping $E : W \to V(n, q)$.*

2. *The subset $C = E(W)$ of size M in $V(n, q)$ is a q-ary (n, M)* block code, *and its elements are* codewords. *The* rate *of the code is $(\log_q M)/n$. The* redundancy *of the code is $r = n - \log_q M$.*

3. *The block code C is a* linear $[n, k]$ *code if $W = V(k, q)$ and C is a subspace of $V(n, q)$. Otherwise C is* nonlinear *(note: even if $M = q^k$).*

4. *A $k \times n$ matrix whose rows form a basis for a linear $[n, k]$ code C is called a* generator matrix *for C.*

5. *The* dual *of a code* C *is the set* $C^{\perp}$ *of elements in* $V(n,q)$ *orthogonal to all codewords of* C, *that is,* $C^{\perp} = \{v \in V(n,q) : v \cdot c = 0, \; \forall \, c \in C\}$, *where* $v \cdot c = \sum_{i=1}^{n} v_i c_i$. *A code* C *is* self-orthogonal *if* $C \subseteq C^{\perp}$ *and* self-dual *if* $C = C^{\perp}$.

The rate of the code measures the transmission cost per symbol of encoding M message words as length n codewords. Two of the goals in constructing a good code are to maximise the number M of message words which can be encoded and to maximise the rate (minimise n, given M). But it is the redundancy which permits error detection and correction mechanisms to be built into the code, as it allows us to distribute codewords within $V(n,q)$ so as to maximise their distance from each other under a discrete metric known as *Hamming distance*.

DEFINITION 3.10 *Let* $\mathbf{v}, \mathbf{w} \in V(n,q)$.

1. *The* Hamming weight $w(\mathbf{v})$ *of* $\mathbf{v}$ *is the number of nonzero symbols in* $\mathbf{v}$.

2. *The* Hamming distance $d(\mathbf{v}, \mathbf{w})$ *between* $\mathbf{v}$ *and* $\mathbf{w}$ *is the number of positions in which they differ:* $d(\mathbf{v}, \mathbf{w}) = w(\mathbf{v} - \mathbf{w})$.

3. *The* (minimum) distance $d = d(C)$ *of a code* C *is the minimum of all the Hamming distances between distinct codewords.*

4. *The* parameters *of* C *are* (n, M, d) — *but written* $[n, k, d]$ *when* C *is linear.*

5. *The* weight enumerator *of* C *is the two-variable polynomial* $W_C(x, y) = \sum_{i=0}^{n} A_i x^{n-i} y^i$, *where* A_i *is the number of codewords in* C *of weight* i.

The weight enumerator determines the probability that a transmitted error will not be detected [237, p. 21], so this can provide a good engineering reason to prefer one code over another code with the same parameters. Even for codes with the same parameters and weight enumerators, one may be preferred: a linear code with a generator matrix of the form $[I_k \; A]$ has much simpler coding and decoding algorithms than one without.

Nonetheless, it is often sufficient theoretically to identify a code up to its equivalence class. Two q-ary codes C and C' of length n are *equivalent* [237, p. 40] if there exist permutations $\pi_1, \pi_2, \ldots, \pi_n$ of $GF(q)$ and a permutation σ of the coordinate positions $1, \ldots, n$ such that, if $(v_1, v_2, \ldots, v_n) \in C$, then $\sigma(\pi_1(v_1), \pi_2(v_2), \ldots, \pi_n(v_n)) \in C'$. If both codes are linear, then the π_i are restricted to compositions of scalar multiples and field automorphisms, and if the π_i all equal the identity, C and C' are *permutation equivalent*. Equivalent codes have the same parameters and weight enumerators.

When all codewords $\mathbf{c}$ in a code C are equally likely to be transmitted, the probability of decoder error (decoding a received word as a codeword different from the transmitted codeword) is minimised under a *maximum likelihood decoding* scheme. In such a scheme, a received word $\mathbf{r}$ is decoded as one of the codewords which was most likely to have been sent, that is, as a codeword $\mathbf{c} = \mathbf{c}_i$ which maximises the conditional probability $p(\mathbf{r}|\mathbf{c})$ over all $\mathbf{c} \in C$. For the q-ary symmetric memoryless channel, this is a codeword c_i for which $d(r, c_i)$ is a minimum, that is, for which the error word $e = r - c_i$ has minimum weight.

If, in a code with distance d, an error pattern converts one codeword to another codeword, the error will be undetectable, so at least d symbols must have been altered in transmission. Any fewer errors will be detected. If a received word is equidistant from two codewords, it cannot be decoded unambiguously.

LEMMA 3.11 *An (n, M, d)-code can simultaneously detect t_d and correct t_c errors, where $t_d + 2t_c \leq d - 1$, and can therefore detect any error pattern up to $t = d - 1$ errors or correct any error pattern up to $t = \lfloor (d-1)/2 \rfloor$ errors.*

In designing good codes, the twin goals of optimising parameters M and n are fundamentally at odds with the goal of maximising error correction capacity by maximising d. The resulting optimisation problems are often expressed through bounds on one parameter in terms of the others. In particular, the parameters n and d are fixed and the optimisation problem is to bound M both above and below. Given n and d, the maximum number of codewords in any q-ary code of length n and distance d is denoted $A_q(n, d)$, and the *Singleton bound* is

$$A_q(n, d) \leq q^{n-d+1}. \tag{3.19}$$

Linear codes meeting the Singleton bound are called *maximum distance separable* (MDS) and are relatively rare (see [263, Theorem 1.3.8, Corollary 1.10.15, Section 4.2.2]). One popular MDS family, the Reed-Solomon codes, is known.

If, when fixing n and d, we require high distance relative to length, say $d \geq (q-1)n/q$, the upper bound on M is *much* lower than the Singleton bound.

LEMMA 3.12 *(Plotkin bound) Suppose $d \geq (q-1)n/q$.*

1. *If $n < qd/(q-1)$, then $A_q(n, d) \leq qd/[qd - (q-1)n]$.*

2. *If $n = qd/(q-1)$, then $A_q(n, d) \leq qn$.*

Proof. For part 1, count distances between pairs of codewords in two ways (see [263, Section 4.2.3]). For part 2, combine the result $A_q(n, d) \leq qA_q(n-1, d)$ [263, Section 4.2.2], with part 1 applied to length $n - 1$. □

No further elements of coding theory will be presented here. The literature on error-correcting codes is extensive and the interested reader is referred to the 'bible' MacWilliams and Sloane [237] and the more recent — and encyclopaedic — *Handbook of Coding Theory* [263].

3.2.2 Hadamard codes

The first practical binary codes were Hamming's optimal single-error-correcting codes, discovered in the late 1940s. Within a few years, Golay had discovered the perfect $[23, 12, 7]$ binary triple error-correcting code $\mathcal{G}_{23}$. Addition of a parity-check bit to every codeword of $\mathcal{G}_{23}$ determines $\mathcal{G}_{24}$, the $[24, 12, 8]$ *extended Golay code*. Together with the two ternary Golay codes — the perfect $[11, 6, 5]$ ternary

Golay code $\mathcal{G}_{11}$ and the $[12, 6, 6]$ extended ternary Golay code $\mathcal{G}_{12}$ — they have such remarkable practical and theoretical properties that they are regarded by many as the most important and elegant of all codes (cf. [237, 321]).

Discovery of the Reed-Muller codes followed in 1954. These are more flexible than the Hamming and Golay codes because of their capacity to correct varying numbers of errors per codeword. Until 1977 a coset of the first-order Reed-Muller code of length 32 provided error control on all of the Mariner deep-space missions flown by the USA, and was the main downlink code returning digital images of the surface of Mars ([263, p. 2126]; for a picture, see [237, Fig. 14.7]). Golay codes were similarly applied in the Voyager missions, returning clear colour pictures of Jupiter and Saturn until 1981 (see [263, Ch. 25]).

The Reed-Muller codes lost their prominence in the space program with the adoption of convolutional codes and sequential decoders. They do not perform as well as long BCH and Reed-Solomon codes, but they have the benefit of an extremely fast maximum likelihood decoding algorithm. Wicker [321] points out that there is renewed interest in using Reed-Muller codes in optical communications because of this.

As we will shortly see, both the binary Golay codes and the first-order Reed-Muller codes can be derived from Hadamard matrices. First, however, we introduce the three codes derived from any Hadamard matrix which are usually referred to as Hadamard codes [237]. We will subsequently (Definition 3.15) call them *Class A Hadamard codes*.

DEFINITION 3.13 *Let H be a normalised Hadamard matrix of order n. The '$\log_{(-1)}$' mapping $1 = (-1)^0 \mapsto 0, -1 = (-1)^1 \mapsto 1$ applied to H defines the* binary normalised Hadamard matrix A_n. *The* binary Hadamard codes *are*

1. $\mathcal{A}_n$ — *the $(n - 1, n, n/2)$ code consisting of the rows of A_n with the first column (of 0s) deleted;*

2. $\mathcal{B}_n$ — *the $(n - 1, 2n, n/2 - 1)$ code consisting of A_n and the complements of its codewords; and*

3. $\mathcal{C}_n$ — *the $(n, 2n, n/2)$ code consisting of the rows of A_n and their complements.*

Clearly $\mathcal{A}_n$ meets the binary Plotkin bound of Lemma 3.12.1 and $\mathcal{C}_n$ meets the binary Plotkin bound of Lemma 3.12.2. In fact, Levenshtein showed that Hadamard codes $\mathcal{A}_m$ can be suitably concatenated with shortened codes $\mathcal{A}'_n$ (found by taking a cross-section of $\mathcal{A}_n$ [237, 1.9, Example VI]) to provide binary codes of any length, meeting the Plotkin bound. For a proof, see [237, 2.3, Theorem 8].

THEOREM 3.14 *(Levenshtein) If the Hadamard Conjecture holds, then for any n and for any $d \geq n/2$, there exists an $(n, A_2(n, d), d)$ code meeting the binary Plotkin bound.*

If the Sylvester Hadamard matrix $\mathcal{S}_n$ is used in Definition 3.13, the resulting Hadamard codes are all linear (by Lemma 3.6.2). In fact, $\mathcal{A}_{2^n}$ is the *simplex* code of dimension n, $\mathcal{B}_{2^n}$ is the punctured first-order Reed-Muller code of dimension $n+1$ and, as we will see, $\mathcal{C}_{2^n}$ is the first-order Reed-Muller code $\mathcal{R}(1, n)$ of dimension $n + 1$.

However, the Hadamard codes resulting from other constructions for Hadamard matrices are usually nonlinear, because the binary rank of A_n (equally, of $\mathcal{A}_n$) is $< \log_2 n$.

The binary normalised Hadamard matrix A_{2^m} derived from a square $(2^m - 1, 2^{m-1} - 1, 2^{m-2} - 1)$-design by (2.9) has binary rank at least m. (By [263, Theorem 8.6 proof], the span of the rows of the incidence matrix B of such a Hadamard design contains $\mathbf{1}$, and its binary rank is bounded below by $m + 1$. If $\mathrm{rank}_2 B = m + 1$, $\mathrm{rank}_2(J - B) = m$. Under the '$\log_{(-1)}$' mapping, the core B' of the normalised Hadamard matrix (2.9) derived from B maps to $J - B$.) This minimum binary rank m is achieved by the binary normalised Hadamard matrix A_{2^m} derived from a Sylvester Hadamard matrix $\mathcal{S}_m$ and by that derived from the Singer $(2m - 1, 2^{m-1} - 1, 2^{m-2} - 1)$-difference set, by Lemma 2.14.

The binary normalised Hadamard matrix A_{p+1} derived from a Paley Type I Hadamard matrix P_p, where p is prime, has binary rank $(p - 1)/2$.

More generally, by a result due to Klemm [14, Theorem 2.4.2], if 2 divides n, the binary normalised Hadamard matrix A_{4n} derived from a square $(4n - 1, 2n - 1, n - 1)$-design by (2.9) has binary rank at most $2n$; moreover, if 4 does not divide n, then $\mathrm{rank}_2(A_{4n}) = 2n$.

Otherwise, there is still very little known about the binary rank of A_n.

Research Problem 9 *Let* $n \geq 12$. *For each of the other families of Hadamard matrices of order n given in Chapter 2, determine the rank over $GF(2)$ of the corresponding binary Hadamard matrix A_n.*

If a Hadamard code is nonlinear, there are at least two ways to linearise it, either by taking the linear span of the codewords, or by forming the code with generator matrix $[I_n\ A_n]$. In the latter case, the dimension of the code is increased to n, but this is achieved at the expense of doubling the code length to $2n$.

To distinguish between the nature of the codes constructed from Hadamard matrices, we introduce here a simple extension of the definition of Hadamard code, classified by construction technique. This classification will be enlarged to include q-ary codes in Chapter 4.4.4.

DEFINITION 3.15 *A Class A binary Hadamard code* is a Hadamard code as given in Definition 3.13.

A Class B binary Hadamard code is a linear code not in Class A, derived from some rows of a Hadamard matrix H.

A Class C binary Hadamard code is a linear code with generator matrix $[I\ A]$ for some binary matrix A associated with H.

Binary Hadamard codes can themselves be used to construct quantum error-correcting codes, which encode states in 2^n-dimensional complex Hilbert space. Quantum codes will not be covered here (see [24, Chapter XIII.5.7, pp. 918–919]).

Each class of Hadamard codes contains very well-known binary codes.

3.2.2.1 Class A Hadamard codes

The first-order Reed-Muller codes $\mathcal{R}(1,n)$ are examples of Class A binary Hadamard codes. They are defined as follows from the linear and affine Boolean functions $f : V(n,2) \to GF(2)$.

DEFINITION 3.16 *For each* $f : V(n,2) \to GF(2)$ *let* **f** *denote its* truth table *(length 2^n string of values $f(v)$, $v = (v_n, v_{n-1}, \ldots, v_1) \in V(n,2)$ in lexicographical order). For $i = 1, \ldots, n$, let $\mathbf{v}_i$ be the truth table of the Boolean function which projects coordinate v_i of $V(n,2)$ and let $\mathbf{1}$ be the truth table of the Boolean function with constant value 1. The* first-order Reed-Muller code $\mathcal{R}(1,n)$ *of length 2^n is the set of 2^{n+1} codewords*

$$\mathcal{R}(1,n) = \{a_0\mathbf{1} + a_1\mathbf{v}_1 + \cdots + a_n\mathbf{v}_n, \ a_i \in GF(2)\}. \tag{3.20}$$

A generator matrix for $\mathcal{R}(1,n)$ consists of the rows $\{\mathbf{1}, \mathbf{v}_i, 1 \le i \le n\}$. Now, any *linear* Boolean function is of the form $L_a(v) = \langle a, v \rangle$ for a fixed $a \in V(n,2)$, since we can write it as $L_a(v_n, v_{n-1}, \ldots, v_1) = \sum_{i=1}^{n} a_i v_i$, where v_i is regarded as the Boolean function with truth table $\mathbf{v}_i$. In other words, the linear and affine Boolean functions together are the codewords of $\mathcal{R}(1,n)$ and by (3.9) and Definition 3.13, this is the code $\mathcal{C}_{2^n}$ derived from the Sylvester Hadamard matrix $\mathcal{S}_n$.

A decoder for a first-order Reed-Muller code uses the FHT (3.17) to compute the correlation between the (± 1) versions of a received vector **r** and each of the codewords $\mathbf{c} = a_1\mathbf{v}_1 + \cdots + a_n\mathbf{v}_n$. This is because if $d(\mathbf{r}, \mathbf{c})$ is a minimum then the correlation has maximum magnitude. As described in Section 3.5.1 below, the coordinate **c** at which the transform vector has maximum absolute value is located. If the transform coefficient there is positive, **r** is decoded as the nearest codeword **c**, and if it is negative, as its complement $\mathbf{1} + \mathbf{c}$. For more details see [237, Chaps. 13, 14] or [263, Chaps. 1.13, 25.3].

3.2.2.2 Class B Hadamard codes

The Paley Type I Hadamard matrix P_q defines only nonlinear Class A Hadamard codes for $q > 8$, so the linear codes generated by the rows of these Class A codes are Class B.

Example 3.2.1 *The binary* quadratic residue (QR) codes $\mathcal{Q}$ *are binary Class B Hadamard codes. If p is prime, $p \equiv -1 \bmod 8$ and 2 is a quadratic residue mod p, the corresponding QR code is a cyclic linear code with parameters $[p, (p+1)/2, d \ge \sqrt{p}]$. A generator matrix is given by the $p \times p$ binary circulant matrix with top row having 0 exactly in the $\bmod p$ quadratic residue indices, and with the all-1s row $\mathbf{1}$ appended [237, Equation (23), p. 488]. The circulant matrix is the matrix $\mathcal{A}_{p+1}$ obtained from the Paley Type I Hadamard matrix P_p, but with its all-0s top row removed.*

For instance the perfect binary $[23, 12, 7]$ Golay code $\mathcal{G}_{23}$ is Class B. A generator matrix is given by the 23×23 binary circulant matrix with top row equal to the second row of $\mathcal{A}_{24}$

$$10000101001100110101111$$

and with 1 appended.

Research Problem 10 *Determine the dimension and distance of the Class B binary Hadamard codes generated by the rows of the Class A Hadamard codes defined by the Paley Type I Hadamard matrices P_{p^m}, where $p^m \equiv 3 \bmod 4$, $m \geq 2$ and p is an odd prime.*

3.2.2.3 Class C Hadamard codes

Class C binary Hadamard codes are often self-orthogonal or self-dual, an important characteristic of many of the best codes known, so this class is a good source of extremal self-dual codes, and has been investigated by several authors, notably Tonchev, Harada and Kimura (see [263, Chapter 15.7]) and Rao (= Baliga) [16, 17].

Example 3.2.2   *The binary extended Golay code $\mathcal{G}_{24}$ is an example of a Class C Hadamard code, since it has a generator matrix of the form $[I_{12}\ A_{12}]$, where A_{12} is a binary 12×12 Hadamard matrix. Specifically, if we negate all rows of the Paley Hadamard matrix P_{11} shown in (2.5) except the first and rotate the first column to the right-hand end, the binary version of this matrix is an A_{12} (see [237, p. 500]).*

Alternatively, by invoking the self-duality of $\mathcal{G}_{24}$ it is also possible to obtain the equivalent generator matrix of [237, Fig. 2.13]. A third construction of $\mathcal{G}_{24}$ of this kind appears in [311] (cited in [321, Fig. 6-2]). If we negate all rows of P_{11} except the first, and swap columns 3–7 with columns 12–8 (that is, mirror about the vertical axis), we obtain the equivalent symmetric Hadamard matrix with back-circulant core

$$
\left[
\begin{array}{c|ccccccccccc}
1 & -1 & -1 & -1 & -1 & -1 & -1 & -1 & -1 & -1 & -1 & -1 \\
\hline
-1 & -1 & -1 & 1 & -1 & -1 & -1 & 1 & 1 & 1 & -1 & 1 \\
-1 & -1 & 1 & -1 & -1 & -1 & 1 & 1 & 1 & -1 & 1 & -1 \\
-1 & 1 & -1 & -1 & -1 & 1 & 1 & 1 & -1 & 1 & -1 & -1 \\
-1 & -1 & -1 & -1 & 1 & 1 & 1 & -1 & 1 & -1 & -1 & 1 \\
-1 & -1 & -1 & 1 & 1 & 1 & -1 & 1 & -1 & -1 & 1 & -1 \\
-1 & -1 & 1 & 1 & 1 & -1 & 1 & -1 & -1 & 1 & -1 & -1 \\
-1 & 1 & 1 & 1 & -1 & 1 & -1 & -1 & 1 & -1 & -1 & -1 \\
-1 & 1 & 1 & -1 & 1 & -1 & -1 & 1 & -1 & -1 & -1 & 1 \\
-1 & 1 & -1 & 1 & -1 & -1 & 1 & -1 & -1 & -1 & 1 & 1 \\
-1 & -1 & 1 & -1 & -1 & 1 & -1 & -1 & -1 & 1 & 1 & 1 \\
-1 & 1 & -1 & -1 & 1 & -1 & -1 & -1 & 1 & 1 & 1 & -1 \\
\end{array}
\right], \quad (3.21)
$$

of which the binary version is an A_{12}.

3.3 SIGNAL MODULATION AND SEPARATION: HADAMARD CODES

This application of Hadamard matrices, by contrast with the error-correcting codes of the previous section, is one which operates on a population scale. Everyone will know someone who is using it, or will — perhaps unwittingly — be using it themselves.

Often in transmission over a communications channel, such as a microwave link or optical fibre, a high-frequency *carrier* waveform is modulated by a lower frequency process such as phase shifting. The phase shifting may represent data symbols and the purpose of the transmission be the transfer of information to a detector. In *code division multiple access (CDMA)* communications the purpose of modulation may be twofold, to separate signals for multiple users and to carry information for each user, with different processes used for each purpose.

The receiver compares a locally generated ideal model of the carrier signal with an incoming signal in order to extract the transmitted information. More generally, when a receiver is trying to decide which one of a set of signals was sent by a particular user, over a noisy channel, it may compare the received signal with locally generated ideal models of each of the possible transmitted signals.

For channels subject to Gaussian noise, the optimal decision process is to calculate the *correlation* between reference and incoming signals and decide that the signal giving the highest value of the correlation corresponds to the signal that was actually sent. See Golomb [127] for an excellent introduction to these ideas.

DEFINITION 3.17 *Let V be a vector space equipped with an inner product $\langle,\rangle$. The* (normalised) correlation *of nonzero vectors $\mathbf{x}$ and $\mathbf{y}$ in V is*

$$C(\mathbf{x}, \mathbf{y}) = \langle \mathbf{x}, \mathbf{y}\rangle/(|\mathbf{x}||\mathbf{y}|) = \langle \mathbf{x}/|\mathbf{x}|, \mathbf{y}/|\mathbf{y}|\rangle.$$

If $C(\mathbf{x}, \mathbf{y}) = \langle \mathbf{x}, \mathbf{y}\rangle = 0$ the vectors $\mathbf{x}$ and $\mathbf{y}$ are orthogonal.

A set $\{\mathbf{x}_i, 1 \leq i \leq n\}$ of n signals, each of length $n - 1$ with real-valued components, is *maximally uncorrelated* if and only if (cf. [127, 1.6])

$$C(\mathbf{x}_i, \mathbf{x}_j) = -1/(n-1), \ 1 \leq i \neq j \leq n.$$

Maximally uncorrelated signal sets are called *simplex codes*.

To maximise the demodulation rate, it is necessary to maintain a fixed ratio between the the frequency with which phase shifts occur (the *chip rate*) and the frequency of the carrier, with phase changes allowed every m, say, periods of the carrier and at no other time. In this *coherent* case, when a received signal of one chip duration is compared with another, maximum distinguishability occurs between signals which are out of phase by π radians, and the simplex codes form the optimal signal sets.

If the modulation is noncoherent, it cannot be assumed that a negative correlation between signals means they are distinguishable, and it is common for the receiver to perform an 'envelope detection' of the modulation pattern and discard the correlation information contained in the carrier (see Figure 3.4). In the noncoherent case, maximum distinguishability occurs when two signals are orthogonal, that is, their correlation is 0. A set of n pairwise orthogonal signals is called an *orthogonal code*. The signal set consisting of an orthogonal set and its negatives is called a *biorthogonal code* and is only useful for coherent detection.

Example 3.3.1 *Let H be a Hadamard matrix of order n. After multiplication by $\frac{1}{\sqrt{n}}$, its rows form an orthogonal signal set of size n taking 'binary' component values $\pm\frac{1}{\sqrt{n}}$. The Class A binary Hadamard codes $\mathcal{A}_n$ and $\mathcal{C}_n$ of Definition 3.13*

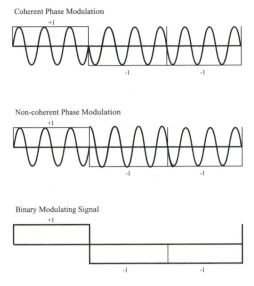

Figure 3.4 Modulation of carrier signal

similarly correspond to simplex and biorthogonal signal sets of sizes n and $2n$, respectively.

3.3.1 CDMA for mobile, wireless and optical communications

When a wireless digital communications system has many users, their signals have been kept separate by multiplexing, traditionally by placing them either in different frequency bands (FDMA) or in different transmission time slots (TDMA). In current implementations TDMA allows 3 to 8 users per frequency band.

But bandwidth is a finite resource, and the enormous global enthusiasm for mobile telephony (cellphones) over the past decade has promoted CDMA as an alternative system providing higher quality voice transmission. Mobile phones typically have very low power, both for reasons of size and economy and to meet community standards on electromagnetic emissions. Weak, unsynchronised, noncoherent signals from many mobile phones which arrive simultaneously at the local base station must be distinguishable from each other. Equally, the base station must be able to distinguish between the phones to which it is transmitting.

By using CDMA, bandwidth can be conserved. Multiple *users* — transmitter-receiver pairs — can communicate simultaneously over the same frequency band, with each of the M users assigned a distinct periodic *signature (or code) sequence* $x_i(1), \ldots, x_i(n)$, $i = 1, \ldots, M$. When user i transmits the signal for symbol a, it is modulated by one period of the signature sequence, and what is sent are the n signals corresponding to $ax_i(1), \ldots, ax_i(n)$. Figure 3.5 depicts transmission of the same symbols using different signatures.

The optimal detector in a CDMA system is a multi-user detector [312]. A cor-

base transmitting station

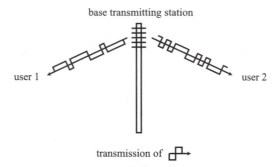

user 1   user 2

transmission of

Figure 3.5 Base station transmission of same symbols to different users

relation detector can make suboptimal decisions on the data of all users (treating signals of other users as interference) which both minimise interference between signals and time self-synchronise individual signals, by choosing signals with low 'odd-correlation' values. Designing signal sets with this property is hard, however, and it is usual instead to try to find signal sets with minimum periodic *correlation* and then test their suitability. See [148, Section 2] for details of this analysis.

Wireless CDMA was introduced commercially in 1995, and is one of the world's fastest-growing wireless technologies, with a consumer base in the hundreds of millions. In 1999, broadband CDMA was selected for integration into the 'third-generation' (3G) wireless standard IMT-2000.

Hadamard codes provide families of zero-correlation signature and transmission rate control sequences for CDMA. For example, the QUALCOMM® cdma2000 high rate packet data system uses the rows of a Walsh Hadamard matrix (3.7) of order 64 (called a *Walsh cover*) to identify each user on the downlink from base station to mobile phone or wireless access point [318, Chapter 4]. On the uplink, the mobile phone uses both a biorthogonal Hadamard code of length 8 and a Walsh Hadamard matrix of order 8 to select channels for forward transmission by the base station.

Paralleling the growth in wireless communications has been the increasing use of optical fibre for high bit-rate communication links, especially for local area networks (LANs). Optical code division multiple access (OCDMA) systems can give each user asynchronous access to the network, without strict wavelength controls, and with graceful degradation in performance as the number of users increases.

Early OCDMA schemes did not code the phase of the signal, so new code classes were developed, including prime codes and optical orthogonal codes (OOC). (An OOC is equivalent to a 'strictly cyclic' $t - (v, k, 1)$ partial design (see, for example, [55]), so they intersect with the Hadamard designs only trivially.) However, these codes generally have much poorer correlation properties than their radio domain counterparts [298], and in recent years spectral amplitude coding has attracted more interest.

In spectral amplitude coding schemes, infrared light of n distinct wavelengths $\lambda_1, \ldots, \lambda_n$ can be reflected down the fibre. To modulate an information bit 1, the 1s in user i's signature sequence $x_i(1), \ldots, x_i(n)$ select which of the wavelengths

$\lambda_1, \ldots, \lambda_n$ will be reflected. To modulate bit 0, the complementary signature sequence selects wavelengths. Hadamard codes using $\mathcal{S}_n$ are shown to perform well as signature sequences in this domain too [171, 298].

3.3.2 3-D holographic memory for data storage and retrieval

The rapidly expanding need for mass memory systems to archive and backup data has created fascinating new applications of Hadamard codes to 3-D (3-dimensional) optical memory. Here the signal is transmitted by *beams* — waveforms in the light frequencies.

In 3-D optical memories, data are arranged in 2-D pages. A whole page can be written or read in a single access operation, so parallel access is possible. The interference pattern between an object beam carrying a 2-D data page and a coherent reference beam is distributed throughout the holographic medium, with multiple pages stored in the same volume by multiplexing.

The focus of this page storage application has been to store thousands of high-resolution images in one voxel (3-D volume element) of holographic material to enable rapid parallel access to 2-D information — for example, image retrieval in real time. Although the possibility of writing dynamic holograms in photorefractive materials was first proposed over 25 years ago, a key constraint to development has been the storage capacity of the holographic material. In photoreactive materials, the local refractive index of the medium is changed by a spatial variation of the incident light intensity. Advances in growth and preparation of photoreactive materials such as iron-doped lithium niobate crystals, and in optical device technology such as spatial light modulators (SLMs) and detector arrays, have made realisation of this idea feasible.

There are several multiplexing techniques in use, but *phase-code multiplexing*, first proposed in the early 1990s, has several advantages over the others, such as simplicity, compactness, high light efficiency, fast access and fixed wavelength.

In phase-code multiplexing, each reference beam consists of a set of n plane wavefronts with a unique phase distribution across its component waves. The phase-code of the reference beam represents the address of the stored data page. To store m pages of data, m phase-codes are used to encode the reference beam. Each data page is retrieved by illuminating the holographic medium with a beam coded for that page. Partial retrieval (with weaker intensity) of pages with different phase-codes is referred to as *cross-talk*. As in the microwave signal case, to avoid cross-talk, the phase-codes must be orthogonal so that the correlation between different reference beams is zero. The maximum number of images that can be stored and reconstructed without, theoretically, any cross-talk is $m = n$.

This work is still in the development stage and is very nicely explained in the survey by X. Y. Yang and Jutamulia [330]; see also [320]. Hadamard matrices of order n have been the sole source of orthogonal phase-codes for this purpose.

At present, storage density close to the theoretical bound has been achieved. Performance is primarily limited by the constrained number of pixels n of currently available SLMs, which is typically not a power of 2. Using a Sylvester Hadamard matrix $\mathcal{S}_6$ is inefficient when the SLM has 100 pixels, as only 64 data pages can be stored. Similarly, using only 100 phase-codes (rows) of $\mathcal{S}_7$ is inefficient, even though the cross-talk noise-to-signal ratio is much reduced, as the reference beams must contain 28 more plane wave components than necessary [320, p. 15].

For these reasons, experimental implementations [320], [330] use Williamson Hadamard matrices with symmetric circulant components as phase-codes, rather than the ubiquitous Sylvester Hadamard matrices. The orders used are 36 and 100, and the component matrices A, B, C and D (see Lemma 2.10) in the Williamson Hadamard matrices have top rows which are symmetric apart from the first element.

A second novel application of phase-code multiplexing for holographic memory is to store beam-steering information for optical (laser) scanners [275]. This application generally needs much lower holographic storage capacity, essentially because it reverses the roles of reference and signal beams, requiring the holographic recording of a set of n 3-D scan reference beams (spherical wavefronts) with n signal beams which are Hadamard-coded. Each 3-D scan beam is recalled by imposing its signature code on the input laser beam. The proof-of-concept experiment in [275] uses $\mathcal{S}_4$ (rather than $\mathcal{S}_3$, to reduce cross-talk) to code 8 signal beams representing the basic voxel of a 3-D scan.

3.4 SIGNAL CORRELATION: PERFECT SEQUENCES AND ARRAYS

Instead of separating signals, the intention of phase shifting may be to permit accurate timing or synchronisation of signals.

When the intention of a transmission is to determine a time interval, or a point in time, very accurately, such as in radar or sonar or in signal synchronisation, a signal sequence $x(1), \ldots, x(n)$ is compared with time-shifts $x(1+t), \ldots, x(n+t)$ of itself or of another signal sequence $y(1+t), \ldots, y(n+t)$.

The ratio of the *in-phase* or *on-peak* correlation, when $t = 0$ and signals are synchronised, and the maximum value of the *out-of-phase* or *off-peak* correlation, when $t \neq 0$, is a measure of the accuracy of this calculation in a noisy environment [127, 3.2].

DEFINITION 3.18 *Let* $\mathbf{x} = x(1), \ldots, x(n)$ *and* $\mathbf{y} = y(1), \ldots, y(n)$ *be complex-valued sequences. The unnormalised aperiodic (cross)correlation function of* $\mathbf{x}$ *and* $\mathbf{y}$ *is*

$$N_{\mathbf{x},\mathbf{y}}(t) = \sum_{i=1}^{n-t} x(i)\overline{y(i+t)}$$

and if $\mathbf{x}$ *and* $\mathbf{y}$ *have period* n, *the unnormalised periodic (cross)correlation function of* $\mathbf{x}$ *and* $\mathbf{y}$ *is*

$$C_{\mathbf{x},\mathbf{y}}(t) = \sum_{i=1}^{n} x(i)\overline{y(i+t)}.$$

If $\mathbf{x} = \mathbf{y}$, *the terms aperiodic (or* finite*) autocorrelation* $N_{\mathbf{x}}(t)$ *and periodic autocorrelation* $C_{\mathbf{x}}(t)$ *are used, respectively.*

Note that $\sum_{i=1+(n-t)}^{n} x(i)\overline{y(i+t)} = \sum_{j=1}^{t} x(j+(n-t))\overline{y(j+n)}$, so for periodic sequences, $N_{\mathbf{x},\mathbf{y}}(t) + N_{\overline{\mathbf{y}},\overline{\mathbf{x}}}(n-t) = C_{\mathbf{x},\mathbf{y}}(t)$ and $N_{\mathbf{x}}(t) + N_{\overline{\mathbf{x}}}(n-t) = C_{\mathbf{x}}(t)$. It is common to focus on periodic rather than aperiodic correlation values in the periodic case.

Different applications have imposed distinct criteria on what constitutes a sufficiently 'low' value for one of these correlation functions.

3.4.1 Timing and synchronisation: Perfect binary sequences

In the aperiodic case, Barker [20] looked for sequences of ± 1 with ideal aperiodic autocorrelation. He asked: for which lengths n do sequences $\mathbf{x}$ consisting of ± 1 exist, with $N_{\mathbf{x}}(t) \in \{-1, 0, 1\}$ for each $1 \leq t \leq n-1$? Such *Barker sequences* exist for $n = 1, 2, 3, 4, 5, 7, 11, 13$ and for no other odd n. If a Barker sequence of even length n exists, then so does a circulant Menon Hadamard matrix of order n (see Definition 2.20). Schmidt, combining his spectacular results (Theorem 2.21) on nonexistence of circulant Hadamard matrices with the fact that n cannot have a prime divisor $p \equiv 3 \bmod 4$, gets the following result (by computer search).

THEOREM 3.19 [281, Theorem 6.4] *There is no Barker sequence of length n for* $13 < n \leq 4 \cdot 10^{12}$.

Research Problem 11 *The Barker Sequence Conjecture. Show that there is no Barker sequence of length > 13.*

Whilst this appears to be the end of the story for ideal binary aperiodic autocorrelation, in the search for binary periodic sequences with low correlation the outlook is not quite so bleak. A fine coverage of this area (to 1998) by Helleseth and Kumar appears in [148].

It is easy to show [193, Corollary 1.2], by comparison with the corresponding $(0, 1)$ sequence, that a periodic (± 1) sequence $\mathbf{x}$ of length v has $C_{\mathbf{x}}(t) \equiv v \bmod 4$ for all t.

In the special case that the off-peak autocorrelation function takes only one value γ, the sequence is called 2-*level*. The best possible value is $\gamma = 0$, in which case the sequence has *ideal autocorrelation* and is known as a *perfect binary sequence*. However, the near-ideal nonzero values $\gamma = \pm 1, \pm 2$ are also possible, and 2-level sequences of period v achieving any of these 5 values of γ are all now called [193] perfect binary sequences.

DEFINITION 3.20 [193] *A periodic 2-level (± 1) sequence with constant off-peak periodic autocorrelation γ equal to one of the five values $0, \pm 1, \pm 2$ is called a perfect binary sequence.*

A comprehensive survey (to 1999) of perfect sequences and almost perfect sequences (that is, sequences which are 2-level apart from exactly one exceptional autocorrelation value), appears in [193]. Perfect binary sequences are equivalent to cyclic difference sets (for a proof see [193, Lemma 1.3]).

LEMMA 3.21 *A perfect binary sequence of period* v, *with* k *entries* $+1$ *per period and 2-level autocorrelation function with off-peak value* γ, *is equivalent to a* (v, k, λ)-*difference set in a cyclic group, where* $\gamma = v - 4(k - \lambda)$.

The ideal case $\gamma = 0$ corresponds to a cyclic Menon-Hadamard difference set (cf. Theorem 2.21). We are in the familiar situation of the circulant Hadamard conjecture and can expect no nontrivial ideal sequences to exist. There is also only a trivial example in the $\gamma = -2$ case, and only one known example in each of cases $\gamma = 1, 2$.

The case $\gamma = -1$ corresponds to the cyclic Hadamard $(4n - 1, 2n - 1, n - 1)$-difference sets. As noted in Section 3.1.4, all known examples belong to one of the three parameter families of Example 2.1.1, where $v = 4n - 1$ is either a prime p, a 'twin prime' $p(p + 2)$ or $v = 2^m - 1$.

If we are willing to accept less than optimal correlation, or if a signal sequence is not restricted to a binary alphabet, or if we allow less restrictive ideas of correlation, such as dyadic correlation (Lemma 3.6.5) or signal array correlation, we can design signal sets with very good correlation performance. We will discuss this topic in Section 3.4.2 below and again in Chapter 4 and Chapter 7.

3.4.2 Signal array correlation: Perfect binary arrays

Some signals are naturally modelled not as a sequence of values but as an array of values in 2 or 3, or even more, dimensions. Multi-element radiating and receiving systems, such as antenna arrays, aperture synthesis systems and optical, X-ray and gamma-ray telescopes, are widely employed in radio science and astronomy.

Analysis and synthesis, synchronisation and error correction coding techniques for signal arrays should take this dimensionality into account, rather than recoding the array of signal values as a signal sequence. Examples are coded aperture imaging, optical image alignment and image coding [186].

One early application of perfect binary *sequences* was to the design of coded apertures used in X-ray astrophysical telescopes, for X-ray imaging. Because X-ray sources are so weak, multi-element receivers are used to improve sensitivity. The plane of the receiver passes radiation through a number of points. Originally the points were chosen randomly, but properly designed coded masks or apertures, with the points chosen on a rectangular grid, increase the signal-to-noise ratio and reduce sidelobe interference (cross-talk) dramatically. The first suitable designs, called *uniformly redundant arrays*, were proposed by Fenimore and Cannon in 1978, and masks of this type are used in both X-ray and gamma ray telescopes [208]. The uniformly redundant arrays they proposed are essentially the perfect binary sequences with $\gamma = -1$ given by the twin prime difference sets (2.11).

The (negation of the) perfect binary sequence defined from a twin prime cyclic Hadamard difference set is naturally a subset of $\mathbb{Z}_p \times \mathbb{Z}_{p+2} \cong \mathbb{Z}_{p(p+2)}$, and so may be written as a 2-D periodic (± 1) array $B = [b(i, j)]_{0 \le i \le p-1, 0 \le j \le p+1}$. The *array autocorrelation* of B, defined by the correlation function

$$C(s, t) = \sum_{i=0}^{p-1} \sum_{j=0}^{p+1} b(i, j) b(i + s, j + t),$$

satisfies $C(0,0) = p(p+2)$ and $C(s,t) = -1$ for all other (s,t). The mask or code for the receiver is the $(0,1)$ array $A = (B + J)/2$ obtained from B by replacing -1 by 0. The aperture is in the form of a $p(p+2)$ grid which is transparent at coordinates coded 1 and opaque at coordinates coded 0. A detailed analysis appears in [193], and a general survey of applications of difference sets in multi-element systems in [208].

Example 3.4.1 *An 11×13 uniformly redundant array for an X-ray telescope mask is given, using (2.11), by the grid with pixel (grid element) $(0,0)$ in the bottom left-hand corner, where $\circ$ denotes a transparent pixel:*

○		○			○	○	○	○			○	
○	○		○	○					○	○		○
○		○		○	○	○	○			○		
○		○		○	○	○	○			○		
○		○		○	○	○	○			○		
○	○		○	○					○	○		○
○	○		○	○					○	○		○
○	○		○	○					○	○		○
○		○		○	○	○	○			○		
○	○		○	○					○	○		○
○												

This corresponds to a $(143, 71, 35)$ cyclic difference set.

Obviously, the 'twin prime' array B, while extremely useful, is not optimal as a 2-D array with respect to periodic array autocorrelation. Optimal array autocorrelation would have $C(0,0) = p(p + 2)$ and $C(s,t) = 0$ for all other (s,t). An m-dimensional (± 1) array with ideal array autocorrelation is called *perfect*, and, in stark contrast to the paucity of ideally correlated binary sequences, they are not hard to find.

DEFINITION 3.22 *An m-dimensional array $A = [a(i_0,\dots,i_{m-1})]$ with entries $a(i_0,\dots,i_{m-1}) = \pm 1$ for $0 \le i_k \le s_k - 1, 0 \le k \le m - 1$ is called an $s_0 \times s_1 \times \cdots \times s_{m-1}$ perfect binary array (PBA) and denoted by $PBA(s_0,\dots,s_{m-1})$ if, for all $j_0,\dots,j_{m-1}$,*

$$\sum_{i_0=0}^{s_0-1} \cdots \sum_{i_{m-1}=0}^{s_{m-1}-1} a(i_0,\dots,i_{m-1})\, a(i_0 + j_0,\dots,i_{m-1} + j_{m-1})$$

$$= \begin{cases} \prod_{i=0}^{m-1} s_i, & j_0 = j_1 = \dots = j_{m-1} = 0, \\ 0 & \text{otherwise,} \end{cases} \tag{3.22}$$

where the index $i_k + j_k$ is reduced $\bmod$ s_k. *(We assume that $\exists\, i : s_i \ne 1$.)*

As with signal sequences (cf. Lemma 3.6), the *energy* of the $PBA(s_0,\dots,s_{m-1})$ is the sum of the squares of the signal values, that is, its volume $s = \prod_{i=0}^{m-1} s_i$. Furthermore, it must equal $4u^2$ for some $u \ge 1$, since it is well known that Menon-Hadamard difference sets over abelian groups and nontrivial PBAs are equivalent.

Theorem 3.23 [186, Theorem 3.1] *Set* $G = \mathbb{Z}_{s_0} \times \cdots \times \mathbb{Z}_{s_{m-1}}$. *An m-dimensional* (± 1) *array* $A = [a(i_0, \ldots, i_{m-1})]$ *is a PBA*$(s_0, \ldots, s_{m-1})$ *if and only if* $D = \{g \in G : a(g) = -1\}$ *is a Menon-Hadamard difference set in G.*

Proof. Set $s = \prod_{i=0}^{m-1} s_i$. For each $g \in G$, let $C_A(g) = \sum_{h \in G} a(h) \, a(h + g)$ and let $\lambda(g)$ denote the number of solutions $(h', h) \in D \times D$ to the equation $h' - h = g$. If $|D| = k$, the number of occurrences of summand $(-1)(-1)$ in $C_A(g)$ is $\lambda(g)$, and the number of occurrences of summand $(1)(-1)$, $(-1)(1)$ and $(1)(1)$, respectively, is $k - \lambda(g)$, $k - \lambda(g)$ and $s - 2k + \lambda(g)$. In other words, $C_A(g) = s - 4(k - \lambda(g))$ for all $g \neq 0 = (0, \ldots, 0)$.

Therefore, A is a $PBA(s_0, \ldots, s_{m-1})$ if and only if $C_A(0) = s$ and $\lambda(g) = k - s/4$ for all $g \neq 0$. By Definition 2.8, A is a $PBA(s_0, \ldots, s_{m-1})$ if and only if D is a $(s, k, k - s/4)$-difference set in G. If D is a $(4u^2, 2u^2 \pm u, u^2 \pm u)$-difference set we are finished.

If A is a PBA it remains to show $s = 4u^2$ for some u. If $g \neq 0$, we know $C_A(g) = 0$ and, as this is a sum of s terms ± 1, s is even. In addition, $s = C_A(0) = \sum_{g \in G} C_A(g) = \sum_{g \in G}(\sum_{h \in G} a(h)a(h + g)) = (\sum_{h \in G} a(h))^2$. □

COROLLARY 3.24 *By Theorem 2.22, for any $a, b \geq 0$ and any odd number m, PBAs with energy $4 \cdot 2^{2a} \cdot 3^{2b} \cdot m^4$ exist.*

The equivalence of PBAs and abelian group developed Hadamard matrices is also apparent. If M is group developed over $G = \mathbb{Z}_{s_0} \times \cdots \times \mathbb{Z}_{s_{m-1}}$ by the mapping $\phi : G \to \{\pm 1\}$, and its rows and columns are indexed by the elements $g \in G$ in lexicographical (mixed-radix) order, the entry in the gth row and h^{th} column is $\phi(g + h)$. Thus, the inner product of the kth row and the $(g + k)^{th}$ row is $\sum_{h \in G} \phi(h) \, \phi(g + h)$.

LEMMA 3.25 [164] *The top row of a Hadamard matrix which is group developed over $\mathbb{Z}_{s_0} \times \cdots \times \mathbb{Z}_{s_{m-1}}$ is a PBA$(s_0, \ldots, s_{m-1})$, and vice versa.*

For example, there is a $\mathbb{Z}_3 \times \mathbb{Z}_3 \times \mathbb{Z}_4$-developed Hadamard matrix of order 36. With lexicographical ordering of the group elements, its top row

$$\begin{bmatrix} 1 & 1 & 1 & 1 & 1 & -1 & -1 & -1 & 1 & -1 & -1 & -1 \\ 1 & 1 & -1 & -1 & 1 & -1 & 1 & -1 & 1 & -1 & -1 & 1 \\ -1 & 1 & -1 & -1 & -1 & -1 & -1 & 1 & -1 & -1 & 1 & -1 \end{bmatrix} \quad (3.23)$$

is a $PBA(3,3,4)$. Equally, using the isomorphism $\mathbb{Z}_3 \times \mathbb{Z}_3 \times \mathbb{Z}_4 \cong \mathbb{Z}_3 \times \mathbb{Z}_{12}$ and the corresponding reordering, it is a $PBA(3, 12)$.

2-D perfect binary arrays $PBA(s, t)$ originated in the engineering community in 1968 (Calabro and Wolf [41]) and for them, the connection with Menon-Hadamard difference sets was known by 1979 [51].

Kopilovich and Sodin [209, 208] have since championed the use of 2-dimensional $PBA(s, t)$ and other 'generalised 2-D difference sets' in astrophysics, both for plane antenna arrays and for aperture synthesis systems which cover a rectangular domain of spatial frequencies. Systems based on PBAs have *filling* or *transparency coefficient* $\beta = k/v = (2u^2 \pm u)/4u^2 \approx 0.5$.

They point out, in the former case, that the sidelobe (cross-talk) levels (SLL) decrease with increasing number of elements k with little change in β, and that these arrays have lower SLL than other proposed antenna designs. In the latter case, it is known that the telescope masks are sidelobe free and are optimal when $\beta < 0.5$, and they point out that the possibility of choosing higher numbers k of 'open' pixels improves the resolution of the telescope over present implementations. As a toy illustration, the 11×13 uniformly redundant array (or $(143, 71, 35)$-difference set) of Example 3.4.1 has $\beta = k/v = 71/143 \approx 0.4965$, whereas the 12×12 PBA with $u = 6$ (or $(144, 66, 30)$-difference set) of [207, Fig. 2] has very similar dimensions but fewer active elements $k = 66$ and lower $\beta = 66/144 \approx 0.4583$. So a PBA-based antenna array would be cheaper to make while a PBA-based telescope mask would have higher sensitivity, respectively, than the twin-prime version.

It appears that the list of $PBA(s, t)$ known by 1992 and appearing in [50, 187] has not been enlarged. The PBAs corresponding to the new constructions of Corollary 3.24 are all at least 4-D. More recent surveys appear in [74, 295].

That is, it is known only that for any $d \geq 1$, $PBA(2^d, 2^d)$, $PBA(2^d, 2^{d+2})$, $PBA(2^d \cdot 3, 2^d \cdot 3)$ and $PBA(2^d \cdot 3, 2^{d+2} \cdot 3)$ exist. The smallest undecided cases [187, p. 253] are $\{s, t\} = \{18, 18\}$ and $\{9, 36\}$.

Research Problem 12 *For which pairs of integers (s, t) which are not relatively prime, and with $st = 4 \cdot 2^{2a} \cdot 3^{2b} \cdot m^4$ for some $a, b \geq 0$ and odd number m, do $PBA(s, t)$ exist?*

3.5 CRYPTOGRAPHY: NONLINEAR FUNCTIONS

The last application of Hadamard matrices we discuss is to disguise characteristics of the data sequence. Encryption of data, for millennia the province of government, diplomacy and the military, now pervades society at all levels. Its use scales from the personal (safeguarding your PIN when an EFTPOS transaction is authorised at the supermarket, or your credit card number for an internet purchase) through the community (protection of medical and tax records) to national and global exchanges (fund reconciliation by national reserve banks).

At its simplest, the aim of cryptography is to maintain confidentiality of information transmitted over an insecure channel. The model is illustrated in Figure 3.6. Alice wants to send a message to Bob which only Bob is entitled to read, over a channel which is not secure. An evesdropper Eve may try to intercept the message and read or modify it. To avoid this, Alice and Bob agree to use a cryptographic system, or *cryptosystem*, which consists of encryption and decryption algorithms. If messages are to be sent frequently, it is practical to reuse the algorithms, but incorporate secret material (the key) which can be changed at regular intervals.

Alice encrypts her message (*plaintext p*) using the encryption algorithm E with her key K_A and transmits the *ciphertext* $z = E(p, K_A)$ to Bob. Bob uses the decryption algorithm D with his key K_B to recover the message $p = D(z, K_B)$. Of necessity, the encryption algorithm must be an invertible function, whatever messages and keys are used, that is, $D(E(p, K_A), K_B) = p$ for all p, K_A, K_B.

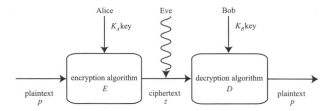

Figure 3.6 Model of transmission over an insecure channel

Cryptosystems are of two main types, *private key* or symmetric systems, where the parties know each other and have disclosed information about their private keys, and *public key* or asymmetric systems, where it is not necessary that the parties know each other and they each have two keys obtained through a trusted authority, a public key published by the authority, and a private key they do not disclose to anyone. The best-known private key systems are DES, the Data Encryption Standard, and its successor Rijndael, the AES (Advanced Encryption Standard) competition winner. These algorithms are suited to fast high volume data transmissions. The best-known public key system is RSA, which is slower but does not require the parties to exchange key information, and is suitable for key distribution and digital signature schemes. An encyclopaedic coverage of cryptography in general appears in [244].

One of the main design features essential to any cryptographic algorithm is confusion — the relationship between any cipher bit and all plaintext bits should appear random. Highly nonlinear functions are important algorithm components for this purpose. In particular, they are used to construct keystream generators for stream ciphers, S-boxes for block ciphers, components of hash algorithms and authentication codes.

Designing for different applications has, of course, engendered different notions of what 'nonlinearity' means and when a function has ideal nonlinearity. Golomb [126] famously stated three conditions which, when satisfied, qualify a binary sequence to be called *pseudonoise*, that is, to have some good statistical properties which are also characteristic of random sequences.

DEFINITION 3.26 *(Golomb's randomness postulates, cf. [244, 5.4.3]) A periodic* $(0, 1)$ *sequence* $\mathbf{x}$ *is a* pseudonoise *sequence if*

1. *in one period of* $\mathbf{x}$*, the number of* 1s *differs from the number of* 0s *by at most* 1;

2. *in one period of* $\mathbf{x}$*, at least half the runs have length* 1*, at least one quarter have length* 2*, at least one eighth have length* 3*, and so on, whilst there are at least* 2 *runs of a given length. Moreover, for each of these lengths, there are (almost) equally many runs of* 0s *as of* 1s;

3. *the unnormalised periodic autocorrelation function* $C_{\mathbf{x}}$ *of the corresponding* (± 1) *sequence is 2-level.*

The last of the three we have already met in connection with perfect binary sequences (Definition 3.20). Whilst important, these properties by no means exhaust the wish-list of desirable design features.

Keystream generators, for instance, attempt to produce a periodic sequence from a relatively short key or seed. They are generally regarded as good if they have long period, good statistical properties, large linear complexity, confusion with respect to relating any keystream bit to all the seed bits, diffusion — the dissipation of redundancies in the keystream into long-range statistics — and a high degree of nonlinearity in equations involving seed bits [36].

Diffusion is traditionally provided by transposition of bits, but in modern systems, such as in Rijndael, more general linear transformations are used. One such family of linear transformations is the *Pseudo-Hadamard Transform (PHT)*, defined recursively over a ring of order 2^n, usually $\mathbb{Z}_{2^n}$ or $GF(2^n)$, by (cf. [302])

$$\text{PHT}_1 = \begin{bmatrix} 1 & 1 \\ 1 & 2 \end{bmatrix}, \ \text{PHT}_t = \otimes^t \text{PHT}_1, \ t \geq 2. \tag{3.24}$$

Here PHT_1 acts on pairs of bit strings of equal length $< n$ and an integer stands for its binary representation, or the corresponding coefficient polynomial in $GF(2)[x]$. The PHT is used, for example, in round functions within the block cipher SAFER and its variants. SAFER+ is implemented in the security features of Bluetooth, a protocol used worldwide for short-range fast communications. Bluetooth technology is used in a large set of wired and wireless devices: mobile phones, PDAs, desktop and mobile PCs, printers, digital cameras, and dozens of others. A recent attack on Bluetooth [289] depends on algebraic representation of the round function containing the PHT.

S-boxes are functions (often permutations) in DES-like block encryption algorithms, whose principal aim is confusion. They have been the subject of intensive study and since the success of the linear cryptanalysis techniques of Matsui [241] have been expected to be capable of resisting linear attacks by having a high degree of nonlinearity in equations relating output and input bits. Since the success of the differential cryptanalysis techniques of Biham and Shamir [26], S-box functions have similarly been expected to be capable of resisting differential attacks by having a high degree of uniformity in the distribution of output differences for each input difference.

In the next two subsections, we look at the relationship of Hadamard matrices to the design of S-boxes. S-box functions are typically mappings $f : V(n, q) \to V(m, q)$, where we assume $n \geq m$, but they can be more general mappings between arbitrary finite groups. We will deal with this general case in Chapter 9 (Sections 9.2.1 and 9.5, see also Definition 7.34), but we begin with the binary case $q = 2$. For a comprehensive coverage of the cryptographic properties of binary-based functions, see Carlet [46].

3.5.1 Binary bent functions and maximally nonlinear functions

First, how do we design functions resistant to linear attacks?

To start with, let us consider Boolean functions $f : V(n, 2) \to GF(2)$ and establish some notation. For convenience in moving between $(0, 1)$ and (± 1) sequences of output values, define

$$F(v) = (-1)^{f(v)}, \quad f(v) = \log_{(-1)} F(v), \quad v \in V(n,2). \tag{3.25}$$

Define the *weight* $w(f)$ of f to be $w(\mathbf{f})$, where $\mathbf{f}$ denotes the truth table of f (Definition 3.16).

DEFINITION 3.27 *A Boolean function* $f : V(n,2) \to GF(2)$ *is* balanced *if* $w(f)$ $= 2^{n-1}$.

For each $v \neq 0 \in V(n,2)$, define the *directional derivative* $(\Delta f)_v$ *of* f *in direction* v to be

$$(\Delta f)_v(u) = f(u+v) + f(u), u \in V(n,2) \tag{3.26}$$

Recall that any linear Boolean function is of the form $L_u(v) = \langle u, v \rangle$ for a fixed u, since we can write it as $L_u(v_1, v_2, \ldots, v_n) = \sum_{i=1}^{n} u_i v_i$, and v_i may be regarded as the Boolean function with truth table $\mathbf{v}_i$ which projects the ith coordinate of $V(n,2)$.

By (3.9) and (3.11), the Walsh-Hadamard Transform $\widehat{F}$ of F is

$$\widehat{F}(u) = \sum_{v \in V(n,2)} (-1)^{\langle u,v \rangle} F(v) = \sum_{v \in V(n,2)} (-1)^{L_u(v)+f(v)}, \ u \in V(n,2).$$
$$\tag{3.27}$$

In the first analysis, if $\widehat{F}(0) = 0$, that is, f is balanced, then it cannot be approximated by a constant function.

We interpret (3.27) — which is the number of times f and L_u are equal minus the number of times they differ — as a measure of how well f may be approximated by the linear function L_u. By computing the WHT of F and searching for the transform coefficient which has the greatest absolute value, we can identify any likely linear approximations to f.

By Parseval's Theorem (3.13) we know that

$$\sum_{v \in V(n,2)} \widehat{F}(v)^2 = 2^n \sum_{v \in V(n,2)} F(v)^2 = 2^{2n}, \tag{3.28}$$

so that, if some of the transform coefficients are smaller than average in absolute value, especially if some are 0, then others must be larger. If a maximum absolute value of $\widehat{F}$ occurs at u, then either L_u is the best linear approximation of F (when $\widehat{F}(u) > 0$) or its complement, the affine function $1 + L_u$, is as good as, or better than, the best linear approximation (when $\widehat{F}(u) < 0$).

Example 3.5.1 *Order the elements of* $V(3,2)$ *lexicographically, viz* $000, 001,$ $010, 011, 100, 101, 110, 111,$ *and let* $\mathbf{f} = [1, 0, 0, 1, 0, 1, 1, 1]$, *so* $\mathbf{F} = [-1, 1, 1, -1,$ $1, -1, -1, -1]$. *By Example 3.1.1 using (3.11),* $\widehat{\mathbf{F}} = [-2, 2, 2, -2, 2, -2, -2, -6]$. *The highest magnitude coefficient is* $\widehat{\mathbf{F}}(111) = -6$, *so* $L_u(v) = v_1 + v_2 + v_3$ *and the affine function* $1 + \mathbf{v}_1 + \mathbf{v}_2 + \mathbf{v}_3 = [1, 0, 0, 1, 0, 1, 1, 0]$ *best approximates* f.

Thus, the maximum absolute value of the WHT coefficients of f can serve as a quantitative measure of the linearity of f. If a Boolean function is equally like each linear function and each of their complements, so that no approximation is better than any other, the function is called *bent*, probably because it is as far from being linear as possible. Bent functions were introduced in 1976 by Rothaus [278] as those functions f whose WHT coefficients are all equal in absolute value.

DEFINITION 3.28 *A Boolean function $f : V(n,2) \rightarrow GF(2)$ is bent if $|\widehat{F}(u)|$ is constant, for all $u \in V(n,2)$.*

It follows from (3.28) that a bent function f must be a function of an even number of Boolean variables.

Several known families of bent functions — the Maiorana-McFarland, nondegenerate quadratic functions and partial spread families — are contained, up to equivalence, in a family of bent functions discovered by Dobbertin [100]. It is not known whether these included families exhaust Dobbertin's family [326]. The existence of another family, proposed by Dillon in terms of trace functions, was proved by Lachaud and Wolfmann in 1987. See Wolfmann [326] for descriptions of these families.

Example 3.5.2 *Let $n = 2k$ and let the function $f : V(n,2) \rightarrow GF(2)$ be quadratic, that is, have the form*

$$f(v_1,\ldots,v_n) = \sum_{1 \leq i < j \leq n} b_{ij} v_i v_j, \ b_{ij} \in GF(2).$$

Equivalently, $f \not\equiv 0$ is quadratic if and only $f(0) = 0$ and ∂f is a nonzero bilinear form, where

$$\partial f(u,v) = f(u+v) + f(u) + f(v).$$

If f is nondegenerate, that is, $\{v \in V(n,2) : \partial f(u,v) = 0, \ \forall u\} = \{0\}$, then f is bent.

For instance [237, p. 429], the quadratic functions

$$\sum_{1 \leq i \leq k} v_{2i-1} v_{2i} \quad \text{and} \quad \sum_{1 \leq i < j \leq 2k} v_i v_j$$

are bent functions of $2k$ variables. The first known bent functions which do not have quadratic terms are homogeneous functions of degree 3 in $V(6,2)$, found by exhaustive search [271], and subsequently, in $V(8,2), V(10,2)$ and $V(12,2)$, found by Charnes et al. using invariant theory [52].

From the simple observation [237, Lemma 14.3.2],

$$\sum_{u \in V(n,2)} \widehat{F}(u)\widehat{F}(u+v) = 0, \ \forall v \neq 0,$$

a bent function is equivalent to a Hadamard matrix. By (3.28) it also follows that f is bent if and only if $\hat{f}$, defined by $2^{-n/2}\widehat{F}(u) = (-1)^{\hat{f}(u)}$, is bent. But then, by the same argument, $\hat{f}$ is bent if and only if

$$\sum_{u \in V(n,2)} F(u)F(u+v) = 0, \ \forall v \neq 0,$$

which is equivalent to the condition that for each $v \neq 0$, the directional derivative $(\Delta f)_v$ of f in direction v (3.26) is balanced.

LEMMA 3.29 [237, Section 14.5] *Suppose* $n = 2k$ *and* $f : V(n, 2) \to GF(2)$. *The following are equivalent.*

1. f *is bent;*

2. $|\widehat{F}(u)| = 2^k$ *for all* $u \in V(n, 2)$;

3. *the* $2^n \times 2^n$ *matrix* $[2^{-k}\widehat{F}(u + v)]$ *is a Hadamard matrix;*

4. *the* $2^n \times 2^n$ *matrix* $[F(u + v)]$ *is a Hadamard matrix;*

5. $w((\Delta f)_v) = 2^{n-1}$ *for all* $v \neq 0$.

The Hadamard matrix $[F(u + v)]$ in Lemma 3.29 must be a Menon Hadamard matrix, since it is group developed from the function $F : \mathbb{Z}_2^n \cong V(n, 2) \to \{\pm 1\}$. By Lemma 2.19 it is equivalent to a Menon-Hadamard difference set.

COROLLARY 3.30 *Let* $n = 2k$. *Then the following are equivalent.*

1. $f : V(n, 2) \to GF(2)$ *is bent;*

2. *the* $\mathbb{Z}_2^n$-*developed matrix* $[(-1)^{f(u+v)}]$ *is a Menon Hadamard matrix;*

3. *the set* $\{v \in V(n, 2) : f(v) = 1\}$ *is a* $(2^{2k}, 2^{2k-1} \pm 2^{k-1}, 2^{2k-2} \pm 2^{k-1})$-*difference set in the abelian group* $V(n, 2) \cong \mathbb{Z}_2^n$. *These Menon-Hadamard difference sets are usually called* elementary Hadamard difference sets.

In binary-based cryptographic systems, a cryptographic block function such as an S-box function is modelled by a mapping $f : V(n, 2) \to V(m, 2)$. Typically we assume $n \geq m$, though S-boxes with $n < m$ are known.

Under a linear attack on $f : V(n, 2) \to V(m, 2)$, the aim is to uncover linear relations between output bits and input bits which have other than the average number of solutions. Thus, we try to take advantage of sets $L_f(a, b) = \{u \in V(n, 2) : \langle a, u \rangle + \langle b, f(u) \rangle = 0\}$, where $a \in V(n, 2), b \neq 0 \in V(m, 2)$ and $|L_f(a, b)| \neq 2^{n-1}$. The resistance of f to linear attack can be measured [49] by

$$\Lambda_f = \max_{b \neq 0, a} ||L_f(a, b)| - 2^{n-1}|;$$

the lower this value is, the more resistant f will be to linear attack. This translates to nonlinearity conditions on the WHT spectrum of f which correspond to those for Boolean functions.

DEFINITION 3.31 [49, 101] *Let* $f : GF(2^n) \to GF(2^m)$. *For each* $b \neq 0 \in GF(2^m)$, *define the* component $f_b : GF(2^n) \to GF(2)$ *of* f *to be* $f_b(u) = \langle b, f(u) \rangle$, *so that* $F_b(u) = (-1)^{\langle b, f(u) \rangle}$ *and, by* (3.27),

$$\widehat{F_b}(u) = \sum_{v \in V(n, 2)} (-1)^{\langle b, f(u) \rangle + \langle u, v \rangle}.$$

The WHT spectrum *of* f *is*

$$\mathcal{W}(f) = \{\widehat{F_b}(u), \ u \in GF(2^n), \ b \neq 0 \in GF(2^m)\}.$$

Define $\mathcal{L}(f) = \max\{|\widehat{F_b}(u)|, \ u \in GF(2^n), \ b \neq 0 \in GF(2^m)\}$.

1. *If n is even, define f to be a* (vectorial) bent function *if f_b is bent, for all $b \neq 0 \in GF(2^m)$.*

2. *If $n = m$, define f to be* maximally nonlinear *if it attains the minimum possible value for $\mathcal{L}(f)$.*

See also Dobbertin [101] for a concise coverage of maximally nonlinear functions.

Example 3.5.3   *Let $f : GF(2^n) \rightarrow GF(2^n)$ be the inversion function $f(x) = x^{2^n-2}$. Then f is maximally nonlinear if and only if n is even.*

Sidel'nikov's bound tells us that $\mathcal{L}(f) \geq 2^{n/2}$ (see [49, 1.3,3.2] and [294]). For even $n = m$, it is an open question to determine the minimal achievable amount of linearity, and for power functions it is conjectured to be $2^{n/2+1}$ [101].

Research Problem 13   *For n even, determine whether or not*

$$\min\{\mathcal{L}(f), f : GF(2^n) \rightarrow GF(2^n)\} = 2^{n/2+1}.$$

If $n = m = 2k+1$ is odd, the minimum achievable amount of linearity is known to equal the lower bound 2^{k+1}. In this case a maximally nonlinear power function is also called *almost bent*.

THEOREM 3.32 [49, Note 2, p. 363] *Let $f : GF(2^n) \rightarrow GF(2^n)$. If $n = 2k + 1$ is odd, f is maximally nonlinear if and only if $\mathcal{W}(f) = \{0, \pm 2^{k+1}\}$.*

There are very few maximally nonlinear functions known. Until recently, up to *affine equivalence* $g \sim f$ (that is, there exist $L \in \mathrm{Aut}(V(n,2))$, **c** constant, such that $g(x) = f(L(x) + \mathbf{c}))$ and to the addition of an affine function, all the known examples were power functions. Up to inverses, and excluding equivalent powers, the known families of binary-based maximally nonlinear power functions for odd n are listed in Table 3.1. For n odd, a family which is affinely inequivalent to any power function appears in [39, Theorem 1]. In the next subsection we will see that maximally nonlinear functions with n odd also resist differential attacks. Thus the almost bent functions are optimal in a cryptographic context, providing maximum possible resistance to both linear and differential cryptanalysis.

3.5.2 Perfect and almost perfect nonlinear functions

It is suspected that some design criteria for the DES S-boxes were not made public precisely because knowledge of the differential attack would have been exposed.

The differential attack introduced by Biham and Shamir [26] has had a substantial effect on the design of block encryption algorithms, requiring the establishment of theoretical measures of resistance to differential attacks and the construction of resistant functions, as well as study of the links with other design criteria.

Under a differential attack on $f : V(n, 2) \rightarrow V(m, 2)$, the aim is to uncover differences between input vectors for which the differences in the resulting output vectors have other than the average number of solutions. Thus, we try to take

Case	Exponent	Maximally Nonlinear?
f_1	$d = 2^k + 1, \ (k,n) = 1, \ 1 \le k \le \frac{n-1}{2}$	yes, n odd
f_2	$d = 2^{2k} - 2^k + 1, \ (k,n) = 1 \ 2 \le k \le \frac{n-1}{2}$	yes, n odd
f_3	$d = 2^{4s} + 2^{3s} + 2^{2s} + 2^s - 1, \ n = 5s$	no
f_4	$d = 2^m + 3, \ n = 2m + 1$	yes
f_5	$d = 2^m + 2^{m/2} - 1, \ n = 2m + 1, \ m$ even $\\ d = 2^m + 2^{(3m+1)/2} - 1, \ n = 2m + 1, \ m$ odd	yes
f_6	$d = 2^n - 2, \ n = 2m + 1$	no (yes, n even)

Table 3.1 APN power functions $f(x) = x^d$ on $GF(2^n)$

advantage of frequencies $n_f(a,b) = |\{u \in V(n,2) : f(a+u) + f(u) = b\}|$, where $a \ne 0 \in V(n,2)$, $b \in V(m,2)$ and $n_f(a,b) \ne 2^{n-m}$. The resistance of f to differential attack can be measured [252] by

$$\Delta_f = \max_{a \ne 0 \in V(n,2), \ b \in V(m,2)} n_f(a,b).$$

The functions least susceptible to differential attack (that is, in the analogous position to bent functions for linear attack) are those for which $n_f(a,b) = 2^{n-m}$ for all $a \ne 0$ and b. That is, for each $a \ne 0$, the *directional derivative* $(\Delta f)_a(u) = f(a+u) + f(u)$ *of f in direction a* (cf. (3.26)) is uniformly distributed. Such functions, introduced by Nyberg [251, Definition 3.1] in 1991, are called *perfect nonlinear* (PN).

If $m = 1$ we see that f is perfect nonlinear if and only if $[F(u+v)]$ is a $\mathbb{Z}_2^n$-developed Menon Hadamard matrix, if and only if (by Lemma 3.29) f is bent. Similarly, if $m \ge 2$, Nyberg shows that perfect nonlinear functions coincide with vectorial bent functions (Definition 3.31.1), and consequently can exist only if $n \ge 2m$ and n is even (for proof, see Chabaud and Vaudenay [49, Theorems 2, 3]).

COROLLARY 3.33 [251] *If $n \ge 2m$ is even, then $f : V(n,2) \to V(m,2)$ is PN if and only if f is (vectorially) bent.*

Unfortunately for the S-box designer, this means that, in the most useful case $n = m$, there is no binary-based permutation ideally resistant to differential attack. Since solutions to $f(a+u) + f(u) = b$ come in pairs $\{u_0, u_0 + a\}$, the best that can be achieved in this case is $\Delta_f = 2$. A function $f : V(n,2) \to V(n,2)$ satisfying $\Delta_f = 2$ is termed *almost perfect nonlinear* (APN).

When n is odd, APN and maximally nonlinear functions are closely related.

COROLLARY 3.34 *Let* $n = 2m + 1$ *and* $f : V(n, 2) \to V(n, 2)$.

1. [49, Theorem 4] *If* f *is maximally nonlinear, then* f *is APN.*

2. [42, 43] *If* f *is APN, then* f *is maximally nonlinear if and only if all the values of the WHT spectrum of* f *are divisible by* 2^{m+1}.

At present, rather more APN functions than maximally nonlinear functions are known. Table 3.1 lists the six known families of binary-based APN power functions. For proofs and the provenance of these functions, see Carlet [46] and Dobbertin [101] and the references cited there.

Dobbertin [101, p. 1272] had conjectured that there are no other infinite families of almost bent or APN functions, up to affine equivalence, but this conjecture does not hold. For n even, a family of APN functions affinely inequivalent to any power function appears in [39, Theorem 2].

In Chapters 8.2 and 9.2 we will argue that affine equivalence is too strong an equivalence relation to represent properly the similarity of functions for cryptographic purposes, and that a weaker *affine bundle equivalence* (Definitions 8.18 and 8.22) is more appropriate.

We will return to the APN functions in Chapters 9.5.1 and 9.5.2.

Chapter Four

Generalised Hadamard Matrices

It takes no great stretch of the imagination to ask what happens when the entries of a Hadamard matrix are allowed to have more values than ± 1. For instance, Hadamard's original interest in matrices with entries from the unit disc makes an extension to complex entries on the unit circle wholly natural.

But what makes a matrix intrinsically 'Hadamard'? Is it a property inherent in the invertibility of Hadamard matrices, as square matrices with entries from a field? Apart from its independent interest, this characteristic drives the applications of Hadamard matrices as digital signal transforms. Is it a property inherent in the balanced numbers of matches and mismatches between corresponding entries in each pair of distinct rows (or columns)? This characteristic drives the applications of Hadamard matrices as error-correcting codes and experimental designs. Is it a property inherent in the orthogonality or zero correlation of distinct rows in a Hadamard matrix? This characteristic drives the applications in cryptography and signal separation.

As always, the spiral process — through which mathematical theory informs practical application, which drives product development and then forces improved performance, which needs innovative modelling techniques and, in turn, inspires new theoretical approaches — has fired our interest in these questions. The applications just mentioned, and illustrated in the previous Chapter, have each imposed distinct directions on subsequent research, and indeed moulded different conceptions of how to generalise Hadamard matrices.

In fact, there are many reasonable ways of generalising Hadamard matrices. Interest in Hadamard's original problem of finding $v \times v$ matrices with real entries $\{\pm 1\}$ and maximum possible determinant — for which the Hadamard matrices are solutions in the case $v \equiv 0 \bmod 4$ — continues to this day. The aim is to find matrices with maximal determinant in each of the classes $v \equiv 0, 1, 2$ and $3 \bmod 4$ and use them to construct D-optimal experimental designs. However, we shall not pursue the question of maximal determinants here; further information may be found, for example, at Will Orrick's website [255] and the references cited there.

Other generalisations relax the requirement that the matrix be square (difference matrices) or allow entries from more general sets (conference and weighing matrices and orthogonal designs). The latter ideas overlap the ones we investigate, but will not be especially developed here. They, too, benefit from the cocyclic approach, expounded by de Launey and Flannery in [87].

This Chapter covers two major formulations of generalised Hadamard matrices, due to Butson and to Drake, and their applications to multiphase (mostly quaternary) signal alphabets. Drake's formulation is the principal interest of this monograph, and is the subject of most of Part 2.

The first two Sections look at unitary matrices with complex entries on the unit circle, summarising our knowledge of construction techniques and equivalence classes for these *complex Hadamard matrices*. Section 4.3 introduces matrices with entries from a finite group which satisfy row balance criteria mimicking those of Hadamard matrices. Jungnickel showed these *generalised Hadamard matrices* are equivalent to class regular semiregular divisible designs, of which those with regular action can be constructed from semiregular relative difference sets.

Applications of both types of generalised Hadamard matrix to spectroscopy, coding, separation and correlation of multiphase signals and data sequences are covered in Section 4.4. Quaternary alphabets $\{\pm 1, \pm i\}$ receive most attention. A unification of the two types — the *Generalised Butson Hadamard (GBH)* matrices, together with the *Generalised Hadamard Transforms (GHT)* they define — is presented in Section 4.5.

4.1 BUTSON MATRICES

We begin with a discussion of the earliest generalisation of Hadamard matrices, to matrices with entries which are complex m^{th} roots of unity. Butson [40] introduced these in 1962, essentially taking the familiar extension of orthogonal real matrices to unitary complex matrices and applying it to the discrete case. These matrices display the invertibility and orthogonality characteristics of Hadamard matrices, but not necessarily their pairwise row and column balancing characteristics.

Because both the terms 'generalised Hadamard matrix' and 'complex Hadamard matrix' have become identified with other families of matrices, we call the matrices Butson introduced by his name.

DEFINITION 4.1 *For m a positive integer, let H be an $n \times n$ matrix, all of whose entries are complex m^{th} roots of unity. Let $\overline{H}$ denote the* conjugate *of H, that is, the entries of $\overline{H}$ are the complex conjugates of the entries of H. If $H(\overline{H})^{\top} = nI_n$, then H is a* Butson matrix *of order n, denoted $BH(m, n)$.*[1]

Assume for simplicity that $n \geq 2$. The elementary properties of Butson matrices mimic Lemma 2.2. (The second property follows because H is invertible over $\mathbb{C}$. The third property follows since $\overline{(A \otimes B)}^{\top} = (\overline{A})^{\top} \otimes (\overline{B})^{\top}$.)

LEMMA 4.2 *Let H be a $BH(m, n)$ and let $d = (m, m')$ denote the greatest common divisor of m and m'.*

1. *If α is an $(m')^{th}$ root of unity, then αH is a $BH(mm'/d, n)$.*

2. *The transpose $H^{\top}$ of H is a $BH(m, n)$.*

3. *If H' is a $BH(m', n')$, then the tensor product $H' \otimes H$ is a $BH(mm'/d, nn')$.*

---

[1]Other authors use m-GHM or $BH(n, m)$.

The Hadamard matrices are the $BH(2,n)$. Probably the next most familiar Butson matrices are the symmetric $BH(n,n)$ arising as the matrices of the length n Discrete Fourier Transform (3.6) and its inverse.

Example 4.1.1   *Let $\omega = \exp(-2\pi i/n) \in \mathbb{C}$, where $i = \sqrt{-1}$. The matrix of the n-point Discrete Fourier Transform (DFT) is*

$$\mathcal{F}_n = [\,\omega^{jk}\,]_{0 \leq j,k \leq n-1} \tag{4.1}$$

and the matrix of the Inverse Discrete Fourier Transform (IDFT) is

$$\overline{\mathcal{F}_n} = [\,\omega^{-jk}\,]_{0 \leq j,k \leq n-1}.$$

Both $\mathcal{F}_n$ and $\overline{\mathcal{F}_n}$ are Butson matrices $BH(n,n)$.

More generally, a Fourier Transform of a complex-valued function of any finite group G may be defined in terms of the matrix representations of G (see [239]).

In the abelian case, the irreducible representations (characters) are all linear and the corresponding transform matrix is a Butson matrix. Recall that an irreducible *character* of a finite abelian group C of exponent m is any group homomorphism from C to the multiplicative group $D = \langle e^{2i\pi/m} \rangle \subset \mathbb{C}$ of all complex m^{th} roots of unity. An example with $m = 2$ is the quadratic character of Chapter 2.1.2.

The *character group* $\widehat{C} = \mathrm{Hom}(C,D)$ of all irreducible characters of C is isomorphic to C.

DEFINITION 4.3 *Let C be a finite abelian group of order w and exponent m and fix an ordering $C = \{c_1, \ldots, c_w\}$. Fix an isomorphism $\chi : C \to \widehat{C}$ and denote the image of $c \in C$ by χ_c. The* Fourier Transform (FT) *of a complex-valued function $\varphi : C \to \mathbb{C}$ is the function $\widehat{\varphi} : C \to \mathbb{C}$ given by*

$$\widehat{\varphi}(c_k) = \sum_{\ell=1}^{w} \varphi(c_\ell)\chi_{c_k}(c_\ell), \ 1 \leq k \leq w, \tag{4.2}$$

and the Inverse Fourier Transform (IFT) *of $\widehat{\varphi}$ is*

$$\varphi(c_\ell) = w^{-1} \sum_{k=1}^{w} \widehat{\varphi}(c_k)\overline{\chi_{c_k}(c_\ell)}, \ 1 \leq \ell \leq w.$$

When C is cyclic, the FT is the usual DFT of (3.6). When C is an elementary abelian 2-group $\mathbb{Z}_2^n$, the FT is the Walsh-Hadamard Transform of (3.11), since any homomorphism is represented by a linear Boolean function $L_{\mathbf{u}} : \mathbb{Z}_2^n \to \mathbb{Z}_2$ with $L_{\mathbf{u}}(\mathbf{v}) = \langle \mathbf{u}, \mathbf{v} \rangle$ and vice versa, from (3.27).

Analogues of Parseval's Theorem and the Convolution Theorem (cf. Lemma 3.6) hold for the Fourier Transform. See [185] and [239] for more details.

Example 4.1.2   *Let $C = \{c_1, \ldots, c_w\}$ be a finite abelian group of exponent m. The matrix $\mathcal{F}_C = [\,\chi_{c_k}(c_\ell)\,]_{1 \leq k,\ell \leq w}$ of the FT and the matrix $(\overline{\mathcal{F}_C})^\top = [\,\overline{\chi_{c_\ell}(c_k)}\,]_{1 \leq k,\ell \leq w}$ of the IFT are $BH(m,w)$.*

For instance, when $C = \mathbb{Z}_p^n$ for prime p, the Fourier Transform matrix is a p-ary version of the Sylvester Hadamard matrix. Matsufuji and Suehiro [240, Theorem

1] show that it factorises similarly into sparse matrices, from which a p-ary FFT, corresponding to Theorem 3.7, is derived. They propose implementation of the p-ary FFT as a correlation detector in a synchronous spread spectrum system (cf. Chapter 3.3.1).

Example 4.1.3 *For p prime, let $C = \mathbb{Z}_p^n$ in lexicographical order and $\omega = \exp\left(-2\pi i/p\right)$. Then $\mathcal{F}_C(k,\ell) = \omega^{\sum_{j=0}^{n-1} k_j \ell_j}$, $0 \le k, \ell \le p^n - 1$ and $\mathcal{F}_C = \otimes^n \mathcal{F}_p$.*

For $m = 3$, $BH(3,n)$ are relatively rare (see Example 4.3.3.2). Most investigation of Butson matrices has centred on the quaternary case $m = 4$, discussed in Section 4.2 following. The $BH(4,n)$ are termed 'complex Hadamard matrices' in the combinatorial literature, but to avoid confusion will be called *quaternary complex Hadamard matrices* here. This is because in high-energy and quantum physics, the name 'complex Hadamard matrix' refers to an invertible matrix with unimodular complex entries, a generalisation of $BH(m,n)$ which corresponds to taking the limit $m \to \infty$ in Definition 4.1. Such matrices will be termed *unimodular complex Hadamard matrices* in the sequel, and include all the Butson matrices.

Comparatively little is known about other Butson matrices apart from scattered nonexistence results. A necessary condition for existence, when $m = p^a$ is a prime power, is due to Winterhof [325], and generalises Butson's original result [40] for $a = 1$. Winterhof's consequent nonexistence results were proved for $a = 1$ by de Launey (cf. [57]).

LEMMA 4.4 *1. [325] If p is prime and there exists a $BH(p^a, n)$, then $n = pt$ for some positive integer t.*

2. [40, Theorem 3.5] If p is prime, there exists a $BH(p, 2^j p^k)$ for all $0 \le j \le k$.

3. [325] Suppose $p \equiv 3 \bmod 4$ is prime, $n = p^b r^2 s$ is odd, s is square-free with $(s,p) = 1$, and there exists a prime $q|s$ with quadratic character value $\left(\frac{q}{p}\right) = -1$. Then there is no $BH(p^a, n)$ and no $BH(2p^a, n)$.

The concepts of equivalence and normalisation are easily extended to Butson matrices. Two $n \times n$ matrices M and M' with entries which are complex m^{th} roots of unity are (Hadamard) *equivalent* if one can be obtained from the other by a finite sequence of row permutations, column permutations, multiplication of a row by a complex m^{th} root of unity or multiplication of a column by a complex m^{th} root of unity. Any equivalence operation applied to a Butson matrix gives a Butson matrix. A matrix is *normalised* if its first row and first column consist entirely of 1s, and every matrix with entries which are complex m^{th} roots of unity is equivalent to a normalised matrix.

LEMMA 4.5 *In a normalised Butson matrix, the elements in each noninitial row (and noninitial column) sum to 0.*

The restriction on order given in Lemma 4.4.1 does not apply when m is not a prime power, as demonstrated by Brock in the smallest case.

Example 4.1.4 [37, Theorem 4.4] *Let* $m = 6$ *and let* w *be a complex cube root of unity. Then*

$$\begin{bmatrix} 1 & 1 & 1 & 1 & 1 & 1 & 1 \\ 1 & -w^2 & -1 & -1 & w & -w & -w \\ 1 & -1 & -w^2 & -1 & -w & w & -w \\ 1 & -1 & -1 & -w^2 & -w & -w & w \\ 1 & w & -w & -w & -w^2 & -1 & -1 \\ 1 & -w & w & -w & -1 & -w^2 & -1 \\ 1 & -w & -w & w & -1 & -1 & -w^2 \end{bmatrix}$$

is a normalised $BH(6, 7)$.

From this example it is clear, first, that not all the mth roots of unity need appear in a normalised $BH(m, n)$ and, second, that the mth roots which do appear in the matrix need not appear equally often in each noninitial row (or column). Even if every m^{th} root of unity does appear in a normalised $BH(m, n)$, the DFT matrices $\mathcal{F}_m$ demonstrate that the second property need not hold. The failure of these properties distinguishes Butson matrices from Hadamard matrices and has a corresponding effect on applicability.

However, it is important to identify those Butson matrices, including the Hadamard matrices themselves, which do satisfy these properties.

DEFINITION 4.6 *A* $v \times v$ *matrix with entries from a group* N *is* normalised[2] *if all the entries in the first row and first column are the identity* 1 *of* N.

A normalised $v \times v$ *matrix with entries from a finite group* N *is* row balanced (column balanced) *if every element of* N *appears equally often (necessarily* $v/|N|$ *times) in each noninitial row (column).*

For example, when $m = 3$, or more generally when $m = p$ a prime, normalised Butson matrices are row and column balanced.

If a normalised Butson matrix is row balanced, then any row equivalence operations followed by normalisation will preserve the row balance property, and similarly for column balance.

Research Problem 14 *Show that in an equivalence class of Butson matrices all the normalised matrices are both row and column balanced, or none of them are.*

4.2 COMPLEX HADAMARD MATRICES

As mentioned, the term 'complex Hadamard matrices' refers to a specific family of Butson matrices in combinatorial contexts but to a generalisation of Butson matrices elsewhere. We will cover them in turn.

---

[2]If N is the complex unimodular group, the term *dephased* is often used instead of 'normalised'.

4.2.1 Quaternary complex Hadamard matrices

Quaternary Butson matrices $BH(4, n)$ are commonly called complex Hadamard matrices in the combinatorial literature, and were first isolated for study by Turyn [308]. They must have even order, by Lemma 4.4.1 (or Lemma 4.5).

DEFINITION 4.7 *Let n be even. A quaternary complex Hadamard matrix of order n is an $n \times n$ matrix H with entries from $\{\pm 1, \pm i\}$ such that $H(\overline{H})^\top = nI_n$.*

A quaternary complex Hadamard matrix of order $n = 2$, equivalent on normalisation to the Hadamard matrix $\mathcal{S}_1$, is

$$\mathcal{C}_1 = \begin{bmatrix} 1 & -i \\ 1 & i \end{bmatrix}.$$

A quaternary complex Hadamard matrix of the smallest order $n = 4$ which is not equivalent to a Hadamard matrix is the matrix of the length 4 IDFT (see Example 4.1.1)

$$\overline{\mathcal{F}_4} = \begin{bmatrix} 1 & 1 & 1 & 1 \\ 1 & i & -1 & -i \\ 1 & -1 & 1 & -1 \\ 1 & -i & -1 & i \end{bmatrix}.$$

A quaternary complex Hadamard matrix of the next smallest order $n = 6$ (an order impossible for a Hadamard matrix) is

$$iI_6 + S = \begin{bmatrix} i & 1 & 1 & 1 & 1 & 1 \\ 1 & i & 1 & -1 & -1 & 1 \\ 1 & 1 & i & 1 & -1 & -1 \\ 1 & -1 & 1 & i & 1 & -1 \\ 1 & -1 & -1 & 1 & i & 1 \\ 1 & 1 & -1 & -1 & 1 & i \end{bmatrix}, \tag{4.3}$$

where S is as given for $q = 5$ in Lemma 2.4. This is the smallest example of an infinite family of quaternary complex Hadamard matrices $iI_{q+1} + S$ of order $q + 1$, where $q \equiv 1 \bmod 4$ is an odd prime power and S is as defined in Lemma 2.4. The normalised form of $iI_6 + S$ is

$$\begin{bmatrix} 1 & 1 & 1 & 1 & 1 & 1 \\ 1 & -1 & i & -i & -i & i \\ 1 & i & -1 & i & -i & -i \\ 1 & -i & i & -1 & i & -i \\ 1 & -i & -i & i & -1 & i \\ 1 & i & -i & -i & i & -1 \end{bmatrix}.$$

A quaternary complex Hadamard matrix of order $n = 8$ is

$$\begin{bmatrix} 1 & 1 & 1 & 1 & 1 & 1 & 1 & 1 \\ 1 & i & -i & 1 & -1 & -i & i & -1 \\ 1 & -i & -1 & i & 1 & -i & -1 & i \\ 1 & 1 & i & i & -1 & -1 & -i & -i \\ 1 & -1 & 1 & -1 & 1 & -1 & 1 & -1 \\ 1 & -i & -i & -1 & -1 & i & i & 1 \\ 1 & i & -1 & -i & 1 & i & -1 & -i \\ 1 & -1 & i & -i & -1 & 1 & -i & i \end{bmatrix}. \qquad (4.4)$$

It is found by normalising the back-circulant matrix having first row

$$1 \quad 1 \quad i \quad 1 \quad 1 \quad -1 \quad i \quad -1. \qquad (4.5)$$

From these examples it also is clear that not every normalised quaternary complex Hadamard matrix is row balanced. Necessarily, a row balanced normalised quaternary complex Hadamard matrix $H = [h_{ij}]$ has order a multiple of 4, in which case the image $\phi(H) = [\phi(h_{ij})]$ of H, under the epimorphism $\phi : \langle i \rangle \rightarrow \langle -1 \rangle$ defined by $\phi(i) = -1$, is plainly a Hadamard matrix.

Seberry [315] conjectured that a quaternary complex Hadamard matrix of every even order exists. Lists of orders for which quaternary complex Hadamard matrices are known (to 1992) appear in [288, Table 11.2]; the first gap is at $n = 70$.

Research Problem 15 *The (Quaternary) Complex Hadamard Conjecture. Show that if n is even, a quaternary complex Hadamard matrix of order n exists.*

Note that for a matrix M of order $2w$ with entries from $\{\pm 1, \pm i\}$, there are matrices A, B of order $2w$ with entries from $\{\pm 1\}$ such that $M = \frac{1-i}{2}(A + iB)$, and vice versa. As Turyn [308] points out, and as can be seen from the first part of the following theorem, the Complex Hadamard Conjecture implies the Hadamard Conjecture (Research Problem 1).

THEOREM 4.8 ([245, Lemma 4],[201])

1. *There is a Hadamard matrix of order $4w$ of the form* $\begin{bmatrix} A & B \\ -B & A \end{bmatrix}$ *if and only if there is a quaternary complex Hadamard matrix of order $2w$ of the form $\frac{1-i}{2}(A + iB)$.*

2. *There is a Hadamard matrix of order $4w$ of the form (2.14) if and only if there is a quaternary complex Hadamard matrix of order $2w$ of the form* $\begin{bmatrix} S & T \\ -\overline{T} & \overline{S} \end{bmatrix}.$

Existence of quaternary complex Hadamard matrices is thus deeply entwined with existence of Hadamard matrices.

4.2.2 Unimodular complex Hadamard matrices

Outside combinatorics, a complex Hadamard matrix is defined as an $n \times n$ matrix H, all of whose entries lie on the complex unit circle, such that $H(\overline{H})^\top = nI_n$. To avoid confusion, we will term such a matrix a *unimodular complex Hadamard matrix*. For applications, the scaled unitary matrix $\frac{1}{\sqrt{n}}H$ often replaces H.

Clearly the unimodular complex Hadamard matrices include the Butson matrices; they may be thought of as the limiting case $m \to \infty$ of the $BH(m, n)$. There is an extensive mathematics, physics and engineering literature dealing with these matrices. Here we will touch only on the notions of group development and equivalence, as they apply to unimodular complex Hadamard matrices.

Group developed unimodular complex Hadamard matrices are defined (see Definition 2.17) by the existence of an indexing group G and a function $\phi : G \to \mathbb{C}$ taking values on the complex unit circle such that for every $g \neq 1 \in G$,

$$\sum_{h \in G} \phi(gh)\overline{\phi(h)} = 0. \tag{4.6}$$

The sequence $(\phi(g), \ g \in G)$ is known as a *generalised unimodular perfect sequence* [117], or simply a *unimodular perfect sequence* in the cyclic case $G = \mathbb{Z}_n$. When G is abelian, transformation of a unimodular generalised perfect sequence by the corresponding Fourier Transform (Definition 4.3) yields another unimodular generalised perfect sequence [117, Section IV]. The tensor product of the corresponding group developed matrices may be used to construct perfect sequences of any length from perfect sequences of prime-power lengths, and there are various constructions known for perfect sequences of prime-power length. The most general constructions are due to Mow [248] for perfect roots-of-unity sequences (PRUS) corresponding to circulant Butson matrices, and to Gabidulin [117] in the unimodular case.

Once entries on the complex unit circle other than roots of unity are allowed, the definition of equivalence suitable for Butson matrices must be modified. Two $n \times n$ matrices M and M' with complex entries of modulus 1 are (Hadamard) *equivalent* if one can be obtained from the other by a finite sequence of row permutations, column permutations, or multiplication of a row or of a column by a unimodular complex number.[3] Whilst it is known that for composite orders n, uncountably infinitely many equivalence classes of unimodular complex Hadamard matrices can exist, parameterised by one or more real variables, it was conjectured for many years that for each prime order, only a single equivalence class exists. This was shown to be false by Petrescu [261], who found 1-parameter families for $n = 7$ and 2-parameter families for $n = 13$.

The number of equivalence classes is known only for $n = 1, 2, 3, 5$, being $1, 1, 1, 1$, respectively (the equivalence class of $\mathcal{F}_n$ in each case) and for $n = 4$,

---

[3]For unimodular matrices, equivalence is sometimes weakened to include transposition and conjugation. Compare with Definition 4.12.

where it is uncountably infinite, with equivalence classes represented by

$$
\begin{bmatrix}
1 & 1 & 1 & 1 \\
1 & ie^{ia} & -1 & -ie^{ia} \\
1 & -1 & 1 & -1 \\
1 & -ie^{ia} & -1 & ie^{ia}
\end{bmatrix}, \ a \in [0, \pi), \tag{4.7}
$$

of which case $a = \pi/2$ is equivalent to the Hadamard matrix $S_2 = F_2 \otimes F_2$ and case $a = 0$ is $\overline{F}_4$. The only Butson matrices of order 4 are in these two equivalence classes. The first order for which the number of parameterised families of equivalence classes is unknown is $n = 6$. There are at least three equivalence classes of Butson matrices of order 6: one containing the $BH(6,6)$ F_6, one containing the quaternary $BH(4,6)$ of (4.3) and one containing the ternary $BH(3,6) = [(e^{i2\pi/3})^{\lambda_{jk}}]$ whose 'log-Hadamard' matrix $[\lambda_{jk}]$ appears in Example 4.3.1 below. For $n = 7$, the $BH(6,7)$ of Example 4.1.4 and the $BH(7,7)$ F_7 are inequivalent.

The tensor product of two unimodular complex Hadamard matrices is a unimodular complex Hadamard matrix. In particular, if $(m, m') = 1$, then $F_m \otimes F_{m'}$ is equivalent to $F_{mm'}$ by a permutation of indices corresponding to the isomorphism $\mathbb{Z}_m \times \mathbb{Z}_{m'} \cong \mathbb{Z}_{mm'}$, but, for example, it is known that S_3, $F_2 \otimes F_4$ and F_8 are all inequivalent matrices of order 8.

A list of open problems on equivalence classes of unimodular complex Hadamard matrices appears in the survey [304], from which the above results on equivalence are extracted.

4.3 GENERALISED HADAMARD MATRICES

The matrices which nowadays hold the title 'generalised Hadamard matrices' were introduced by Drake [102], who was initially unaware of Butson's work seventeen years earlier. Drake discovered them in the course of his study of finite geometries and orthogonal arrays. They display two characteristics of Hadamard matrices: the pairwise balancing of distinct rows and (a version of) orthogonality, but not necessarily their invertibility characteristics.

However, they do free us wholly from the bond of the complex unit circle!

DEFINITION 4.9 *Let N be a finite group of order w, written multiplicatively.*

1. *Let w divide v. A $v \times v$ matrix $H = [h_{ij}]$ with entries from N is* row pairwise balanced *if, for all $i \neq j$, the sequence of quotients $h_{ik}h_{jk}^{-1}, 1 \leq k \leq v$ contains each element of N equally often.*

2. *Such a row pairwise balanced matrix H is termed a* generalised Hadamard *matrix of order v over N and denoted $GH(w, v/w)$.*

3. *Equivalently, H is a $GH(w, v/w)$ over N if, in the integral group ring $\mathbb{Z}N$,*

$$
HH^* = vI_v + v/w \left(\sum_{u \in N} u \right) (J_v - I_v), \tag{4.8}
$$

where $H^* = [h_{ij}^*]$ is the transinverse *of H: the transpose of the matrix of inverses of entries in H, that is, $h_{ij}^* = (h_{ji})^{-1}$.*

Drake's original concept required that H satisfy both the row pairwise balanced condition and the corresponding *column pairwise balanced* condition (or equivalently, that $H^\top$ satisfy the row pairwise balanced condition). This is no longer part of the definition of a generalised Hadamard matrix [57, 86], probably because in the common case of abelian N, one condition implies the other (Lemma 4.10). See also the new result for arbitrary N in Chapter 7 (Lemma 7.26.3).

Example 4.3.1 *A normalised $GH(4,1)$ over $\mathbb{Z}_2^2$ and a normalised $GH(3,2)$ over $\mathbb{Z}_3$ (both written additively):*

$$
\begin{bmatrix}
00 & 00 & 00 & 00 \\
00 & 01 & 10 & 11 \\
00 & 10 & 11 & 01 \\
00 & 11 & 01 & 10
\end{bmatrix}
\qquad
\begin{bmatrix}
0 & 0 & 0 & 0 & 0 & 0 \\
0 & 0 & 1 & 1 & 2 & 2 \\
0 & 1 & 2 & 0 & 1 & 2 \\
0 & 1 & 0 & 2 & 2 & 1 \\
0 & 2 & 1 & 2 & 1 & 0 \\
0 & 2 & 2 & 1 & 0 & 1
\end{bmatrix}.
$$

Possibly the most familiar generalised Hadamard matrices are the multiplication tables of $GF(q)$, where N is the underlying additive group. The $GH(4,1)$ of Example 4.3.1 is the smallest case of nonprime order.

Example 4.3.2 *Let $N = \mathbb{Z}_p^n = (GF(p^n), +)$. The matrix $M_\mu = [\mu(g,h)]_{g,h \in N}$, with $\mu(g,h) = gh$ in $GF(p^n)$, is a $GH(p^n, 1)$ over $\mathbb{Z}_p^n$.*

4.3.1 Generalised Hadamard matrix constructions

Some elementary constructions for generalised Hadamard matrices may be predicted from our previous experience, but Lemmas 2.2 and 4.2 do not quite generalise in the case of the transpose.

LEMMA 4.10 *Let H be a $GH(w, v/w)$ over N.*

1. *The matrix of inverses $H^{(-1)} = [h_{ij}^{-1}]$ is a $GH(w, v/w)$ over N.*

2. *If H' is a $GH(w, v'/w)$ over N, the tensor product $H \otimes H'$ is a $GH(w, vv'/w)$ over N.*

3. *If N is abelian, the transpose $H^\top$ of H is a $GH(w, v/w)$ over N.*

Proof. Transpose both sides of (4.8) for part 1; part 2 is straightforward. For part 3, see [37, Theorem 4.1], which corrects the earlier [188]. □

When N is nonabelian the transpose of a $GH(w, v/w)$ is not necessarily also one, though the author knows of no instance in which this is the case. Deep results in Chapter 7 (see Lemma 7.26) greatly extend the set of $GH(w, v/w)$ with transpose which is also a $GH(w, v/w)$.

Research Problem 16 *Find a $GH(w, v/w)$ whose transpose is not a $GH(w, v/w)$ or prove that no such matrix exists.*

The general tensor product (2.3) of generalised Hadamard matrices is again a generalised Hadamard matrix; this result has been extended by No and Song [250].

LEMMA 4.11 [250, Theorem 1] *Let N be an abelian group of order w, let $v = w\lambda$, let $H = [h_{ij}]$ be a $GH(w^v, \lambda')$ over the direct product N^v and let $K = \{K_i, 1 \leq i \leq m = w^v\lambda'\}$, be a set of not necessarily distinct $GH(w, \lambda)$ over N. For $u = (u_1, u_2, \ldots, u_v) \in N^v$ and a $v \times v$ matrix M over N, denote by $u \odot M$ the $v \times v$ matrix whose j^{th} column is u_j times the jth column of M, $1 \leq j \leq v$.*

Then the block matrix $H \odot K$, defined to have rows $[h_{i1} \odot K_i, \ldots, h_{im} \odot K_i]$, $1 \leq i \leq m$, is a $GH(w, w^v\lambda\lambda')$ over N.

Left-multiplying a row or right-multiplying a column of a $GH(w, v/w)$ over N by an element of N will still give a $GH(w, v/w)$, as will permuting rows or columns. Applying a fixed automorphism of N to all the entries of a $GH(w, v/w)$ will still give a $GH(w, v/w)$. (This operation leaves the matrix unchanged when $N = \{\pm 1\} \cong \mathbb{Z}_2$ and corresponds to complex conjugation when $N = \{\pm 1, \pm i\} \cong \mathbb{Z}_4$.)

DEFINITION 4.12 *Two $v \times v$ matrices M and M' with entries in a group N are (Hadamard) equivalent, written $M \sim M'$, if either can be obtained from the other by performing a finite sequence of the following operations:*

1. *(permutation equivalence) permute the rows or the columns;*

2. *right-multiply a column by an element of N;*

3. *left-multiply a row by an element of N;*

4. *replace every entry by its image under a fixed automorphism of N.*

Each equivalence class $[M]$ *therefore contains normalised representatives and either consists entirely of $GH(w, v/w)$ or contains no $GH(w, v/w)$.*

By weakening Definition 4.12.4 and taking the image of a generalised Hadamard matrix H under an epimorphism of N, a new generalised Hadamard matrix is obtained. Applied to Hadamard matrices this construction is degenerate, but in Section 4.2 we have already seen it applied to row balanced complex Hadamard matrices to give Hadamard matrices.

LEMMA 4.13 [102, Proposition 1.8] *Let $H = [h_{ij}]$ be a $GH(w, v/w)$ of order v over N. Let $\phi : N \to N'$ be an epimorphism of groups, with $|N'| = w'$. Then the projection $\phi(H) = [\phi(h_{ij})]$ of H is a $GH(w', v/w')$ of order v over N'.*

By projection (Lemma 4.13) of the $GH(p^n, 1)$ of Example 4.3.2, there exist $GH(p^i, p^j)$ over elementary abelian groups of order p^i for all primes p and integers i and j.

However, examples with nonabelian N are known: de Launey [78] constructs $GH(w, v/w)$ with entries from nonabelian groups of prime power order. No example is known of a $GH(w, v/w)$ for which w is not a prime power.

Research Problem 17 *Does there exist a GH$(w, v/w)$ for which w is not a prime power?*

For a summary of existence and construction results see [86] or the earlier [57, 11.3]. Remarkably few values of $v < 100$ are known for which the existence of GH$(w, v/w)$ for all $w|v$ is settled, even over abelian groups which are direct products of elementary abelian groups (Table 5.13 of [86]). The smallest unsettled case is $v = 12$, with GH$(2, 6)$, GH$(3, 4)$ and GH$(4, 3)$ known to exist but the existence of GH$(6, 2)$ and GH$(12, 1)$ unknown (but presumed not to exist).

Research Problem 18 *Complete Table 5.13 of* [86].

4.3.2 Generalised Hadamard matrices and Butson matrices

There exist Butson matrices which are not generalised Hadamard matrices (for instance, Example 4.1.4) and generalised Hadamard matrices which are not Butson matrices (for instance, the GH$(4, 1)$ of Example 4.3.1), but clearly the intersection of the two types contains at least the BH(p, pt) for p a prime [102, 1.3.iii]. In particular, a BH$(3, 3t)$ must be a GH$(3, t)$.

A normalised matrix which is both a Butson matrix and a generalised Hadamard matrix must be both row and column balanced — compare each row (column) with the initial row (column). It is not known whether these conditions are sufficient to characterise the intersection of the two types.

Research Problem 19 *Must a normalised Butson matrix which is both row and column balanced be a generalised Hadamard matrix?*

One subset of the generalised Hadamard matrices is of particular interest because the matrices in it are invertible, so it contains the intersection of the set of generalised Hadamard matrices and the set of Butson matrices.

DEFINITION 4.14 *Let R be a ring with unity, with characteristic char R not dividing v and with group of units R^*. Let H be a GH$(w, v/w)$ over $N \leq R^*$ (for instance, $R = \mathbb{Z}N$). Then H is* invertible *over R if $HH^* = H^*H = vI_v$.*

If N is abelian and H is a GH$(w, v/w)$ over N, then so is H^*, by Lemma 4.10, so that H is invertible if and only if $\sum_{u \in N} u = 0$ in R. For instance, if $N = \{e^{(2i\pi/w)k}, 0 \leq k < w\}$ in $\mathbb{C}$, then $\sum_{u \in N} u = 0$. Furthermore, if $N = R^*$ is finite abelian and char $R \neq 2$, then N is a disjoint union of two finite sets S and $-S$, so $\sum_{u \in N} u = 0$.

Example 4.3.3 *Examples of GH$(w, v/w)$ invertible over a ring R with unity.*

1. *A BH$(2, v)$ is a GH$(2, v/2)$ and vice versa, invertible over $N = \{\pm 1\} \subset \mathbb{Z}$. It is a Hadamard matrix of order v and vice versa.*

2. *A BH$(3, v)$ is a GH$(3, v/3)$ and vice versa, invertible over $\mathbb{Z}[e^{2i\pi/3}] \subset \mathbb{C}$ (if $N = \langle \beta : \beta^3 = 1 \rangle$ use the isomorphism $\beta \leftrightarrow e^{2i\pi/3}$).*

3. *A normalised GH(4, v/4) over* $N = \{\pm 1, \pm i\} \subset \mathbb{C}$ *is a row and column balanced BH(4, v).*

4. *A normalised GH(w, v/w) over* $N = \{e^{(2i\pi/w)k}, 0 \leq k < w\} \subset \mathbb{C}$ *is a row and column balanced BH(w, v).*

5. *A GH(q − 1, v/(q − 1)) over* $N = GF(q)^*$ *is invertible over GF(q).*

6. *If R is commutative,* $|R^*| = w$ *and char* $R \neq 2$*, a GH(w, v/w) over* R^* *is invertible over R.*

4.3.3 Generalised Hadamard matrices and class regular divisible designs

A Hadamard matrix exists if and only if a Hadamard design exists (Lemma 2.7). For $w \geq 2$, there is a corresponding result which characterises generalised Hadamard matrices as a particular class of *divisible* designs.

A square *divisible* (v, w, k, λ)*-design* is a pair $\mathcal{D} = (P, B)$ consisting of a set P of vw points and a set B of vw blocks, each containing k points. The point set is partitioned into v point classes of w points each, such that two points in distinct point classes are both contained in precisely λ blocks, and no block contains distinct points in the same point class. Since the design is square, each point is contained in precisely k blocks.

When $k = v$, these designs are termed *semiregular* and are characterised by the existence of an incidence matrix A for which

$$AA^\top = vI_{vw} - (v/w)J_w \otimes I_v + (v/w)J_{vw}, \tag{4.9}$$

where J_n is the order n matrix containing only 1s (cf. [24, I.7.6] or [266, p. 3]). When $k = v \geq 3$, the class of square divisible (v, w, v, λ)-designs coincides with the class of *transversal designs* $TD_\lambda(v, w)$ [24, Proposition I.7.3]. The following definition was introduced by Jungnickel [189, §6], adopting the terminology for transversal designs.

DEFINITION 4.15 *A square divisible* (v, w, v, λ)*-design is* class regular *with respect to* N *if it admits an automorphism group* N *that acts regularly on each point class.*

The equivalence of generalised Hadamard matrices and class regular semiregular divisible designs was proved in a seminal paper by Jungnickel.

THEOREM 4.16 (Jungnickel [189, 6.5, 6.8]) *The existence of a* $v \times v$ *generalised Hadamard matrix GH(w, v/w) with entries in* N *is equivalent to that of a divisible* $(v, w, v, v/w)$*-design, class regular with respect to* N.

Proof. Given the design, select one point p_i from each of the v point classes $P_i, 1 \leq i \leq v$, and let $B_1, \ldots, B_v$ be the v distinct blocks incident with p_1. For $i, j \in \{1, \ldots, v\}$, block B_j meets point class P_i in precisely one point, say b_{ij}. Since N acts regularly on P_i, there is a unique $h_{ij} \in N$ such that $p_i^{h_{ij}} = b_{ij}$. For

$i \neq k \in \{1, \dots, v\}$ and for $h \in N$, the set $\{j : h_{ij} h_{kj}^{-1} = h\}$ has size v/w. Hence the matrix $[h_{ij}]$ is a $GH(w, v/w)$.

Conversely, if $M = [m_{ij}]$ is the $GH(w, v/w)$, give the point set of the design as the union of the point classes $P_i = \{(u, i) : u \in N\}, 1 \leq i \leq v$ and the blocks as $B_{ui} = \{(m_{ij}u, j) : 1 \leq j \leq v\}, u \in N, 1 \leq i \leq v$. $\square$

Ordinary representation theory provides a suitable incidence matrix for this design. Let $\mathcal{R} : N \to M_w(R)$ be the regular representation of N in the algebra $M_w(R)$ of $w \times w$ matrices with entries in a commutative ring with unity R. That is, for each $u \in N$, the $(0,1)$ matrix $\mathcal{R}(u)$ is indexed by the elements $u_i, 1 \leq i \leq w$, of N and $\mathcal{R}(u)_{kl} = 1$ if and only if $u_k u_l^{-1} = u$. Note that $\mathcal{R}(u)$ may be obtained from the multiplication table of N by swapping the column indexed by u_l for the column indexed by u_l^{-1} and in the resulting table, replacing $u_k u_l^{-1}$ by 1 whenever it equals u and by 0 otherwise. Consequently, $\mathcal{R}(u)$ is N-invariant.

LEMMA 4.17 [260, Lemma 3.1] *Suppose M is a $v \times v$ generalised Hadamard matrix $GH(w, v/w)$ over N. Replace each entry m_{ij} of M by its representation matrix $\mathcal{R}(m_{ij})$, and call the resulting $(0, 1)$ matrix A. Then A is the incidence matrix of a divisible $(v, w, v, v/w)$-design, class regular with respect to N.*

Proof. Since M is a generalised Hadamard matrix, $\sum_{j=1}^{v} m_{ij} m_{ij}^{-1} = v1$ and $\sum_{j=1}^{v} m_{ij} m_{kj}^{-1} = (v/w) \sum_{u \in N} u$ whenever $i \neq k$, in $\mathbb{Z}N$. Recall that $\mathcal{R}$ is a group homomorphism, with $\mathcal{R}(u)^{\top} = \mathcal{R}(u^{-1}) = \mathcal{R}(u)^{-1}$.

Thus $\sum_{j=1}^{v} \mathcal{R}(m_{ij}) \mathcal{R}(m_{ij})^{\top} = vI_w$ and $\sum_{j=1}^{v} \mathcal{R}(m_{ij}) \mathcal{R}(m_{kj})^{\top} = (v/w)J_w$ whenever $i \neq k$, and $AA^{\top} = [(v/w)J_w] \otimes (J_v - I_v) + (vI_w) \otimes I_v = vI_{vw} - (v/w)J_w \otimes I_v + (v/w)J_{vw}$, as required by (4.9). $\square$

4.3.4 Group developed $GH(w, v/w)$ and semiregular relative difference sets

We know from Lemma 2.19 that a group developed Hadamard matrix exists if and only if a Menon-Hadamard difference set exists. Naturally, we ask if there is a construction for divisible (v, w, k, λ)-designs which parallels the construction (see Theorem 2.9) for (v, k, λ)-designs from (v, k, λ)-difference sets.

Again, we have Jungnickel to thank for the answer: he uses the (normal) relative (v, w, k, λ)-difference sets introduced by Elliot and Butson [106] in 1966. We now provide a few relevant details of the topic, but it is an area with an extensive literature, and for more depth the reader is referred to Pott's monograph [266] and survey [267].

DEFINITION 4.18 *A relative (v, w, k, λ)-difference set $((v, w, k, \lambda)$-RDS), in a finite group E of order vw relative to a normal subgroup N of order w, is a k-element subset $R = \{r_1, \dots, r_k\}$ of E such that the sequence of quotients*

$$r_i r_j^{-1}, \quad r_i, r_j \in R, \ i \neq j \tag{4.10}$$

lists each element of $E \backslash N$ exactly λ times and lists no element from N. The subgroup N is called the forbidden *subgroup. An RDS R is said to be* normalised *if R contains the identity of E, and* central *if N lies in the centre of E. An RDS is called cyclic, abelian, metabelian, etc. if E has this property.*

Equally, R is a (v, w, k, λ)-RDS in E if and only if, in $\mathbb{Z}E$,

$$\sum_{r \in R} \sum_{s \in R} rs^{-1} = k.1 + \lambda \sum_{g \in E \setminus N} g. \tag{4.11}$$

If we employ the shorthand (beloved of researchers in difference sets) whereby a subset $X \subseteq G$ of a group G is identified with its sum $X = \sum_{x \in X} x$ in the group ring $\mathbb{Z}G$, then (4.11) is abbreviated as

$$RR^{(-1)} = k + \lambda E - \lambda N. \tag{4.12}$$

The value $w = 1$ corresponds to the case of an ordinary (v, k, λ)-difference set, and is usually excluded. Similarly, we generally assume that $v > 1$ and $k > 1$. Counting the quotients in two ways, we get the fundamental equation

$$k(k - 1) = \lambda w(v - 1). \tag{4.13}$$

The following result, often referred to as *projection* of RDS, highlights the close connection between RDSs and ordinary difference sets.

LEMMA 4.19 [106, Theorem 2.1] *Suppose that R is a (v, w, k, λ)-RDS in E relative to N. If $\rho : E \to H$ is an epimorphism with kernel K of order u contained in N, then $\rho(R)$ is a $(v, w/u, k, \lambda u)$-RDS in H relative to $\rho(N)$. In particular, there is always an ordinary $(v, k, \lambda w)$-difference set in E/N.*

An *extension*[4] of N by G is a short exact sequence of groups

$$1 \to N \overset{\imath}{\to} E \overset{\pi}{\to} G \to 1 , \tag{4.14}$$

that is, $\imath : N \rightarrowtail E$ is a monomorphism and $\pi : E \twoheadrightarrow G$ is an epimorphism satisfying $\ker(\pi) = \operatorname{im}(\imath)$. We will also call E an *extension* of N by G, or an *extension group*. The group N is the *kernel* of the extension and G is the *quotient*. Then N is isomorphic to the normal subgroup $\imath(N)$ of E and $G \cong E/\imath(N)$. A mapping $t : G \to E$ from G onto a transversal of $\imath(N)$ in E, that is, such that $\pi \circ t = \operatorname{id}_G$, is called a *section* of π.

For any normal subgroup N of E, there is always an extension

$$1 \to N \to E \overset{\eta}{\to} E/N \to 1 ,$$

where $\eta : E \to E/N$ is the natural quotient map. According to Lemma 4.19, any RDS R in E relative to N has an associated *underlying* ordinary difference set $D = \eta(R)$ in E/N. Relative difference sets are classified according to the type of their underlying ordinary difference sets: RDSs with underlying (v, v, v)-difference sets are called *semiregular*, while all other RDSs are called *regular*.

A semiregular RDS in E relative to N has parameters $(v, w, v, v/w)$ and is a complete transversal of N in E.

THEOREM 4.20 (cf. [189, Theorem 2.7]) *Let E be a finite group of order vw with a normal subgroup N of order w, and let $R \subseteq E$ be a k-subset of E. Then*

---

[4]Some authors call this an extension of G by N.

1. R is a (v, w, k, λ)-RDS in E relative to N if and only if

2. $\mathrm{dev}(R) = \big(E, \{Re : e \in E\}\big)$ is a (v, w, k, λ)-divisible design with point class partition $\{Ne : e \in E\}$, regular group E, where E acts on points by $h^e = he$ and on blocks by $(Rh)^e = Rhe$ for all $h, e \in E$, and is class regular with respect to N.

Moreover, any (v, w, k, λ)-divisible design $\mathcal{D}$ with regular group E and class regular with respect to N is isomorphic to $\mathrm{dev}(R)$ for a suitable (v, w, k, λ)-RDS R in E relative to N.

In [106] Elliott and Butson called an RDS R an 'extension' of its underlying difference set $\eta(R)$. More recently the term 'lifting' has been adopted by several authors (see [11], for example). More generally, we define a lifting of an ordinary difference set as follows.

DEFINITION 4.21 *Let $N \overset{\imath}{\rightarrowtail} E \overset{\pi}{\twoheadrightarrow} G$ be an extension of N by G, where N and G are finite groups. A* lifting *(in E) of an ordinary difference set $D \subseteq G$ is any RDS R in E relative to $\imath(N)$ such that $\pi(R) = D$.*

In the literature it has been traditional to call an RDS in E relative to N *splitting* if $E \cong N \times G$, so that any splitting RDS with N abelian is necessarily central. However, the perspective of Part 2 allows us to extend this definition. The following definition coincides with the traditional definition in the central case and provides a more general interpretation for splitting RDSs in the noncentral case as well.

DEFINITION 4.22 *An RDS R in E relative to N is a* splitting *RDS if E splits over N, that is, if there is a subgroup $H \leq E$ with $E = NH$ and $N \cap H = \{1\}$ (or equivalently, if E is isomorphic to a semidirect product $N \rtimes E/N$ of N by E/N).*

Two (v, w, k, λ)-RDS R and R' in a group E are *equivalent* (Pott [267, p. 198]) if there exist $\alpha \in \mathrm{Aut}(E)$ and $d, e \in E$ such that

$$R = d \cdot \alpha(R') \cdot e, \tag{4.15}$$

and *isomorphic* if $d = e = 1$ in (4.15).

The promised result of Jungnickel for G-developed generalised Hadamard matrices is as follows.

COROLLARY 4.23 (Jungnickel [189, 7.4]) *The existence of the following is equivalent:*

1. *a G-developed $GH(w, v/w)$ over N;*

2. *a relative $(v, w, v, v/w)$-difference set in $N \times G$, relative to $N \times \{1\}$;*

3. *a divisible $(v, w, v, v/w)$-design, class regular with respect to $N \times \{1\}$, with regular group $N \times G$.*

An obvious question to ask is whether any group developed $GH(w, v/w)$ exist. A table listing positive results for $v \leq 50$ and N a direct product of elementary abelian groups appears in de Launey [81, Table 2]. In particular, there exist $\mathbb{Z}_p$-developed $GH(p, 1)$ and $\mathbb{Z}_{p^2}$-developed $GH(p, p)$ over $\mathbb{Z}_p$, and Ma and Pott [232] determine necessary conditions for the existence of some group developed $GH(p^a, p^b)$, for p an odd prime.

However there are *no* $\mathbb{Z}_2^2$-developed $GH(4, 1)$ over $\mathbb{Z}_2^2$ [81, Table 2]. There also exist $GH(2^m, 2^m)$ which are not G-developed for any G (see Pott [267, Theorem 5.6], and apply Corollary 4.23).

We will see the first matrix of Example 4.3.1, while not normalised group developed, has a $\mathbb{Z}_2^2$-cocyclic development over $\mathbb{Z}_2^2$ (see Example 6.2.7), but the second matrix has neither construction (see Example 7.4.2).

4.4 APPLICATIONS OF COMPLEX AND GENERALISED HADAMARD MATRICES

Unimodular complex Hadamard matrices have important applications in quantum optics, high-energy physics, construction of *-subalgebras in finite von Neumann algebras and in investigation of Fuglede's conjecture. They also play a crucial rôle in quantum information theory, for construction of teleportation schemes or dense codes [304]. Such applications in mathematics and physics are not described here.

Circulant unimodular complex Hadamard matrices (or at least the perfect sequences which are formed by any row) have been widely adopted in linear system parameter identification, real-time channel evaluation, synchronisation, timing measurements, spread spectrum multiple access and 2-D signal processing [117].

When $GH(w, v/w)$, whether group developed or not, do exist, their transform, coding, spreading and correlation applications mimic those of Hadamard matrices. Similarly, semiregular RDSs have applications in signal processing precisely because of their excellent array correlation properties. For example, PBAs (Definition 3.22) are equivalent to splitting $(4u^2, 2, 4u^2, 2u^2)$-RDSs ([189], see Corollary 7.33). Jedwab's generalised perfect binary arrays (GPBAs) (see [186] for the definition) are equivalent to certain abelian $(4t, 2, 4t, 2t)$-RDSs ([186], see Lemma 7.38). Kumar's ideal matrices [213] for FDMA communications systems are the 2-D characteristic functions of splitting $(v, v, v, 1)$-RDSs in $\mathbb{Z}_v^2$, but are rare — see the examples of planar functions in Chapter 9.2.1.

In this Section, some of these extensions of Hadamard matrix applications to digital signals and data sequences, outlined in Chapter 3, are covered. We will begin with quaternary alphabets.

4.4.1 Quaternary complex Hadamard transforms

A quaternary complex Hadamard matrix transform for signals of length $n = 2^t$, the *complex BIFORE Transform (CBT)*, was introduced by Ahmed and Rao in 1970 (see [105, 10.2]). Measured by the number of nth roots of unity involved in any computation, the CBT has higher complexity than the WHT but much lower than the DFT. The transform matrices, which we denote $\mathcal{C}_t$, $t \geq 1$, have a recursive construction from the WHT and the basic 2×2 quaternary complex Hadamard matrix $\mathcal{C}_1$:

$$\mathcal{C}_2 = \begin{bmatrix} \mathcal{S}_1 & \mathcal{S}_1 \\ \mathcal{C}_1 & -\mathcal{C}_1 \end{bmatrix}, \quad \mathcal{C}_t = \begin{bmatrix} \mathcal{C}_{t-1} & \mathcal{C}_{t-1} \\ \mathcal{C}_1 \otimes \mathcal{S}_{t-2} & -\mathcal{C}_1 \otimes \mathcal{S}_{t-2} \end{bmatrix}, \quad t \geq 3. \quad (4.16)$$

Like the WHT (Theorem 3.7) and DFT, the CBT is fast to implement because the transform matrix factorises into sparse block diagonal matrices with blocks constructed from tensor products of $\mathcal{S}_1$, $\mathcal{C}_1$ and identity matrices [105, pp. 366-370].

A second family of quaternary complex Hadamard transforms for signals of length $n = 2^t$, which we denote $\mathcal{C}'_t$, has been developed by Rahardja and Falkowski [272, 273] originally for application to classification of switching (Boolean) functions. Such classification is an important problem in computer-aided design of logic circuits. Spectral methods using the WHT have proved cumbersome even for functions with a small number of variables. Using any one of 32 alternatives, for example $\mathcal{C}'_1 = \mathcal{C}_1$ or $\mathcal{C}'_1 = \begin{bmatrix} 1 & i \\ -i & -1 \end{bmatrix}$ (all equivalent to $\mathcal{C}_1$ and of course to $\mathcal{S}_1$), construction mimics that of the WHT, or more generally that of Example 4.1.3, with $\mathcal{C}'_t = \mathcal{C}'_1 \otimes \mathcal{C}'_{t-1}$ for $t \geq 2$. Unlike the CBT, the corresponding fast algorithm has a constant geometry: only one 'butterfly' stage has to be implemented and the processed data can be fed back to the input to be processed by the same circuitry. This transform is said to give a more efficient classification scheme for switching functions than the standard technique using the WHT, because only half the spectrum is needed for spectral analysis [273, Theorem 5].

4.4.2 Perfect quaternary sequences and arrays

When signal or data sequences are represented by 4 symbols, usually $\{1, i, -1, -i\}$ or their 'log $_i$' images $\{0, 1, 2, 3\} \cong \mathbb{Z}_4$, or modulated as 4-phase signals, as in QPSK (quadrature phase-shift keying) communications systems, they are called *quaternary* sequences. We have the same design and performance questions to ask as for binary sequences. It is not surprising that their answers may involve quaternary complex Hadamard matrices — or alternatively, their 'log $_i$' images over $\mathbb{Z}_4$.

When are quaternary sequences ideally correlated, and how easy is it to find ideally correlated sequences? The answer for quaternary sequences is nearly as restrictive as in the binary case, though, as we will see, this is really not surprising, by virtue of their connections with quaternary complex Hadamard matrices, and hence with real Hadamard matrices.

A quaternary sequence $\mathbf{x}$ of period n is *perfect* if its out-of-phase periodic autocorrelation (see Definition 3.18 and (4.6)) is always 0, and, analogously with the binary case, this is equivalent to the existence of a circulant quaternary complex Hadamard matrix, or equally, a $\mathbb{Z}_n$-developed quaternary complex Hadamard matrix. Turyn [308] proved that there exist circulant quaternary complex Hadamard matrices of orders 2, 4, 8 and 16, for instance, those with top rows

$$1\,i, \quad -1\,1\,1\,1, \quad 1\,1\,i -1\,1 -1\,i\,1 \; or \; 1\,1\,i\,1\,1 -1\,i -1,$$

[8, Example 2], see (4.5) and

$$1\ 1\ -i\ 1\ i\ -1\ 1\ -1\ 1\ -1\ -i\ -1\ i\ 1\ 1\ 1,$$

respectively, but that none exist with periods $n = 2^t, t > 4$ or $n = 2p^t$, for p an odd prime. It is conjectured that no other orders are possible, and Arasu et al. [9] adapt Schmidt's nonexistence results (Theorem 2.21) for circulant Hadamard matrices in support.

THEOREM 4.24 [9, Theorem 2.24] *Let Π be any finite set of primes. Then there are only finitely many circulant quaternary complex Hadamard matrices of order n, where all prime divisors of n are in Π.*

They confirm the conjecture for all but 11 orders $n \leq 1000$. They are $n = 260$, 340, 442, 468, 520, 580, 680, 754, 820, 884 and 890.

Research Problem 20 *Prove that no circulant quaternary complex Hadamard matrix of order $2t > 16$ exists.*

The frustration of searching for ideal sequences finally evaporates when we allow quaternary 2-D arrays rather than 1-D sequences of signals. Now the limitation, which seems apparent even in the case of 2-D binary arrays, does not exist.

DEFINITION 4.25 *An m-dimensional array $A = [a(i_0, \ldots, i_{m-1})]$ with entries $a(i_0, \ldots, i_{m-1}) \in \{\pm 1, \pm i\}$ for $0 \leq i_k \leq s_k - 1, 0 \leq k \leq m - 1$ is called an $s_0 \times s_1 \times \cdots \times s_{m-1}$ perfect quaternary array (PQA) of energy $\prod_{i=0}^{m-1} s_i$ and denoted by $PQA(s_0, \ldots, s_{m-1})$ if, for all $j_0, \ldots, j_{m-1}$,*

$$\sum_{i_0=0}^{s_0-1} \cdots \sum_{i_{m-1}=0}^{s_{m-1}-1} a(i_0, \ldots, i_{m-1}) \overline{a(i_0 + j_0, \ldots, i_{m-1} + j_{m-1})}$$

$$= \begin{cases} \prod_{i=0}^{m-1} s_i, & j_0 = j_1 = \ldots = j_{m-1} = 0, \\ 0 & \text{otherwise,} \end{cases} \quad (4.17)$$

where the index $i_k + j_k$ is reduced $\mod s_k$. Equally, the mapping $a : \mathbb{Z}_{s_0} \times \cdots \times \mathbb{Z}_{s_{m-1}} \to \{\pm 1, \pm i\}$ defining A is termed a PQA. (We assume that $\exists i : s_i \neq 1$.)

Arasu and de Launey [8] prove that many primes can divide the energy of a 2-D perfect quaternary array, in marked contrast to our best knowledge in the binary case. For some of these dimensions ($PQA(14, 14)$ and $PQA(28, 28)$, for instance) it is known that no PBA can exist.

THEOREM 4.26 [8, Theorem 7] *Let $l \geq 0, n, m$ be integers, let $g_0, g_1, \ldots, g_k$ be a nondecreasing sequence of positive integers and let $p_0, p_1, \ldots, p_k \equiv 3 \mod 4$ be a sequence of primes such that $p_0 = 2^{g_0} 3^{2l} - 1$ and $p_i = 2^{g_i} 3^{2l} p_0^2 \ldots p_{i-1}^2 - 1$ for $i > 0$. Then there exists a 2-D $PQA(2^n 3^l p_0 \ldots p_k, 2^m 3^l p_0 \ldots p_k)$ whenever $-4 \leq n - m \leq 4$ and $n + m \geq g_k - 1$.*

The argument giving the equivalence (Lemma 3.25) between Hadamard matrices which are group developed over $\mathbb{Z}_{s_0} \times \cdots \times \mathbb{Z}_{s_{m-1}}$ and $PBA(s_0, \ldots, s_{m-1})$ carries over with no difficulty to quaternary complex Hadamard matrices and PQAs.

LEMMA 4.27 *The top row of a quaternary complex Hadamard matrix which is group developed over $\mathbb{Z}_{s_0} \times \cdots \times \mathbb{Z}_{s_{m-1}}$ is a $PQA(s_0, \ldots, s_{m-1})$, and vice versa.*

There have been several attempts to generalise the idea of a perfect array, notably by Jedwab and Hughes. Jedwab's GPBAs are shown by Hughes [175] to include the PQAs as a particular case.

LEMMA 4.28 [175, Theorem 2.1] *Let $G = \mathbb{Z}_{s_0} \times \cdots \times \mathbb{Z}_{s_{m-1}}$. If $\varphi : G \rightarrow \{\pm 1, \pm i\}$ and $\phi : \mathbb{Z}_2 \times G \rightarrow \{\pm 1\}$ are related by*

$$\varphi(g) = \frac{1-i}{2}(\phi(0,g) + i\phi(1,g)), \ g \in G, \qquad (4.18)$$

then φ is a $PQA(s_0, \ldots, s_{m-1})$ if and only if ϕ is a $GPBA(2, s_0, \ldots, s_{m-1})$ of type $(1, 0, \ldots, 0)$.

By Lemma 4.27, if a group developed $GH(4, v/4)$ exists, its top row must be a particular kind of PQA. Hughes [175] calls a PQA *flat* if it is the top row of a group developed $GH(4, v/4)$, and proves that a PQA is flat if and only if its square is a PBA.

LEMMA 4.29 [175, Theorem 3.2] *Given a PQA φ, or equivalently a GPBA ϕ related by (4.18), then φ is flat if and only if $\varphi(g)^2 = \phi(0,g)\phi(1,g)$ is a PBA.*

As will be shown in Part 2 (see Chapter 7.4 and Lemma 7.37), the behaviour of PQAs can be better understood through their representation as relative difference sets and cocyclic matrices.

4.4.3 Quaternary error-correcting codes

The preferred assignment of the 4 possible phases in QPSK to two information bits is the Gray map $a + 2b \mapsto (b, a + b)$, for $a, b \in \{0, 1\}$ (see (3.8)), so that adjacent phases differ by only 1 bit. For error-correcting codes, this Gray map representation has the advantage that when a quaternary sequence is transmitted across an additive white Gaussian noise (AWGN) channel, the errors most likely to occur are those which, after demodulation, result in only a single bit error.

Of the binary error-correcting codes, linear codes are by far the most important, since they are easier to construct, encode and decode than nonlinear codes. However, the discovery, around 1970, of binary nonlinear codes having at least twice as many codewords as any linear code with the same length and minimum distance, means that some of the best possible codes are nonlinear. Examples are the Nordstrom-Robinson $(16, 256, 6)$ code, the Preparata codes, the Kerdock codes and the Goethals codes.

It was also realised that the weight enumerator of the extended Preparata code is the MacWilliams transform of that of the Kerdock code of the same length, though they are not dual to each other. Similarly, the Goethals codes are formally self-dual. This relationship was one of the great mysteries of coding theory, whose explanation by Nechaev [249] and Hammons et al. [139] in terms of $\mathbb{Z}_4$ and the Gray map caused great excitement in the coding world. The crux of this explanation is that there exist 'linear' error-correcting codes with alphabet $\mathbb{Z}_4$ whose binary images under the Gray map are these good nonlinear binary codes. Dual $\mathbb{Z}_4$-linear codes map to binary codes whose weight enumerators satisfy the MacWilliams identity. This remarkable breakthrough led to a great outpouring of new coding theory for codes over alphabets more general than fields.

A clear account of quaternary error-correcting codes appears in Wan [317]. The definitions for quaternary error-correcting block codes mimic those of Definition 3.9, with the rôle of vector space $V(n, q)$ taken by the free $\mathbb{Z}_4$-module $\mathbb{Z}_4^n$. In particular, a $\mathbb{Z}_4$-code is *linear* if it is a $\mathbb{Z}_4$-submodule of $\mathbb{Z}_4^n$, or equally, if it is a subgroup of $\mathbb{Z}_4^n$. The *dual* of a linear code C is the set $C^\perp$ of elements in $\mathbb{Z}_4^n$ orthogonal to all codewords of C, that is, $C^\perp = \{v \in \mathbb{Z}_4^n : v \cdot c = 0, \ \forall\, c \in C\}$, where $v \cdot c = \sum_{i=1}^n v_i c_i$. A $k \times n$ matrix over $\mathbb{Z}_4$ whose rows generate a linear code C but for which no proper subset of rows generates C is called a *generator matrix* for C.

For reference, two basic results on $\mathbb{Z}_4$-linear codes are recorded (for example, see [317, Proposition 1.1, Theorem 3.7]).

LEMMA 4.30   *1. Any $\mathbb{Z}_4$-linear code C containing some nonzero codewords is permutation equivalent to a $\mathbb{Z}_4$-linear code with a generator matrix of the form*

$$\begin{pmatrix} I_{k_1} & A & B \\ 0 & 2I_{k_2} & 2A' \end{pmatrix},$$

where A and A' are $\mathbb{Z}_2$-matrices and B is a $\mathbb{Z}_4$-matrix. Then C is an abelian group isomorphic to $\mathbb{Z}_4^{k_1} \times \mathbb{Z}_2^{k_2}$ containing $2^{2k_1+k_2}$ codewords, and C is a free $\mathbb{Z}_4$-module if and only if $k_2 = 0$.

2. Let C and $C^\perp$ be dual $\mathbb{Z}_4$-linear codes and let $\phi(C)$ and $\phi(C^\perp)$ be their binary images under the Gray map. Then the weight enumerators of $\phi(C)$ and $\phi(C^\perp)$ are related by the binary MacWilliams identity

$$W_{\phi(C^\perp)}(x, y) = |\phi(C)|^{-1} W_{\phi(C)}(x + y, x - y).$$

Nechaev [249] originally showed that it is possible to permute the coordinates of the Kerdock code (punctured in two coordinates) to obtain a binary cyclic code, using a permutation derived from a $\mathbb{Z}_4$-linear code. The $\mathbb{Z}_4$-linear code can be constructed from a family of low cross-correlation $\mathbb{Z}_4$-sequences (discovered by Solé [300], Boztaş et al. [34]) known as Family $\mathcal{A}$, and its Gray map image is the Kerdock code [139]. The same construction works for other sequence families. This idea has been exploited in the reverse direction, to obtain families of low cross-correlation $\mathbb{Z}_4$-sequences from binary cyclic codes which are permuted images of $\mathbb{Z}_4$-linear codes (see [263, 21.8]). They are being implemented: Family $\mathcal{A}$ forms part of IMT-2000, the CDMA standard for new 3G (third-generation) wireless systems.

Derivation of these high performance quaternary codes and low correlation $\mathbb{Z}_4$-sequence families from generalised Hadamard matrices is deferred to Chapter 9 (Example 9.1.2).

4.4.4 Generalised Hadamard matrices and Hadamard codes

It is a short step conceptually from binary and quaternary block codes to block codes over arbitrary alphabets, though to date the only symbol alphabets implemented are the finite fields $GF(q)$ and the finite ring $\mathbb{Z}_4$.

If we suppose the code alphabet is a finite group N of order w, and make the obvious modifications to Definitions 3.9 and 3.10 so that a code of length n is a subset of N^n, many of the code construction techniques for Hadamard matrices and designs (cf. [14, p. 41, §7]) apply to generalised Hadamard matrices. The Hamming distance between any two distinct rows of a $GH(w, v/w)$ is $v(w-1)/w$, and the Hamming weight of any noninitial row of a normalised $GH(w, v/w)$ is also $v(w-1)/w$, so we can construct codes with high distance relative to length and codes with constant weight. Nonbinary constant-weight codes may be used to construct spherical codes. Unless N has a commutative ring structure, a definition of linearity modelled on $\mathbb{Z}_4$-linearity (Section 4.4.3) makes no sense, and the notion of *additivity* is substituted.

DEFINITION 4.31 *Let N be a finite group of order w. A code $C \subset N^n$ is* additive *if it is a subgroup of N^n. If N is the additive group of a commutative ring R, then C is R-*linear *if C is an R-submodule of R^n.*

The Plotkin bound for w-ary codes with high distance relative to length is $A_w(n, d) \leq wd/(wd - n(w-1))$ if $wd > n(w-1)$ and $A_w(n, d) \leq wn$ if $wd = n(w-1)$. (The proof of Lemma 3.12 translates directly.)

Mackenzie and Seberry prove the w-ary analogue of Levenshtein's Theorem 3.14, provided sufficiently many generalised Hadamard matrices exist [234, Theorem 12]. They also derive w-ary analogues of the Hadamard code constructions of Definition 3.13, as well as an extra construction when $v = w$ [234, Lemma 4]. (For $w = 2$ this extra construction is the $[3, 2, 2]$ even weight code.) Following Definition 3.15 we will call them *Class A w-ary Hadamard codes*.

DEFINITION 4.32 *Let H be a normalised $GH(w, v/w)$ over N. The* Class A w-ary Hadamard codes *are*

1. $\mathcal{A}_v$ — *the w-ary $(v - 1, v, v(w-1)/w)$ code consisting of the rows of H with the first column deleted, which meets the Plotkin bound $A_w(n, d) = wd/(w-1)$ since $wd - n(w-1) = w - 1$;*

2. $\mathcal{B}_v$ — *the w-ary $(v-1, vw, v(w-1)/w-1)$ code consisting of the translates $u\mathcal{A}_v$, $u \in N$;*

3. $\mathcal{C}_v$ — *the w-ary $(v, vw, v(w-1)/w)$ code consisting of the rows of the translates uH, $u \in N$, which meets the Plotkin bound $A_w(n, d) = wn$ since $wd = n(w-1)$; and*

4. $\mathcal{D}_v$ — when $v = w$, the w-ary $(w + 1, w^2, w)$ code consisting of the rows of $(uH)\mathbf{c}$, for all $u \in N$ and any fixed noninitial column $\mathbf{c}$ of H, which meets the Plotkin bound $A_w(n, d) = wd$ since $wd - n(w - 1) = 1$.

Typically, these codes are not additive or linear but there are obvious analogues of the linearisation methods used in the binary case.

One popular technique for constructing $GF(p)$-linear codes for p a prime is to take the $GF(p)$-span of the rows of the incidence matrix of a design. Bounds on the $GF(p)$-dimension of such a code were obtained by Klemm [14, Theorem 2.4.2]. If A is the incidence matrix (Lemma 4.17) of the divisible design $\mathcal{D}$ corresponding to a $GH(w, v/w)$ over N, these bounds can be partially generalised to the code spanned by the rows of A.

COROLLARY 4.33 *Let A be the incidence matrix of a divisible $(v, w, v, v/w)$-design. Then for any prime p,*

$$
\begin{aligned}
vw - (v - 1) &\leq \operatorname{rank}_p A \leq vw - \tfrac{1}{2}(v - 1), & (p, vw) = 1, \\
1 &\leq \operatorname{rank}_p A \leq \tfrac{1}{2}(vw + 1), & (p, vw) \neq 1.
\end{aligned}
$$

Proof. Pott [266, Lemma 1.1.4] shows that the eigenvalues of $AA^\top$ in (4.9) are v^2 with multiplicity 1; v with multiplicity $(vw - v)$; and 0 with multiplicity $(v - 1)$. It follows that $\operatorname{rank}_p(AA^\top) \leq vw - v + 1$ for any p. Then apply Pott's analysis [266, pp. 158–159] to this case. $\square$

As for the binary case, we extend the coinage *Hadamard code* to codes derived from generalised Hadamard matrices, and classify them according to the construction technique used.

DEFINITION 4.34 *Let H be a normalised $GH(w, v/w)$ over N, and let R be a commutative ring with unity.*

A Class A w-ary Hadamard code is as given in Definition 4.32.

A Class B w-ary Hadamard code is an R-linear code not in Class A, derived from some rows of H.

A Class C w-ary Hadamard code is an R-linear code with generator matrix $[I \ A]$ for some matrix A associated with H.

Instances of well-known codes in these classes appear in Chapter 9.1.3.3.

4.5 UNIFICATION: GENERALISED BUTSON HADAMARD MATRICES AND TRANSFORMS

The two principal notions of generalisation for Hadamard matrices are very simply reconciled. We simultaneously extend the entries permissible for a Butson matrix from m^{th} roots of unity in $\mathbb{C}$ to elements in a subgroup of units in a ring with unity, and require a generalised Hadamard matrix over N to be invertible (Definition 4.14), for example over the integral group ring $\mathbb{Z}N$.

DEFINITION 4.35 *Suppose R is a ring with unity 1, group of units R^* and that char R does not divide v. A square matrix M of order $v \geq 2$, with entries from a subgroup $N \leq R^*$ is a Generalised Butson Hadamard (GBH) matrix, denoted GBH(N, v), if M is unitary[5], that is*
$$MM^* = M^*M = vI_v,$$
where M^* is the transpose of the matrix of inverse elements of M: $m_{ij}^* = (m_{ji})^{-1}$. (Write GBH(w, v) if N is finite of order w and its structure is irrelevant.)

Usually we restrict to finite N in Definition 4.35 both for simplicity and because this is the working environment for discrete applications. This condition is not necessary. Interest in and application of matrices with entries from the complex unimodular group has continued since Hadamard's day. Sylvester worked with unrestricted entries from $\mathbb{C}^*$. Craigen and Woodford [70] define "power Hadamard matrices" and work with the ring of formal Laurent polynomials $R = \mathbb{Q}[x, x^{-1}]$ and $N = \langle x \rangle \cong \mathbb{Z}$ (but allow $MM^* = vI_v \bmod f(x)$ for some Laurent polynomial $f(x)$ of degree > 0) to explore all the above matrix types.

GBH matrices include all unimodular complex Hadamard matrices and all invertible generalised Hadamard matrices. Pseudo-Hadamard matrices (3.24), however, are not GBH matrices, since their inverses are not of the required form.

For unimodular complex Hadamard matrices and for invertible generalised Hadamard matrices over an abelian group N, $HH^* = vI_v \Rightarrow H^*H = vI_v$. The largest known class of generalised Hadamard matrices for which $HH^* = vI_v \Rightarrow H^*H = vI_v$ is the subset of invertible matrices in the set of matrices described in Lemma 7.26.3, a new result found by the group extension techniques of Part 2.

It it not known if the implication is true for all GBH matrices.

Research Problem 21 *Suppose R is a ring with unity 1, group of units R^*, and that char $R \nmid v$. If a square matrix M of order $v \geq 2$, with entries from a subgroup $N \leq R^*$ satisfies $MM^* = vI_v$, does $M^*M = vI_v$?*

A GBH matrix is always equivalent to a *normalised* GBH matrix, which has first row and column consisting of all 1s. By taking the inner product of any noninitial row of a normalised GBH matrix M with the all-1s first column of M^*, we see that the sum of the entries in any row of M, apart from the first, must equal 0, and similarly for rows of M^* (columns of the matrix of inverses $M^{(-1)} = [m_{ij}^{-1}]$). The tensor product of two GBH matrices over the same group N is a GBH matrix over N. For instance, in Example 4.1.3, $\otimes^n \mathcal{F}_p$ is a GBH matrix over the group of complex p^{th} roots of unity.

The next subsection describes a new construction for GBH matrices of even order and additional internal structure, introduced in [160] and improved in [166].

4.5.1 The jacket matrix construction

Lee introduced generalisations both of the WHT [220] and of the even length DFT [219] under the name *reverse jacket transforms (RJT)*, so-called because the uni-

---

[5]This is a slight, but common, abuse of notation; more accurately, the scaled matrix $\widehat{M} = \sqrt{v}^{-1} M$ is unitary: $\widehat{M}\widehat{M}^* = I_v$, but this requires $\sqrt{v} \in R^*$.

tary matrix representing the transform has a border ('jacket') and centre which switch some elements under inversion. This family includes the 'centre-weighted Hadamard transforms' with a border of $\{\pm 1\}$ and real centre entries of absolute value greater than 1, which weight the mid-band frequencies of the signal more.

In [160] the author names those GBH matrices which have a border of $\{\pm 1\}$, including the centre-weighted Hadamard transform matrices, *jacket matrices*. In [166] the maximum width of a border of $\{\pm 1\}$ is proposed as a new parameter for classifying transforms.

Throughout this subsection, let G be an indexing set of order v (sometimes G is a group such as $\mathbb{Z}_v$ or $\mathbb{Z}_2^t$ but G may be nonabelian or the group structure may be irrelevant).

DEFINITION 4.36 *Let R be a ring with unity* 1. *A normalised GBH(N, v) matrix K indexed by $G = \{1, \ldots, v\}$ with entries from $N \leq R^*$ is a* jacket matrix *if it is permutation equivalent to a matrix of the form*

$$\widetilde{K} = \begin{bmatrix} 1 & 1 & \ldots & 1 & 1 \\ 1 & * & \ldots & * & \pm 1 \\ \vdots & \vdots & \vdots & \vdots & \vdots \\ 1 & * & \ldots & * & \pm 1 \\ 1 & \pm 1 & \ldots & \pm 1 & \pm 1 \end{bmatrix}, \qquad (4.19)$$

where the central entries $*$ are from N. The jacket width *of K is $m \geq 1$ if K is permutation equivalent to a jacket matrix $\widetilde{K}$ in which rows $1, \ldots, m, v - m + 1, \ldots, v$ and columns $1, \ldots, m, v-m+1, \ldots, v$ all consist of ± 1 and m is maximal for this property. If K is not permutation equivalent to any jacket matrix, it has jacket width 0.*

Since all noninitial (± 1) rows and columns of a jacket matrix sum to 0 in R, the order v of a jacket matrix K must be even.

COROLLARY 4.37 *If v is odd, any normalised GBH(N, v) matrix has jacket width 0.*

If K is a jacket matrix of order $2n$ and jacket width $m \geq 1$, it follows that K^* is itself a jacket matrix of jacket width m. If also $2n \in R^*$, then K has an *inverse* $K^{-1} = (2n)^{-1} K^*$ over R. If $(2n)^{\frac{1}{2}} \in R^*$, then the scaled matrix $\widehat{K} = (2n)^{-\frac{1}{2}} K$ satisfies $\widehat{K}\widehat{K}^* = I_{2n}$ so is unitary in the usual sense.

Example 4.5.1   *The matrix $\mathcal{C}_t$, $t \geq 2$, of the CBT of length 2^t (4.16) is a jacket matrix.*

Proof. For $t \geq 2$, $\mathcal{C}_t$ is normalised and the first two rows of $\mathcal{C}_t$ consist of 2^{t-1} copies of $\mathcal{S}_1$. By induction the $(2^{t-1} + 1)^{st}$ column of $\mathcal{C}_t$ is $[\mathbf{1} \; -\mathbf{1}]^\top$, where $\mathbf{1}$ has length 2^{t-1}. Rotating the second row to the bottom and the $(2^{t-1} + 1)^{st}$ column to the right of $\mathcal{C}_t$ produces a matrix of form (4.19). (For $t = 1$, $\mathcal{C}_1 \sim \mathcal{S}_1$.) $\square$

Example 4.5.2   *The matrix $\mathcal{F}_{2n}$ of the DFT of length $2n$ (3.6) is a jacket matrix of width* 1.

Proof. [221, Theorem 1, Definition 1] Let $\mathcal{F}_{2n} = [\,\omega^{jk}\,]_{0 \le j,k \le 2n-1}$, where $\omega = e^{-\pi i/n}$, $n \ge 1$. Represent the indices in mixed radix notation $j = j_1 n + j_0 = (j_1, j_0)$, with index set $G = \mathbb{Z}_2 \times \mathbb{Z}_n$. The permutation $(j_1, j_0) \mapsto (j_1, (1-j_1)j_0 + (n-1-j_0)j_1)$ leaves the first n indices unchanged and reverses the order of the last n indices. Under this permutation on rows and columns, $\mathcal{F}_{2n}$ is equivalent to

$$\mathcal{K}_n(\omega) = [\omega^{\{(1-j_1)j_0 + (n-1-j_0)j_1 + j_1 n\}\{(1-k_1)k_0 + (n-1-k_0)k_1 + k_1 n\}}], \qquad (4.20)$$

the matrix of Lee's complex RJT, which is of width ≥ 1. But for a fixed j, $\omega^{jk} = \pm 1$ for all $0 \le k \le 2n-1$, if and only if $\omega^j = \pm 1$, if and only if either $j = 0$ (the initial row and column) or $j = n$. Thus the width is exactly 1. $\qquad \square$

At the other extreme, the WHT matrix $\mathcal{S}_t$ of order 2^t, with jacket width 2^{t-1}, is 'all jacket'. In fact this holds for any Hadamard matrix.

Example 4.5.3 *A jacket matrix K of order $2n$ has* maximum *width n if and only if K is a normalised Hadamard matrix. If so, either $K = \mathcal{S}_1$ or n is even.*

Note that, if K, K' are jacket matrices indexed by G, G' of orders $2n, 2n'$, respectively, with entries from R^*, then the tensor product $K \otimes K'$ is a jacket matrix indexed by $G \times G'$ of order $4nn'$, with entries from R^*, since the border condition is easily seen to be satisfied. In fact, a tensor product of jacket matrices is a jacket matrix of width ≥ 2.

THEOREM 4.38 *If K_i is a normalised GBH(N, v_i) matrix of jacket width m_i with entries from R^*, for $i = 1, 2$, then $K_1 \otimes K_2$ is a jacket matrix of width at least $2m_1 m_2$, if $m_1 m_2 \ge 1$, and of width at least m_j, $(j \ne i)$, if $m_i = 0$.*

Proof. Let K_1 have order $2n$ and K_2 have order $2n'$. Permute K_i to $\widetilde{K}_i$, so $K_1 \otimes K_2$ is permutation equivalent to $\widetilde{K}_1 \otimes \widetilde{K}_2$. Let $i \in \{1, \ldots, m_1, 2n - m_1 + 1, \ldots, 2n\}$ be an index of an all-(± 1)s row in $\widetilde{K}_1$. The corresponding i^{th} block row in $\widetilde{K}_1 \otimes \widetilde{K}_2$ consists of $2n$ copies of $\pm 1 \widetilde{K}_2$, so each row indexed $\{2, \ldots, m_2, 2n' - m_2 + 1, \ldots, 2n'\}$ of each copy consists of all-(± 1)s, contributing $2m_2 - 1$ all-(± 1) rows to the ith block row of $\widetilde{K}_1 \otimes \widetilde{K}_2$. If $i = 1$ the top row is all 1s and if $i > 1$ the top row consists of n all-1s subrows and n all-(-1)s subrows. Those in the top m_1 block rows of $\widetilde{K}_1 \otimes \widetilde{K}_2$ may be permuted to occupy the top $2m_1 m_2$ rows and those in the bottom m_1 block rows to the bottom $2m_1 m_2$ rows. A similar process applies for columns. If $m_i = 0$ the argument easily adapts to show $K_1 \otimes K_2$ has width at least m_j $(j \ne i)$. $\qquad \square$

If a jacket matrix may be decomposed as a tensor product of two smaller jacket matrices, the decomposition may be repeated until no further tensor product decomposition is possible.

DEFINITION 4.39 *A jacket matrix of length $2n$ is a primary jacket matrix K_n if it is minimal with respect to tensor product, that is, there are no jacket matrices K, K' such that K_n is permutation equivalent to $K \otimes K'$.*

Clearly, any jacket matrix of width 1, such as the even length DFTs of Example 4.5.2, is primary, but by Example 4.5.3 this sufficient condition is not necessary for a jacket matrix to be primary.

COROLLARY 4.40 *A jacket matrix of width* 1 *is primary. A jacket matrix of order* $2n \geq 4$ *and maximum width* $n = 2k$ *is primary whenever* k *is odd.*

Examples of primary jacket matrices of width 1 for the first four values of n are

$$K_1 = \mathcal{S}_1 = \begin{bmatrix} 1 & 1 \\ 1 & -1 \end{bmatrix},$$

$$K_2(r) = \begin{bmatrix} 1 & 1 & 1 & 1 \\ 1 & -r & r & -1 \\ 1 & r & -r & -1 \\ 1 & -1 & -1 & 1 \end{bmatrix}, \quad r \neq \pm 1 \in R^*, \tag{4.21}$$

$$K_3(\alpha) = \begin{bmatrix} 1 & 1 & 1 & 1 & 1 & 1 \\ 1 & \alpha & \alpha^2 & \alpha^5 & \alpha^4 & -1 \\ 1 & \alpha^2 & \alpha^4 & \alpha^4 & \alpha^2 & 1 \\ 1 & \alpha^5 & \alpha^4 & \alpha & \alpha^2 & -1 \\ 1 & \alpha^4 & \alpha^2 & \alpha^2 & \alpha^4 & 1 \\ 1 & -1 & 1 & -1 & 1 & -1 \end{bmatrix}, \tag{4.22}$$

where α is a primitive 6th root of unity in an integral domain R, and

$$K_4(i) = \begin{bmatrix} 1 & 1 & 1 & 1 & 1 & 1 & 1 & 1 \\ 1 & i & -i & 1 & -1 & i & -i & -1 \\ 1 & -i & -1 & i & i & -1 & -i & 1 \\ 1 & 1 & i & i & -i & -i & -1 & -1 \\ 1 & -1 & i & -i & i & -i & 1 & -1 \\ 1 & i & -1 & -i & -i & -1 & i & 1 \\ 1 & -i & -i & -1 & 1 & i & i & -1 \\ 1 & -1 & 1 & -1 & -1 & 1 & -1 & 1 \end{bmatrix}. \tag{4.23}$$

The matrix K_1 is the unique 2×2 jacket matrix, and $K_2(1) = K_1 \otimes K_1 = \mathcal{S}_2$, so is not primary. The matrix $K_2(r)$ for $r \neq \pm 1 \in \mathbb{R}^*$ is a 'centre-weighted Hadamard transform' (CWHT) matrix [221] and, for $r = i \in \mathbb{C}^*$, is $\mathcal{K}_2(i)$ in (4.20). The matrix $K_3(\alpha)$ with $\alpha = e^{i\pi/3}$ is $\mathcal{K}_3(e^{i\pi/3})$ and with α the fourth power of a primitive root in $GF(25)$ is an 'extended' complex RJT matrix [221, Example 4]. The matrix $K_4(i)$ is permutation equivalent to (4.4).

There are infinite families of jacket matrices (with entries from a fixed N, indexed by orders $2n$) with minimum width 1 and with maximum width n, but what of other widths?

Research Problem 22 *Given* N, *do families (of infinitely many orders* $v = 2n$) *of primary jacket matrices with entries in* N *exist in all possible jacket widths* $w = 1, \ldots, n$ (n *even) and* $w = 1, \ldots, n - 1$ (n *odd)?*

For complex Hadamard matrices, the important question for applications is how many equivalence classes exist for a given n, as N ranges over the unimodular group, and whether they are *isolated* classes [304, Definition 3.1]. Perhaps jacket width is a classifier here too.

Research Problem 23 *For each n, do equivalent unimodular complex Hadamard matrices of order n have the same jacket width?*

COROLLARY 4.41 *A jacket matrix is permutation equivalent to a tensor product of one or more primary jacket matrices. Conversely, any tensor product of primary jacket matrices is a jacket matrix.*

Research Problem 24 *Is the decomposition of a jacket matrix as a permutation equivalent tensor product of primary jacket matrices unique (up to order of the factors)?*

The tensor product of two jacket matrices is a jacket matrix, but by Theorem 4.38, for a tensor product of normalised GBH matrices to be a jacket matrix it is enough that one factor is a jacket matrix.

COROLLARY 4.42 *Let B be a normalised GBH matrix and K a jacket matrix, both with entries in R^*. Then $B \otimes K$ is a jacket matrix.*

This result explains the generation of some primary jacket matrices and is fundamental to the construction of Generalised Hadamard Transforms in Section 4.5.2.

Example 4.5.4 *Let $\beta \neq 1 \in R^*$ satisfy $\beta^2 + \beta + 1 = 0$, so $\beta^3 = 1$. Let*

$$
B_3 = \begin{bmatrix} 1 & 1 & 1 \\ 1 & \beta & \beta^2 \\ 1 & \beta^2 & \beta \end{bmatrix},
$$

so B_3 is a normalised GBH$(3,3)$. Then

$$
B_3 \otimes K_1 \sim \begin{bmatrix}
1 & 1 & 1 & 1 & 1 & 1 \\
1 & \beta & \beta & \beta^2 & \beta^2 & 1 \\
1 & \beta & -\beta & \beta^2 & -\beta^2 & -1 \\
1 & \beta^2 & \beta^2 & \beta & \beta & 1 \\
1 & \beta^2 & -\beta^2 & \beta & -\beta & -1 \\
1 & 1 & -1 & 1 & -1 & -1
\end{bmatrix}.
$$

This jacket matrix relates to the DFT matrix of (4.22) as follows. Permutation (2543) cycling central rows and columns gives the jacket matrix

$$
\begin{bmatrix}
1 & 1 & 1 & 1 & 1 & 1 \\
1 & -\beta & \beta^2 & -\beta^2 & \beta & -1 \\
1 & \beta^2 & \beta & \beta & \beta^2 & 1 \\
1 & -\beta^2 & \beta & -\beta & \beta^2 & -1 \\
1 & \beta & \beta^2 & \beta^2 & \beta & 1 \\
1 & -1 & 1 & -1 & 1 & -1
\end{bmatrix}.
$$

When α is a primitive 6th root of unity in R^* with $\alpha^2 = \beta$ and $\alpha^5 = \gamma$, then this matrix equals $K_3(\gamma)$.

Example 4.5.5 *Let $r \neq \pm 1 \in R^*$. Then $B_3 \otimes K_2(r) \sim K_6(\beta, r) =$*

$$
\begin{bmatrix}
1 & 1 & 1 & 1 & 1 & 1 & 1 & 1 & 1 & 1 & 1 & 1 \\
1 & -r & r & 1 & -r & r & -1 & 1 & -r & r & -1 & -1 \\
1 & r & -r & 1 & r & -r & -1 & 1 & r & -r & -1 & -1 \\
1 & 1 & 1 & \beta & \beta & \beta & \beta & \beta^2 & \beta^2 & \beta^2 & \beta^2 & 1 \\
1 & -r & r & \beta & -r\beta & r\beta & -\beta & \beta^2 & -r\beta^2 & r\beta^2 & -\beta^2 & -1 \\
1 & r & -r & \beta & r\beta & -r\beta & -\beta & \beta^2 & r\beta^2 & -r\beta^2 & -\beta^2 & -1 \\
1 & -1 & -1 & \beta & -\beta & -\beta & \beta & \beta^2 & -\beta^2 & -\beta^2 & \beta^2 & 1 \\
1 & 1 & 1 & \beta^2 & \beta^2 & \beta^2 & \beta^2 & \beta & \beta & \beta & \beta & 1 \\
1 & -r & r & \beta^2 & -r\beta^2 & r\beta^2 & -\beta^2 & \beta & -r\beta & r\beta & -\beta & -1 \\
1 & r & -r & \beta^2 & r\beta^2 & -r\beta^2 & -\beta^2 & \beta & r\beta & -r\beta & -\beta & -1 \\
1 & -1 & -1 & \beta^2 & -\beta^2 & -\beta^2 & \beta^2 & \beta & -\beta & -\beta & \beta & 1 \\
1 & -1 & -1 & 1 & -1 & -1 & 1 & 1 & -1 & -1 & 1 & 1
\end{bmatrix}
$$

is a 12×12 primary jacket matrix by Corollary 4.40, since it has width 1.

Jacket width is defined for all normalised $GBH(N, v)$ and is an invariant of each permutation equivalence class, but it is not known whether the width can change if all equivalence operations are allowed.

Research Problem 25 *Is jacket width an invariant of each equivalence class of $GBH(N, v)$?*

4.5.2 The Generalised Hadamard Transform

There are many discrete signal transforms (3.5) whose transform matrices have entries on the complex unit circle. For instance, the Fourier Transforms (4.2), found by interpreting the Cooley-Tukey Fast Fourier Transform in terms of abelian group characters, include the WHT and DFT. The family of discrete Generalised Transforms $\{(GT)_r, 0 \leq r \leq m - 1\}$ for signals of length $n = 2^m$ [105, 10.2] includes the CBT (complex BIFORE Transform) of (4.16) (as case $r = 1$) as well as the WHT (as case $r = 0$) and the 2^m-point DFT (as case $r = m - 1$).

Both the WHT and DFT are suboptimal discrete unitary transforms, but each has wide application, and efficient, easily implemented fast algorithms exist to compute them.

An optimal discrete unitary transform (in a statistical sense) is the Karhunen-Loève Transform, but it has significant disadvantages in implementation and is seldom used in signal processing [105]. Instead it can be considered as a standard against which performance of other transforms may be evaluated.

When entries outside the complex unit circle are allowed, Lee's reverse jacket transforms are multiphase or multilevel generalisations of the WHT [220] and of the even length DFT [219]. Some of them admit a recursive factorisation into tensor products so represent a fast transform similar to that of the WHT. (The formula [221, Definition 5] is not a generalised transform unifying both the WHT and even-length DFT, as claimed, because it is not unitary.)

In the most general situation, we work in a ring R with unity 1. This includes $\mathbb{R}$, $\mathbb{C}$ and Galois field alphabets $GF(p^a)$, though if we need to distinguish signal values

x and $-x$, the ring must have characteristic $\neq 2$. The family of GBH matrices defines a Generalised Hadamard Transform which includes the RJTs, the $(GT)_r$ and the FTs, so is a truly unifying transform.

DEFINITION 4.43 *Let* **x** *be a signal of length n from a ring R with unity 1, where $n \in R^*$, let $N \leq R^*$ and let B be a* **GBH**(N, n). *A* Generalised Hadamard Transform (GHT) *of* **x** *is*

$$\hat{\mathbf{x}} = B\,\mathbf{x} \qquad (4.24)$$

and an Inverse Generalised Hadamard Transform (IGHT) *of* $\hat{\mathbf{x}}$ *is*

$$\mathbf{x} = n^{-1}B^*\,\hat{\mathbf{x}}. \qquad (4.25)$$

The jacket width w of a $GBH(N, n)$ is a third parameter, after n and N, which we can use to construct and classify signal transforms over varying signal alphabets.

Research Problem 26 *What properties of a complex-valued signal of length n does the jacket width of a GHT with entries from $N = \langle e^{i\,2\pi/m} \rangle$, or more general unimodular complex alphabets, measure?*

Those GHT matrices which are jacket matrices have additional structure, by virtue of their tensor product decomposition into primary jacket matrices and their jacket form, which may particularly suit them to specific applications.

Consider the set of jacket matrices with entries in an integral domain R

$$\{K = (\otimes^{\ell} K_1) \otimes K_2(r)^{\epsilon} \otimes \mathcal{K}_n(\alpha)^{\delta};\ \ell \geq 0, \epsilon, \delta \in \{0, 1\}\}, \qquad (4.26)$$

where $r \neq \pm 1 \in R^*$, α is a primitive $2n^{th}$ root of unity and where by M^0 we mean the 1×1 identity matrix. When $\ell \geq 1, \epsilon = 0, \delta = 0$, this is the WHT of (3.11). When $\ell = 0, \epsilon = 0, \delta = 1$ and $R = \mathbb{C}$ this is equivalent to the $2n$-point DFT of (3.6). When $\epsilon = 1, \delta = 0$ and $R = \mathbb{R}$, this is the CWHT. When $\epsilon = 0, n = 2, \alpha = i \in \mathbb{C}^*, \delta = 1$, or when $\epsilon = 1, r = i \in \mathbb{C}^*, \delta = 0$, this is the complex RJT, and when $\epsilon = 0, \delta = 1$, this is the extended complex RJT.

To summarise: the GHT includes the Walsh-Hadamard, complex BIFORE, Discrete Fourier, Fourier, Generalised, centre-weighted Hadamard, Complex Reverse Jacket and extended Complex Reverse Jacket Transforms. In the jacket case, GHT matrices can be permuted into tensor products of primary GHT matrices. Primary GHT matrices may themselves be tensor products of a GBH matrix which is not a primary jacket matrix, and a primary jacket matrix.

Research Problem 27 *Determine the attributes and performance of the GHT as a discrete signal transform. Determine the relative advantage or disadvantage of the primary jacket tensor form*

$$K_{i_1} \otimes K_{i_2} \otimes \cdots \otimes K_{i_k} \qquad (4.27)$$

where $K_{i_1}, K_{i_2}, \ldots, K_{i_k}$ are primary jacket matrices, over other forms.

For the next two Chapters, we will see no more of Butson, complex Hadamard or generalised Hadamard matrices. Generalised Hadamard matrices are critical to the theoretical advances of Part 2 and are met again in Chapter 7, while the Butson and GBH matrices are revisited in Chapter 9 (Section 9.1.4).

Our attention now returns to arrays with entries from $\{\pm 1\}$.

Chapter Five

Higher Dimensional Hadamard Matrices

As long ago as 1971, Shlichta discovered the existence of higher dimensional (± 1) arrays with a range of orthogonality properties [292]. In particular, he constructed 3-dimensional cubical arrays $A = [a(i_1, i_2, i_3)]$ with the property that any 2-D subarray, obtained by fixing an index in one dimension, is a Hadamard matrix.

By the time of Shlichta's discovery, Hadamard matrices had already been implemented very successfully in a variety of practical applications. The concurrent development, early in the 1970s, of image processing techniques (especially for television), the publication by Harmuth [141] of methods of applying 2- and 3-D Walsh functions for processing signals in several space or time dimensions, and the first work on perfect binary arrays [41], make it all the more surprising that no apparent notice of Shlichta's 3- and 4-D Hadamard matrices was taken.

Shlichta commented on this lack of activity over the intervening period, when he returned to the topic in 1979 [293]. He extended some of his constructions to n dimensions, pointing out that such arrays might have applications to encryption and error-correcting codes. His paper caused a brief flurry of interest, with Hammer and Seberry [136, 137, 138] and Agaian [1] applying his ideas to define and construct higher dimensional orthogonal designs and Butson matrices, respectively.

However, despite its potential, the subject remained, and remains, seriously under-developed. With the notable exception of Y. X. Yang and a little recent work by de Launey, the author and Ma, there is still almost no one investigating higher dimensional Hadamard matrices. Most of the subsequent results in the area are due to Yang, and some are not easily accessible. The only survey is Yang's [334].

Nonetheless, it was in this fallow ground that the theory of cocyclic Hadamard matrices germinated. De Launey, who had been a student of Seberry's, was interested in constructing higher dimensional combinatorial designs in even more generality than for orthogonal designs. By 1990 he had isolated a functional condition on the entries of a pairwise balanced design which would ensure that the design generated a *proper* higher dimensional pairwise balanced design, in any number of dimensions [79]. His discovery is introduced in Section 5.4.

The first three Sections of the Chapter cover earlier constructions, equivalence and applications of proper higher dimensional Hadamard matrices.

Most of the multidimensional arrays we treat in this Chapter have the same size v in each dimension (they are higher dimensional arrays of *order* v, or *hypercube* arrays) but this is not always so. The general case is left for pursuit by the interested reader, as is the case of arrays with entries not restricted to $\{\pm 1\}$.

We begin by defining various subarrays.

DEFINITION 5.1 *Let* $v \geq 2$ *and let* $A = [a(i_1, i_2, \ldots, i_n)]_{1 \leq i_j \leq v, 1 \leq j \leq n}$ *be an n-dimensional array of order v with entries in a field.*

A k-dimensional section of A, for some $0 \leq k \leq n - 1$, *is a subarray consisting of all the elements of A which have a particular set of fixed index values (say,* $i_{j_1}, i_{j_2}, \ldots, i_{j_{n-k}}$) *in the* $n - k$ *fixed dimensions* $j_1, j_2, \ldots, j_{n-k}$, *respectively.*

Two k-dimensional subarrays are parallel *in dimension h if they have their fixed indices* $i_{j_1}, i_{j_2}, \ldots, i_{j_{n-k}}$ *and* $l_{j_1}, l_{j_2}, \ldots, l_{j_{n-k}}$ *in the same dimensions and there exists* $h \in \{j_1, j_2, \ldots, j_{n-k}\}$ *such that* $i_h \neq l_h$.

If k = 2, then the 2-D section (a $v \times v$ *matrix)*

$$[a(i_1, i_2, \ldots, i_{l-1}, x, i_{l+1}, \ldots, i_{m-1}, y, i_{m+1}, \ldots, i_n)]_{1 \leq x, y \leq v},$$

where $i_1, i_2, \ldots, i_{l-1}, i_{l+1}, \ldots, i_{m-1}, i_{m+1}, \ldots, i_n$ *are fixed indices, is called a* plane.

If k = 1, then the 1-D section (a vector of length v)

$$[a(i_1, i_2, \ldots, i_{l-1}, x, i_{l+1}, \ldots, i_n)]_{1 \leq x \leq v},$$

where $i_1, i_2, \ldots, i_{l-1}, i_{l+1}, \ldots, i_n$ *are fixed indices, is called a* row *or (sometimes) a* column.

Shlichta [293] defined an n-dimensional array of order v with entries in $\{\pm 1\}$ to be *Hadamard* if, in each dimension, all its parallel $(n - 1)$-dimensional sections are mutually orthogonal.

DEFINITION 5.2 *Let* $n \geq 2$. *An n-dimensional Hadamard matrix of order v is a* (± 1) *array* $A = [a(i_1, i_2, \ldots, i_n)]_{1 \leq i_j \leq v, 1 \leq j \leq n}$ *such that, for each* $1 \leq l \leq n$, *and for all indices x and y in dimension l,*

$$\sum_{j \neq l} \sum_{1 \leq i_j \leq v} a(i_1, \ldots, x, \ldots, i_j, \ldots, i_n) a(i_1, \ldots, y, \ldots, i_j, \ldots, i_n) = v^{n-1} \delta_{xy}.$$

$$(5.1)$$

An n-dimensional Hadamard matrix A may have stronger orthogonality properties in some dimensions. For each dimension l, let D_l be a set of d dimensions including l and, for each choice c_l of indices in the other $(n - d)$ dimensions, let $A(c_l)$ be the corresponding d-dimensional section of A.

The *propriety* d_l *in dimension* l of an n-dimensional Hadamard matrix A is defined as the *smallest* number of dimensions d, $2 \leq d \leq n$, such that all the parallel $(d - 1)$-dimensional sections of $A(c_l)$ in dimension l are mutually orthogonal, that is, such that for all indices x and y in dimension l, and for all choices of D_l and c_l,

$$\sum_{j \neq l \in D_l} \sum_{1 \leq i_j \leq v} a(i_1, \ldots, x, \ldots, i_j, \ldots, i_n) a(i_1, \ldots, y, \ldots, i_j, \ldots, i_n) = v^{d-1} \delta_{xy}.$$

$$(5.2)$$

Therefore, if an n-dimensional Hadamard matrix A has propriety $\leq d$ in every dimension, then every d-dimensional section is a d-dimensional Hadamard matrix. The minimum such value $d = \min\{d_l, 1 \leq l \leq n\}$ is termed the *propriety* of the n-dimensional matrix.

A *proper n*-dimensional Hadamard matrix of order v, for $n \geq 2$, is one with the minimum possible propriety 2 in every dimension, that is, one in which every plane is a Hadamard matrix (cf. [138]).

DEFINITION 5.3 *An n-dimensional Hadamard matrix* $A = [a(i_1, i_2, \ldots, i_n)]$ *of order v has* propriety *d, where* $2 \leq d \leq n$, *if, for each choice of a set D of d dimensions, for all indices x and y in dimension $l \in D$, and for each choice of fixed indices in the other $n - d$ dimensions,*

$$\sum_{j \neq l \in D} \sum_{1 \leq i_j \leq v} a(i_1, \ldots, x, \ldots, i_j, \ldots, i_n) a(i_1, \ldots, y, \ldots, i_j, \ldots, i_n) = v^{d-1} \delta_{xy},$$

$$(5.3)$$

and this is not true for any number of dimensions less than d.

An n-dimensional Hadamard matrix $A = [a(i_1, i_2, \ldots, i_n)]$ *of order v is* proper *if it has propriety $d = 2$, that is, for each pair of dimensions j, l, for all indices x and y in dimension l, and for each set of fixed indices in the other $n - 2$ dimensions,*

$$\sum_{1 \leq i_j \leq v} a(i_1, \ldots, x, \ldots, i_j, \ldots, i_n) a(i_1, \ldots, y, \ldots, i_j, \ldots, i_n) = v\, \delta_{xy}. \quad (5.4)$$

It follows that a proper n-dimensional Hadamard matrix must have order 2 or a multiple of 4. This restriction does not apply in the improper case (see, for example, the 4-D Hadamard matrix of order 6 of [334, Theorem 6.1.2]), but the order must still be even [334, Theorem 6.1.1].

COROLLARY 5.4 (cf. [293, p. 570]) *If A is an n-dimensional Hadamard matrix of propriety d, then every k-dimensional section of A, $d \leq k < n$, is a k-dimensional Hadamard matrix of propriety d. If A is proper, so is every k-dimensional section, $k \geq 2$.*

5.1 CLASSICAL CONSTRUCTIONS

The tensor product is the oldest general construction for creating higher dimensional Hadamard matrices.

We are already familiar with the tensor product for square matrices, and there is a simple higher dimensional construction technique using it: if $H_1, \ldots, H_n$ are all Hadamard matrices of order v, then $H = H_1 \otimes \cdots \otimes H_n$ is a Hadamard matrix of order v^n. On reindexing H, we obtain a $2n$-dimensional Hadamard matrix H' of order v — but usually not of propriety less than $2n$ — with

$$h'(i_1, j_1, i_2, j_2, \ldots, i_n, j_n) = h_1(i_1, j_1) h_2(i_2, j_2) \ldots h_n(i_n, j_n). \quad (5.5)$$

Shlichta [293, Fig. 3] remarks that 3-D Hadamard matrices of order v^2 can be produced as a triple tensor product of three Hadamard matrices of order v by orienting the factors in three mutually perpendicular directions in space. The construction is detailed in Yang [334, Theorem 3.1.2] and uses the general definition of tensor product [334, Definition 4.2.5] of which we will state only the hypercube case.

DEFINITION 5.5 *If $A = [a(i_0, \ldots, i_{n-1})]$ and $B = [b(j_0, \ldots, j_{n-1})]$ are n-dimensional arrays of orders v and w, respectively, then their tensor product $C = A \otimes B$ is the n-dimensional array of order vw obtained by replacing each entry $a(i_0, \ldots, i_{n-1})$ of A by the n-dimensional subarray $a(i_0, \ldots, i_{n-1})B$. That is, for $0 \leq i_j \leq vw - 1, 0 \leq j \leq n - 1$, the entry $c(i_0, \ldots, i_{n-1})$ in C is*

$$c(i_0, \ldots, i_{n-1}) = a(\lfloor i_0/w \rfloor, \ldots, \lfloor i_{n-1}/w \rfloor) \, b(i_0 \bmod w, \ldots, i_{n-1} \bmod w).$$

In mixed radix notation (Definition 3.3), where $l = l_1 w + l_0 \equiv (l_1, l_0)$ and $0 \leq l_1 \leq v - 1, 0 \leq l_0 \leq w - 1$, this is

$$c(i_0, \ldots, i_{n-1}) = a((i_0)_1, \ldots, (i_{n-1})_1) \, b((i_0)_0, \ldots, (i_{n-1})_0).$$

For instance, 3-D Hadamard matrices of order 2^t can be produced by taking $t - 1$ successive tensor products of 3-D Hadamard matrices of order 2. Examples can be found in [293], where Shlichta notes that the tensor product of two matrices is proper only in those dimensions in which both the parent matrices are proper.

Indeed, this applies to tensor products of n-dimensional Hadamard matrices, for any n and any degree of propriety. (As Shlichta says, 'In Hadamard matrices, as in life, propriety once lost is never regained'.)

LEMMA 5.6 [138, p. 774] *Let A and B be n-dimensional Hadamard matrices of orders v and w, respectively. Then $A \otimes B$ is an n-dimensional Hadamard matrix of order vw. If A has propriety α_k and B has propriety β_k in dimension k, then $A \otimes B$ has propriety $\max\{\alpha_k, \beta_k\}$ in dimension k.*

Apart from the tensor product, there is a multitude of constructions due to Yang, for which the reader is referred to [334]. Even so, it is not known whether there exists an n-dimensional Hadamard matrix of every even order.

Research Problem 28 [334, Question 12, p. 316] *Let $n \geq 4$. Is there an n-dimensional Hadamard matrix of order $2t$ for every $t \geq 1$?*

We will concentrate on constructions of proper higher dimensional Hadamard matrices, since they have the full hierarchy of orthogonality: propriety 2 in every dimension implies propriety d for $2 \leq d \leq n$ in every dimension. The first construction is the tensor product.

COROLLARY 5.7 *Let A and B be n-dimensional proper Hadamard matrices of orders v and w, respectively. Then $A \otimes B$ is an n-dimensional proper Hadamard matrix of order vw.*

Four direct constructions of proper higher dimensional Hadamard matrices follow.

5.1.1 Boolean function construction for order 2

The simplest n-dimensional Hadamard matrices are of order 2. Yang shows they are equivalent to Boolean functions satisfying an additional property. Recall that the *weight* $w(f)$ of a Boolean function f is the number of 1s in its truth table **f**.

DEFINITION 5.8 [333] *If $f(v_1, v_2, \ldots, v_n)$ is a Boolean function of n variables, define n Boolean functions $g_1, g_2, \ldots, g_n$ of $n - 1$ variables by*

$$g_i(v_1, v_2, \ldots, \widehat{v_i}, \ldots, v_n) = f(v_1, v_2, \ldots, v_{i-1}, 0, v_{i+1}, \ldots, v_n) \qquad (5.6)$$
$$+ f(v_1, v_2, \ldots, v_{i-1}, 1, v_{i+1}, \ldots, v_n),$$

where $\widehat{\cdot}$ represents a deleted variable. The function f is H-Boolean if $w(g_i) = 2^{n-2}, 1 \le i \le n$.

For example, if $n = 3$, the Boolean function $f(v_1, v_2, v_3)$ with truth table $\mathbf{f} = [0, 0, 0, 1, 0, 1, 0, 0]$ is H-Boolean, since $\mathbf{g}_1 = [0, 1, 0, 1]$, $\mathbf{g}_2 = [0, 1, 0, 1]$ and $\mathbf{g}_3 = [0, 1, 1, 0]$. It is not bent, since n is odd.

By Definition 5.2, an n-dimensional Hadamard matrix of order 2 is a (± 1) array $A = [a(i_1, \ldots, i_n)]_{0 \le i_j \le 1, 1 \le j \le n}$ in which, for every $1 \le l \le n$ and every choice of indices x and y in dimension l,

$$\sum_{j \ne l} \sum_{i_j=0}^{1} a(i_1, \ldots, x, \ldots, i_j, \ldots, i_n) a(i_1, \ldots, y, \ldots, i_j, \ldots, i_n) = 2^{n-1} \delta_{xy}.$$

$$(5.7)$$

As usual (cf. (3.25)), we have the exponential correspondence which allows us to switch between $(0, 1)$ and (± 1) versions of n-dimensional Boolean functions:

$$a(i_1, i_2, \ldots, i_n) = (-1)^{f(i_1, i_2, \ldots, i_n)}.$$

Therefore, if $x = y$ the left-hand side of (5.7) is 2^{n-1}, and if $x \ne y$ it is

$$\sum_{j \ne l} \sum_{i_j=0}^{1} a(i_1, \ldots, x, \ldots, i_j, \ldots, i_n) a(i_1, \ldots, y, \ldots, i_j, \ldots, i_n)$$

$$= \sum_{j \ne l} \sum_{i_j=0}^{1} (-1)^{g_l(i_1, i_2, \ldots \widehat{i_l}, \ldots, i_n)}$$

by Definition 5.8, and this sum is 0 if and only if g_l has weight 2^{n-2}.

THEOREM 5.9 [333, 334, Theorem 5.1.1] *Let f be a Boolean function of n variables and $A = [(-1)^{f(i_1, i_2, \ldots, i_n)}]_{0 \le i_j \le 1, 1 \le j \le n}$ the corresponding n-dimensional (± 1) array. Then f is H-Boolean if and only if A is an n-dimensional Hadamard matrix.*

Suppose $f : V(n, 2) \to GF(2)$ is a bent Boolean function of n variables. By Lemma 3.29, its directional derivative $f_{\mathbf{u}}$ in the direction of any nonzero vector $\mathbf{u}$ has weight 2^{n-1}, so, in particular, its directional derivative $f_{\mathbf{e}_k}$ in the direction of the standard unit vector $\mathbf{e}_k \in V(n, 2)$ has weight 2^{n-1}, for each $1 \le k \le n$. This means that, for each $1 \le k \le n$,

$$\sum_{j \ne k} \sum_{0 \le v_j \le 1} (-1)^{f(v_1, \ldots, 0, \ldots, v_j, \ldots, v_n)} (-1)^{f(v_1, \ldots, 1, \ldots, v_j, \ldots, v_n)} = 0. \qquad (5.8)$$

By Theorem 5.9, $[(-1)^f]$ is an n-dimensional Hadamard matrix. So, not only does every bent function determine a 2-D Hadamard matrix of order 2^n, it determines an n-dimensional Hadamard matrix of order 2.

Example 5.1.1   [334, Theorem 5.3.8] *Every bent function is H-Boolean.*

Proper higher dimensional Hadamard matrices of order 2 have been completely classified as corresponding to the H-Boolean functions which are sums of an affine Boolean function and a fixed quadratic Boolean function.

THEOREM 5.10 [331, Theorem 4] (or see [334, Theorem 5.1.13]) *Let f and A be as in Theorem 5.9. Then A is proper if and only if there exist $b_j \in GF(2), 0 \leq b_j \leq n$, such that*

$$f(v_1, v_2, \ldots, v_n) = b_0 + \sum_{j=1}^{n} b_j v_j + \sum_{1 \leq j < k \leq n} v_j v_k. \tag{5.9}$$

Consequently there are exactly 2^{n+1} proper n-dimensional Hadamard matrices of order 2. When n is even, the Boolean function $\sum_{1 \leq j < k \leq n} v_j v_k$ is bent, so in this case all the Boolean functions (5.9) are bent.

DEFINITION 5.11 *The H-Boolean proper n-dimensional Hadamard matrices are the 2^{n+1} matrices of order 2 constructed in Theorem 5.10.*

5.1.2 Product construction

A surprisingly simple method for generating a proper n-dimensional Hadamard matrix from any Hadamard matrix is due to Yang [331, Theorem 1] and de Launey [76]. (Proof follows immediately from the definitions.)

If $H = [h(i_1, i_2)]$ is a Hadamard matrix of order v, then $A = [a(i_1, i_2, \ldots, i_n)]$, where

$$a(i_1, i_2, \ldots, i_n) = \prod_{1 \leq j < k \leq n} h(i_j, i_k), \tag{5.10}$$

is a proper n-dimensional Hadamard matrix of order v.

DEFINITION 5.12 *If H is a $v \times v$ Hadamard matrix, then $\mathcal{P}(n, v, H)$ denotes a proper n-dimensional Hadamard matrix of order v constructed from H by the product construction (5.10).*

5.1.3 Group developed construction

Hammer and Seberry [136, Theorem 4] used a different technique to construct proper higher dimensional Hadamard matrices. They showed that if G is an abelian group of order v, and $[\phi(g_i g_j)]_{1 \leq i, j \leq v}$ is a G-developed Hadamard matrix, then setting $a(i_1, i_2, \ldots, i_n) = \phi(g_{i_1} g_{i_2} \cdots g_{i_n})$ determines a proper n-dimensional Hadamard matrix of order v.

In fact, their construction holds for nonabelian groups, too. For any group $G = \{g_1, g_2, \ldots, g_v\}$ of order v and map $\phi : G \to \{\pm 1\}$, if $[\phi(g_i g_j)]_{1 \leq i, j \leq v}$ is a G-developed Hadamard matrix, and

$$a(i_1, i_2, \ldots, i_n) = \phi(g_{i_1} g_{i_2} \cdots g_{i_n}), 1 \leq i_j \leq v, 1 \leq j \leq n, \tag{5.11}$$

then every plane of $A = [a(i_1, i_2, \ldots, i_n)]$ is a Hadamard matrix.

LEMMA 5.13 [164] *The construction* (5.11) *applied to a G-developed Hadamard matrix determines a proper n-dimensional Hadamard matrix A, for any $n \geq 2$. If G is abelian, every plane in A is G-developed.*

Proof. Fix the indices of A in all but dimensions k and l. Assume without loss of generality that $k < l$ and let $h_1 = g_{i_1} g_{i_2} \cdots g_{i_{k-1}}$, $h_2 = g_{i_{k+1}} g_{i_{k+2}} \cdots g_{i_{l-1}}$ and $h_3 = g_{i_{l+1}} g_{i_{l+2}} \cdots g_{i_n}$. In any plane P obtained from A in this manner, the inner product of the two rows in dimension k indexed by group elements a and b is

$$
\sum_{c \in G} \phi(g_{i_1} \ldots a \ldots c \ldots g_{i_n}) \phi(g_{i_1} \ldots b \ldots c \ldots g_{i_n})
$$
$$
= \sum_{c \in G} \phi(h_1 a h_2 . ch_3) \phi(h_1 b h_2 . ch_3)
$$
$$
= \sum_{c \in G} \phi(a'.c) \phi(b'.c) = v\, \delta_{a'b'} = v\, \delta_{ab},
$$

so P is also Hadamard. If G is abelian, the mapping $\phi^* : G \to \{\pm 1\}$ given by $\phi^*(a) = \phi(a . h_1 h_2 h_3)$, $a \in G$, shows that $P(a,c) = \phi(h_1 a h_2 ch_3) = \phi^*(ac)$. $\square$

DEFINITION 5.14 *If there is a Menon Hadamard matrix $[\phi]$, where $\phi : G \to \{\pm 1\}$ is a set mapping and G is a group of order $v = 4u^2$, then $\mathcal{G}(n, v, G, \phi)$ denotes a proper n-dimensional Hadamard matrix of order v constructed from $[\phi]$ by the group developed construction (5.11).*

The product and group developed constructions each have advantages and disadvantages. The product construction requires $O(n^2)$ operations to calculate each entry, while group development requires only $O(n)$ operations. However, the product construction applies to any Hadamard matrix; the group development construction applies only to Menon Hadamard matrices (but is also applicable to other designs). Nonetheless, Theorem 2.22 tells us that group developed higher dimensional Hadamard matrices form a major class of proper higher dimensional Hadamard matrices.

COROLLARY 5.15 *For each $a, b \geq 0$ and odd number m, there is at least one abelian group G such that a $\mathcal{G}(n, 4 \cdot 2^{2a} \cdot 3^{2b} \cdot m^4, G, \phi)$ exists for every $n \geq 2$.*

5.1.4 Perfect binary array construction

In [332] (see [334, Section 6.2.1]), Yang gives two methods to construct an $(n+1)$-dimensional Hadamard matrix of order r from an n-dimensional perfect binary array $PBA(r, \ldots, r)$ of energy r^n (Definition 3.22). However, only construction in a single higher dimension will result, and the higher dimensional Hadamard matrix so constructed will not be proper. Finally, the requirement that $r^n = 4u^2$ significantly restricts the perfect binary arrays to which his method applies.

Corollary 5.15 above improves Yang's results optimally. By Lemma 3.25, any construction of a higher dimensional Hadamard matrix from an abelian group developed Hadamard matrix applies to the corresponding perfect binary array. This includes the Shlichta and 2-D tensor power (5.5) constructions (with factors all equal), product (5.10) and group developed (5.11) constructions. The latter two constructions determine proper n-dimensional Hadamard matrices, and of these the second is faster, and hence is preferable.

LEMMA 5.16 [164, Proposition 3.4] *For each PBA and $n \geq 2$, there exists a proper n-dimensional Hadamard matrix. If a $PBA(s_0, \ldots, s_{m-1})$ of energy $s = \prod_{i=0}^{m-1} s_i$ has elements $\phi(g)$, $g \in G = \mathbb{Z}_{s_0} \times \cdots \times \mathbb{Z}_{s_{m-1}}$ and $n \geq 2$, then a $\mathcal{G}(n, s, G, \phi)$ is given by $a(i_1, i_2, \ldots, i_n) = \phi(g_{i_1} + g_{i_2} + \cdots + g_{i_n}), 1 \leq i_j \leq s, 1 \leq j \leq n$.*

Addition in $\mathbb{Z}_{s_0} \times \cdots \times \mathbb{Z}_{s_{m-1}}$ is fast to implement since it involves $n - 1$ additions mod s_k in coordinate k, for each of the m coordinates.

To illustrate, consider calculation of a proper 3-D Hadamard matrix from the $PBA(3,3,4)$ given in (3.23). The entry $a((0, 1, 0), (1, 0, 2), (2, 1, 3))$, using the product construction, is $\phi((1, 1, 2))\phi((2, 2, 3))\phi((0, 1, 1)) = 1 \cdot -1 \cdot -1 = 1$; while, using the group developed construction, it is $\phi((0, 2, 1)) = -1$. So the constructions determine different 3-D Hadamard matrices.

This completes the listing of classical constructions. Explanation of the cocycle construction is reserved until Section 5.4.

5.2 EQUIVALENCE CLASSES

Ma [229] has investigated the relationship of planes in these proper higher dimensional Hadamard matrices to the Hadamard matrices from which they are constructed.

LEMMA 5.17 [229, Theorems 2.5, 2.6] *Any plane in $\mathcal{P}(n, v, H)$ is equivalent to H. Any plane in $\mathcal{G}(n, v, G, \phi)$ is equivalent to the Menon Hadamard matrix $[\phi]$ from which it is constructed.*

Proof. The first part is proved by induction on the number of dimensions n. The second follows from the proof of Lemma 5.13, since $P(a, c) = \phi(h_1 a h_2.c h_3) = \phi(\sigma(a)\tau(c))$ for suitable permutations σ, τ of G. □

The concept of equivalence between higher dimensional Hadamard matrices was not pinned down until 2002.

DEFINITION 5.18 (Ma [229]) *Let M be an n-dimensional Hadamard matrix of order v. An n-dimensional matrix M' is (Hadamard) equivalent to M if M' can be obtained from M by performing a finite sequence of the following operations:*

1. *(permutation equivalence) Permute parallel $(n - 1)$-dimensional sections, that is, given a dimension k and indices x and y, exchange each corresponding pair of entries indexed by x and y in dimension k.*

2. *Negate any $(n - 1)$-dimensional section, that is, given a dimension k and an index x, multiply all entries with index x in dimension k by -1.*

These operations preserve propriety, and when $n = 2$, they reduce to the usual Definition 2.12 of Hadamard equivalence. Given a proper n-dimensional Hadamard matrix A, Ma shows that, if A' is derived from A under the second operation, then each plane P' in A' is equivalent to the plane P in A with the same fixed indices. For permutation equivalence, he shows that the same result holds if A is constructed by either the product or group developed construction. The latter result also holds (trivially) if A is H-Boolean, since there is only one equivalence class of Hadamard matrices of order 2.

LEMMA 5.19 [229, Theorem 3.3] *If A and A' are equivalent proper n-dimensional Hadamard matrices and A is constructed by the H-Boolean, product or group developed constructions, then every plane P in A is equivalent to the corresponding plane P' in A'.*

Ma then goes on to show, as one would expect, that for any $n \geq 2$, if H and H' are equivalent Hadamard matrices, then the proper n-dimensional Hadamard matrices $\mathcal{P}(n, v, H)$ and $\mathcal{P}(n, v, H')$ are equivalent. He also shows that, if $H = \mathcal{G}(2, v, G, \phi)$ and $H' = \mathcal{G}(2, v, G', \phi')$ are permutation equivalent (only), then $\mathcal{G}(n, v, G, \phi)$ and $\mathcal{G}(n, v, G', \phi')$ are equivalent. It is likely that the result holds in general, but it has not yet been demonstrated.

Research Problem 29 *Show that, if $\mathcal{G}(2, v, G, \phi)$ and $\mathcal{G}(2, v, G', \phi')$ are equivalent, then for any $n \geq 2$, $\mathcal{G}(n, v, G, \phi)$ and $\mathcal{G}(n, v, G', \phi')$ are equivalent.*

Combining these equivalence results with Lemmas 5.19 and 5.17, Ma links the number of equivalence classes of higher dimensional matrices to that of the Hadamard matrices generating them.

COROLLARY 5.20 [229, Corollaries 4.4, 4.8] *The number of equivalence classes of $\mathcal{P}(n, v, H)$ is equal to the number of equivalence classes of $v \times v$ Hadamard matrices. Given G and v, for each $n \geq 2$, the number of equivalence classes of $\mathcal{G}(n, v, G, \phi)$ is at least equal to the number of equivalence classes of $\mathcal{G}(2, v, G, \phi)$.*

If f is as in (5.9) and A is the corresponding proper n-dimensional order 2 matrix, how do the equivalence operations on A translate to equivalence operations on f?

Research Problem 30 *Determine the number of equivalence classes of proper n-dimensional Hadamard matrices of order 2.*

5.3 APPLICATIONS IN SPECTROSCOPY, CODING AND CRYPTOGRAPHY

A very small number of applications of higher dimensional Hadamard matrices is known to the author. This is an immensely fruitful area for future research.

5.3.1 Multidimensional Walsh Hadamard transforms

The earliest applications of higher dimensional Hadamard matrices were as 2-D Walsh-Hadamard Transforms, resulting in construction of fast matrix hardware for digital filtering of television images [23] (and see [141, Sections 2.3.2-3]). It was clear at the time that the advantages of the Walsh-Hadamard Transform would extend to multidimensional signals or data such as images, X-rays or seismic waves. This applies equally to emerging 3-D technologies such as holography (Chapter 3.3.2) and NMR spectroscopy. For example, Kupče and Freeman [214] transform 2-D spectra (obtained by direct NMR excitation of a small protein, agitoxin) one dimension at a time, using a 2-D Walsh-Hadamard Transform (specifically, the tensor product of two Hadamard matrices of order 8) with speed and spectral simplification gains over the 2-D Fast Fourier Transform of the same spectra.

In some communities, multidimensional Hadamard transform spectroscopy refers to ordinary WHT spectroscopy of multidimensional signals using a 1-D mask reconfigured as a 2-D mask (see Chapter 3.1.4). For instance, Hammaker et al. [135, 8.2] use a 127-element cyclic 1-D Hadamard mask to form a 10×10 2-D mask, ignoring the last 27 elements. A received array of measurements is then transformed using a Hadamard matrix of order 128 as usual.

Only if the transform matrix is itself a tensor product of Hadamard matrices, which is not the case in [135], is the technique true multidimensional Hadamard spectroscopy in the sense meant here, of using a higher dimensional Hadamard matrix as transform matrix.

DEFINITION 5.21 [105, 8.9] *Let* $\mathbf{x} = [x(i_1, \ldots, i_n), \ 0 \le i_j \le 2^{l_j - 1}, 1 \le j \le n]$ *be an n-dimensional signal or data sequence. The n-dimensional WHT of* $\mathbf{x}$ *is* $\hat{\mathbf{x}}$, *where for* $0 \le k_j \le 2^{l_j - 1}, 1 \le j \le n$,

$$\hat{x}(k_1, \ldots, k_n) = 2^{-(\sum_{j=1}^{n} l_j)} \sum_{j=1}^{n} \sum_{i_j} x(i_1, \ldots, i_n)(-1)^{\sum_{j=1}^{n} \langle (k_j)_2, (i_j)_2 \rangle}$$

$$= 2^{-(\sum_{j=1}^{n} l_j)} \sum_{j=1}^{n} \sum_{i_j} x(i_1, \ldots, i_n) \prod_{j=1}^{n} \mathcal{S}_{l_j}(k_j, i_j), \qquad (5.12)$$

by (2.4) (or (3.9)), and the inverse *n-dimensional WHT is*

$$x(k_1, \ldots, k_n) = \sum_{j=1}^{n} \sum_{i_j} \hat{x}(i_1, \ldots, i_n)(-1)^{\sum_{j=1}^{n} \langle (k_j)_2, (i_j)_2 \rangle}$$

$$= \sum_{j=1}^{n} \sum_{i_j} \hat{x}(i_1, \ldots, i_n) \prod_{j=1}^{n} \mathcal{S}_{l_j}(k_j, i_j). \qquad (5.13)$$

Multidimensional versions of Parseval's Theorem (3.13) and the Convolution Theorem (3.15) also hold [105].

These transforms may be thought of either as application of a sequence of 1-D Walsh-Hadamard Transforms (3.10), taken one dimension at a time, or as 1-D transforms using a tensor product transform matrix. For instance, the 2-D transform $[\hat{x}(k_1, k_2)] = 2^{-(l_1+l_2)} S_{l_1} [x(k_1, k_2)] S_{l_2}$ is equivalent to the 1-D transform $\hat{\mathbf{x}} = 2^{-(l_1+l_2)} S_{l_1} \otimes S_{l_2} \mathbf{x}$ when the 2-D data are sequenced in lexicographical order.

The corresponding transform matrix, the *2n-dimensional Sylvester Hadamard matrix* $S_{(l_1, l_2, \ldots, l_n)}$, has entries

$$S_{(l_1, l_2, \ldots, l_n)}(k_1, i_1, k_2, i_2, \ldots, k_n, i_n) = \prod_{j=1}^{n} S_{l_j}(k_j, i_j), \qquad (5.14)$$

so it is the ordinary 2-D tensor product $S_{(l_1, l_2, \ldots, l_n)} = S_{l_1} \otimes \cdots \otimes S_{l_n}$ of n 2-D Sylvester Hadamard matrices S_{l_j} (3.9). That is, it is the general version of the construction (5.5), not requiring the same order in each dimension, and is easily shown to be a $2n$-dimensional Hadamard matrix of this more general type.

If $l_1 = l_2 = \cdots = l_n = l$, then $S_{(l,l,\ldots,l)}$ is a $2n$-dimensional Hadamard matrix of order 2^l, so, using the binary representations $(k_j)_{l-1}(k_j)_{l-2} \ldots (k_j)_0$ of the indices k_j,

$$
\begin{aligned}
S_{(l,l,\ldots,l)}&(k_1, i_1, k_2, i_2, \ldots, k_n, i_n) \\
&= (-1)^{\sum_{j=1}^{n} \sum_{m=0}^{l-1} (k_j)_m (i_j)_m} \\
&= (-1)^{\sum_{m=0}^{l-1} \sum_{j=1}^{n} (k_j)_m (i_j)_m} \\
&= \prod_{m=0}^{l-1} S_{(1,1,\ldots,1)}((k_1)_m, (i_1)_m, (k_2)_m, (i_2)_m, \ldots, (k_n)_m, (i_n)_m),
\end{aligned}
$$

it follows that $S_{(l,l,\ldots,l)}$ is the lth tensor power (cf. Definition 5.5) of a $2n$-dimensional Hadamard matrix of order 2 :

$$S_{(l,l,\ldots,l)} = \otimes^l S_{(1,1,\ldots,1)}.$$

The classic picture [141, Fig. 28] of the 3-D Walsh functions with matrix equivalent to $S_{(2,2,2)}$ is still worth viewing. We have already met the basic case $S_{(l)} = \otimes^l S_{(1)} = S_l = \otimes^l S_1$: it specifies the lexicographical ordering seen in (2.4).

5.3.2 Error-correcting array codes

Array error control codes are typically linear block or convolutional codes with codewords formed by attaching check symbols to information symbols arranged in two or more dimensions. The most efficient and widely used array codes are those designed for burst-error correction of single- or multidimensional error or erasure bursts. There are numerous approaches to the problem, and the survey by Blum, Farrell and van Tilborg [29] and the concise update by Farrell [109] are recommended.

The range of array code applications has increased enormously in the past 15 years. The largest application area continues to be tape, disc and optical information storage systems. Others are wireless transmission, network applications, development of practical encoders and decoders, encryption schemes and fault tolerance [109].

Examples of 2-D array codes are block array codes, such as the space-time block codes for communication over Rayleigh fading channels using multiple transmit antennas [305], and convolutional array codes and maximum-rank array codes used to correct criss-cross errors such as those found in memory chip arrays and multi-track magnetic tape recording [277].

When more than one layer of information is to be stored, higher dimensional array codes ought to be considered. For example, in holographic storage, [28] describes a 3-D interleaving scheme to correct 3-D burst errors in a 3-D hologram.

However, none of these examples specifically use 2- or 3-D Hadamard matrices.

The error-correcting potential of n-dimensional Hadamard matrices was first raised by Shlichta in 1979 [293], when he pointed out that the full hierarchy of orthogonalities in a proper matrix meant that a variety of correlation checks could be run, ranging from large-scale comparisons of $(n-1)$-dimensional sections which could detect the occurrence of errors in a particular $(1/v)^{th}$ portion of the entries, to highly localised row checks for locating specific errors. It therefore seems sensible to consider higher dimensional Hadamard matrices for more general targetting problems, such as in fault-tolerant computing.

The first result in this area is by de Launey, who showed that the set of proper n-dimensional Hadamard matrices of order 2 is equivalent to a coset of the first-order Reed-Muller code. A very simple proof of this result uses the characterisation in Theorem 5.10.

LEMMA 5.22 [80, Theorem 2] *Let* $A = [(-1)^f]$ *be an n-dimensional Hadamard matrix of order 2, where f is an H-Boolean function of n variables with truth table* **f**. *Then* $\{\mathbf{f} : A \text{ is proper}\}$ *is a coset of* $\mathcal{R}(1, n)$.

Proof. Set $c(v_1, \ldots, v_n) = \sum_{1 \le j < k \le n} v_j v_k$. By Theorem 5.10 and Definition 3.16,

$$\{\mathbf{f} : A \text{ is proper}\} = \mathbf{c} + \{b_0 \mathbf{1} + b_1 \mathbf{v}_1 + \cdots + b_n \mathbf{v}_n, \; b_i \in GF(2)\} = \mathbf{c} + \mathcal{R}(1, n).$$

$\square$

De Launey remarks that this implies a corrupted proper n-dimensional Hadamard matrix of order 2 may be repaired by any of the methods used to decode $\mathcal{R}(1, n)$, such as the WHT. He goes on to extend Lemma 5.22 and prove that the set of n-dimensional Hadamard matrices of order 2 and propriety d is a union of cosets of the first-order Reed-Muller code [80, Corollary 6].

One immediate option for developing a general theory of n-dimensional Hadamard codes, suggested by the near-linearity of the n-dimensional Hadamard matrices of order 2 and propriety d, is to concentrate on linear codes whose codewords are n-dimensional q-ary arrays. Here we do not restrict to hypercube arrays.

DEFINITION 5.23 *Let* $V(m_1 \times m_2 \times \cdots \times m_n, q)$ *denote the vector space of dimension* $m = \prod_{i=1}^{n} m_i$ *over* $GF(q)$ *consisting of all* $m_1 \times m_2 \times \cdots \times m_n$ *n-dimensional arrays whose entries are elements of* $GF(q)$. *An n-dimensional* $(m_1 \times m_2 \times \cdots \times m_n, M)$ *array code C over* $GF(q)$ *is a subset of size M of* $V(m_1 \times m_2 \times \cdots \times m_n, q)$. *If C is a subspace of* $V(m_1 \times m_2 \times \cdots \times m_n, q)$

of dimension k, *then* C *is an* n-dimensional $[m_1 \times m_2 \times \cdots \times m_n, k]$ linear array code *with* $M = q^k$. *Elements of* C *are called* codewords *of* C.

For instance, if there exists a proper $2n$-dimensional Hadamard matrix of order v, it can be used to construct an n-dimensional binary array code meeting the Plotkin bound (Lemma 3.12). This code is one of the Class A Hadamard block codes C_{v^n} of Definition 3.13, with each codeword configured as an n-dimensional array.

LEMMA 5.24 [230, Lemma 4.3.5] *Let* $A = [a(i_1, i_2, \ldots, i_{2n})]$ *be a proper* $2n$-*dimensional Hadamard matrix of order* v. *Then there exists an* n-*dimensional* $(v \times v \times \cdots \times v, 2v^n, \frac{v^n}{2})$ *binary (Hadamard) array code meeting the Plotkin bound.*

Proof. Given any choice of n dimensions, there are v^n distinct n-dimensional parallel sections of A, each distinct pair of which has Hamming distance $\frac{v^n}{2}$. String out each of the distinct n-dimensional parallel sections of A in lexicographical order to get v^n (± 1) strings of length v^n. The construction of Definition 3.13.3 gives $2v^n$ binary strings (codewords) of length v^n, each of which can be reconfigured as an n-dimensional array.   □

We appear to have nothing new in the way of codes here because, as vector spaces, $V(m_1 \times m_2 \times \cdots \times m_n, q) \cong V(m, q)$, so all the theory of Chapter 3.2 for block codes translates without further effort (or insight).

However, this view ignores the additional structure of the codewords explicitly available to us through their array formation. Farrell [109] points out, for example, that the position of check symbols and the order in which symbols are sequenced in a 1-D version of the codeword are crucial in determining the error control performance of cluster-error-correcting array codes.

Research Problem 31 (cf. [109, Open problem (ii)]) *Create a unified framework for the theory of linear array codes for control of multidimensional cluster errors, when the generator codewords are* n-*dimensional (binary) Hadamard matrices.*

When errors in a $v \times v$-array code are confined to a number t of rows, or columns, or both, the optimal error-correcting code for this *criss-cross* error model is a maximum-rank array code. Such codes meet a Singleton bound for linear array codes in terms of an array-based definition of distance, *covering distance*, which is not the same as Hamming distance. These definitions can be extended to n-dimensional arrays.

DEFINITION 5.25 [277] *Let* $A = [a(i_1, i_2, \ldots, i_n)]_{1 \le i_j \le v, 1 \le j \le n}$ *be an* n-*dimensional array of order* v *over a field. A* cover *of* A *is an* n-*tuple* $(X_1, X_2, \ldots, X_n)$ *not all empty, where each* X_j *is a set of* $(n-1)$-*tuples over* $\{1, 2, \ldots, v\}$ *such that* $a(i_1, i_2, \ldots, i_n) \neq 0$ *implies that* $(i_1 i_2 \cdots i_{j-1} i_{j+1} \cdots i_n) \in X_j$ *for at least one* j. *The* size *of a cover* $(X_1, X_2, \ldots, X_n)$ *is defined to be* $\sum_{j=1}^{n} |X_j|$, *and the* covering weight $w_c(A)$ *of* A *is the minimum size of any cover of* A. *The* covering distance d_c *between two* n-*dimensional arrays* A *and* B *is defined to be* $d_c(A, B) = w_c(A - B)$.

Let $d_h(A, B)$ denote the Hamming distance between A and B. Since a single set X could cover every nonzero entry of $A - B$ and $|X|$ cannot be larger than $d_h(A, B)$, we have $d_h(A, B) \geq d_c(A, B)$.

Example 5.3.1 *Consider the 4×4 arrays over $GF(2)$*

$$A = \begin{bmatrix} 1 & 0 & 0 & 0 \\ 0 & 1 & 0 & 0 \\ 1 & 0 & 1 & 1 \\ 0 & 1 & 0 & 0 \end{bmatrix}, \quad B = \begin{bmatrix} 0 & 0 & 0 & 0 \\ 0 & 1 & 0 & 1 \\ 0 & 1 & 1 & 0 \\ 0 & 0 & 1 & 1 \end{bmatrix}.$$

Two covers of size 3 for A are $(\{1, 3\}, \{2\})$, which covers rows 1 and 3 as well as column 2, and $(\{3\}, \{1, 2\})$, which covers row 3 as well as columns 1 and 2. Since three nonzeroes on the main diagonal of A belong to distinct rows and columns, the covering weight of A must be at least 3. Therefore the covering weight of A is $w_c(A) = 3$. The Hamming weight of A is $w(A) = 6$ [277]. Similarly, the binary Hadamard matrix B has $w_c(B) = 3$, $w(B) = 6$.

Roth states a Singleton bound (cf. (3.19)) for higher dimensional linear codes.

THEOREM 5.26 [277] *Let C be a q-ary n-dimensional $[v \times v \times \cdots \times v, k, d_c]$ linear array code. Then $k \leq v(v^{n-1} - d_c + 1)$.*

Ma extends the Plotkin bound to higher dimensional codes using techniques similar to those of the original proof (Lemma 3.12).

THEOREM 5.27 [230, Theorem 4.3.11] *Let C be a n-dimensional binary $(v \times v \times \cdots \times v, M, d_c)$ array code. If $d_c > v^n/2$, then $M \leq 2d_c/(2d_c - v^n) - 1$ if v is odd, and $M \leq 2d_c/(2d_c - v^n)$ if v is even.*

5.3.3 Cryptography: bent functions and the strict avalanche criterion

The n-dimensional Hadamard matrices of order 2 are characterised by several properties which make the corresponding H-Boolean functions (Theorem 5.9) attractive as sources of highly nonlinear sequences.

Rather than requiring that every directional derivative have weight 2^{n-1}, as with bent functions, less stringent conditions may be imposed on potential cryptographic Boolean functions. Certainly, one of the desirable characteristics of an S-box function is that of completeness: each output bit of the function depends on all the input bits. Webster and Tavares [319] proposed the more stringent *strict avalanche criterion* (SAC) which requires that whenever one input bit is changed, the output bit must change with probability $\frac{1}{2}$. In other words, the directional derivative in the direction of every unit vector must have weight 2^{n-1}; see (5.8). Clearly, every bent function satisfies SAC.

An inductive hierarchy of SAC orders was subsequently introduced by Forré [115], which ranges from SAC (= SAC(0)) to SAC($n - 2$).

DEFINITION 5.28 *A Boolean function f of n variables satisfies SAC if for each unit vector $\mathbf{e}_k, 1 \leq k \leq n$, $w((\Delta f)_{\mathbf{e}_k}) = 2^{n-1}$.*

For $1 \leq m \leq n - 2$, *if* f *satisfies SAC*$(m - 1)$ *then it* satisfies *SAC*(m) *if and only if any function of* $n - m$ *variables obtained from* f *by keeping* m *input bits constant, satisfies SAC.*

We see that the H-Boolean functions are precisely the functions satisfying SAC, and further, that if f is an H-Boolean function of n variables satisfying SAC(m) then all the k-dimensional sections ($k \geq n - m$) of the n-dimensional Hadamard matrix $[(-1)^f]$ are k-dimensional Hadamard matrices of order 2 [334, p. 200]. That is, if f is an H-Boolean function of n variables satisfying SAC(m) then $[(-1)^f]$ has propriety $d \leq n - m$, and vice versa.

Alternatively, the SAC condition may be strengthened by successively increasing the weight of the direction $\mathbf{u}$ in which the directional derivative of f is required to have equal numbers of 0s and 1s. These conditions were introduced under the name *propagation criteria of degree* k [270].

DEFINITION 5.29 *For* $1 \leq k \leq n$, *a Boolean function* f *of* n *variables satisfies the propagation criterion of degree* k *(PC(k)) if* $w((\Delta f)_{\mathbf{u}}) = 2^{n-1}$ *for all* $\mathbf{u} \in V(n, 2)$ *for which* $1 \leq w(\mathbf{u}) \leq k$.

The relationship of these nonlinearity criteria to higher dimensional Hadamard matrices and bent functions may be summarised as follows.

LEMMA 5.30 *Let* f *be a Boolean function of* n *variables and* $A = [(-1)^f]$ *the corresponding* n-*dimensional matrix of order* 2.

1. f *satisfies SAC if and only if* f *satisfies PC(1), if and only if* A *is an* n-*dimensional Hadamard matrix, if and only if* f *is H-Boolean.*

2. f *satisfies SAC(m) if and only if* A *is an* n-*dimensional Hadamard matrix for which every* k ($\geq n - m$)-*dimensional section is a* k-*dimensional Hadamard matrix, that is, if and only if* A *has propriety at most* $n - m$.

3. f *satisfies SAC($n - 2$) if and only if* A *is a proper* n-*dimensional Hadamard matrix, if and only if* f *is of form (5.9).*

4. *Let* n *be even. Then* f *is bent if and only if* f *satisfies PC(n), and then* f *satisfies SAC.*

5.4 THE SECOND LINK: COCYCLIC CONSTRUCTION

In the late 1980s, Warwick de Launey became curious about the group developed construction of higher dimensional designs, which at that time was known for abelian groups (cf. (5.11)), and was applied to classes of combinatorial designs whose defining properties depend on their set of rows. (In this context, *combinatorial designs* are square matrices, not incidence structures. They can be thought of as generalisations of incidence matrices instead. Orthogonal designs form a major class, as do Hadamard matrices; the standard reference here is [123].)

However, other types of combinatorial design, not group developed, such as negacyclic weighing matrices and ω-cyclic generalised weighing matrices, are very useful for generating families of designs, including Hadamard matrices, and de Launey believed an encompassing construction could be found.

In [79], he developed a general notion of combinatorial designs (as $v \times v$ arrays with entries from a set S) whose distinct row pairs conform to certain 'balance' rules. These designs are parameterised by size v and symbol set S, by a set of rules β, and by subgroups Π_r and Π_c of the permutation group of S. The rules are specified only by the requirement that they remain invariant under a collection of group actions on the array. The group actions are essentially the equivalence operations usual to weighing matrices and generalised Hadamard matrices (cf. Definition 4.12): permutations of rows or of columns, and application of an element in Π_r to all the elements of an individual row, or of an element in Π_c to all the elements of an individual column. Within this framework, *proper higher dimensional* $(v, \Pi_r, \Pi_c, \beta, S)$-*designs* are just those for which every 2-D section, or its transpose, is a $(v, \Pi_r, \Pi_c, \beta, S)$-design [79, Definition 2.7].

Intriguingly, for the case $\Pi_r = \Pi_c = \Pi$, de Launey isolated a quasi-associative condition relating pairs of elements in a row with pairs of elements in a column of a (v, Π, β, S)-design, which would guarantee construction of a proper n-dimensional (v, Π, β, S)-design.

This insight lies at the heart of the cocyclic development of designs. For de Launey expressed his condition in terms of an 'abelian extension function' from $G \times G$ to an abelian subgroup C of Π, where G is a group of order v. The author, working with de Launey, recognised his abelian extension function to be a particular kind of 2-D *cocycle*, more commonly seen in the cohomology theory of finite groups, but now appearing in yet another guise.

For finite groups G and C with C abelian, a function $\psi : G \times G \to C$ satisfying

$$\psi(g, h)\psi(gh, k) = \psi(h, k)\psi(g, hk), \quad g, h, k \in G \tag{5.15}$$

is a 2-D cocycle from G to C, with trivial action. It is usually normalised; that is, $\psi(1, 1) = 1.$

Further elaboration, examples and properties of cocycles and cocyclic Hadamard matrices appear (see Chapter 6) in Part 2 — the oak tree grown from this acorn.

Cocyclic development of more general pairwise combinatorial designs, using the same framework, is to be treated in the monograph [87].

We complete Part 1 by giving de Launey's construction [79, Theorem 4.2] for proper higher dimensional Hadamard matrices, in its present form, and relating it to the other constructions.

THEOREM 5.31 [90, Theorem 2.7] *Let $G = \{g_1, \ldots, g_v\}$ be a finite group of order v. Let $\psi : G \times G \to C = \{\pm 1\}$ satisfy (5.15) and let $\phi : G \to \{\pm 1\}$ be a set map. If*

$$[\psi(g_i, g_j)\phi(g_i g_j)]_{1 \leq i, j \leq v} \tag{5.16}$$

is a Hadamard matrix, then for any $n \geq 2$, there is a proper n-dimensional Hada-mard matrix $A = [a(i_1, i_2, \ldots, i_n)]_{1 \leq i_j \leq v, 1 \leq j \leq n}$, where

$$a(i_1, i_2, \ldots, i_n) = \prod_{k=2}^{n} \psi\Big(\prod_{j=1}^{k-1} g_{i_j}, g_{i_k}\Big)\phi(g_{i_1} g_{i_2} \cdots g_{i_n}). \qquad (5.17)$$

Proof. Assume $\phi \equiv 1$. Only proof of this case is required, because of the two possible approaches described below and their reconciliation in (5.19). The case $n = 2$ is the Hadamard matrix $[\psi(g_i, g_j)]_{1 \leq i,j \leq v}$ and the case $n = 3$ illustrates the inductive argument sufficiently for the general case to be left to the reader. Suppose $n = 3$. Fix $j \in \{1, 2\}, \ell = 3$ and let

$$S = \sum_{1 \leq i_j \leq v} \psi(g_{i_1}, g_{i_2})\psi(g_{i_1} g_{i_2}, g_x)\psi(g_{i_1}, g_{i_2})\psi(g_{i_1} g_{i_2}, g_y)$$

$$= \sum_{1 \leq i_j \leq v} \psi(g_{i_1} g_{i_2}, g_x)\psi(g_{i_1} g_{i_2}, g_y) \text{ since } \psi(g_{i_1}, g_{i_2})^2 = 1$$

$$= v\delta_{xy}.$$

Now suppose $j, \ell \in \{1, 2\}$, so that without loss of generality

$$S = \sum_{1 \leq i_1 \leq v} \psi(g_{i_1}, g_x)\psi(g_{i_1} g_x, g_{i_3})\psi(g_{i_1}, g_y)\psi(g_{i_1} g_y, g_{i_3})$$

$$= \sum_{1 \leq i_1 \leq v} \psi(g_x, g_{i_3})\psi(g_{i_1}, g_x g_{i_3})\psi(g_y, g_{i_3})\psi(g_{i_1}, g_y g_{i_3}) \text{ by } (5.15)$$

$$= \psi(g_x, g_{i_3})\psi(g_y, g_{i_3}) \sum_{1 \leq i_1 \leq v} \psi(g_{i_1}, g_x g_{i_3})\psi(g_{i_1}, g_y g_{i_3})$$

$$= v\delta_{x*y*} \text{ (where } g_{x*} = g_x g_{i_3}, g_{y*} = g_y g_{i_3}) = v\delta_{xy}.$$

The other cases are proved similarly. The case for arbitrary n follows by induction and similar arguments, together with the identity

$$\prod_{k=2}^{n} \psi\Big(\prod_{j=1}^{k-1} g_{i_j}, g_{i_k}\Big) = \prod_{k=1}^{n-1} \psi\Big(g_{i_k}, \prod_{j=k+1}^{n} g_{i_j}\Big),$$

which itself follows by induction from (5.15).   □

Construction (5.17) can be interpreted as a generalisation of the group developed construction (5.11) in either of two ways.

First, matrices of the form (5.16) clearly specialise to group developed matrices when ψ is the trivial cocycle which always maps to 1. Second, however, group developed matrices are intrinsically cocyclic: we may assume without loss of generality that $\phi(1) = 1$ and $g_1 = 1$, in which case normalising a group developed matrix $[\phi(g_i g_j)]$ determines a Hadamard equivalent normalised matrix $[\partial\phi(g_i, g_j)]$, where

$$\partial\phi(g_i, g_j) = \phi(g_i)^{-1}\phi(g_j)^{-1}\phi(g_i g_j). \qquad (5.18)$$

The function $\partial\phi$ satisfies (5.15) and is a particular type of cocycle known as a *coboundary*. (A family of coboundaries has already been met in Example 3.5.2.)

Therefore, application of the construction (5.17) to a Menon Hadamard matrix determines two distinct n-dimensional Hadamard matrices.

COROLLARY 5.32 [164, Corollary 4.2] *Let $H = [\varphi(g_i g_j)]_{1 \leq i,j \leq v}$ be a Menon Hadamard matrix over the group $G = \{g_1, \ldots, g_v\}$. Then, for $j = 1, 2$ and any $n \geq 2$, $A_j = [a_j(i_1, i_2, \ldots, i_n)]_{1 \leq i_j \leq v, 1 \leq j \leq n}$, is a proper n-dimensional Hadamard matrix, where*

1. $a_1(i_1, i_2, \ldots, i_n) = \varphi(g_{i_1} g_{i_2} \cdots g_{i_n})$, *and*

2. $a_2(i_1, i_2, \ldots, i_n) = \prod_{k=2}^{n} \partial\varphi(\prod_{j=1}^{k-1} g_{i_j}, g_{i_k})$.

Proof. 1. Set $\psi \equiv 1$ in Theorem 5.31. 2. H is Hadamard equivalent to the coboundary matrix $[\partial\varphi(g_i, g_j)]_{1 \leq i,j \leq v}$, which is therefore Hadamard. Set $\phi \equiv 1$ in Theorem 5.31.   □

Clearly, A_1 is a $\mathcal{G}(n, v, G, \varphi)$ (5.11). Its relationship to A_2 is easily explained from the definitions.

LEMMA 5.33 [164, Lemma 4.3] *Let $H = [\varphi(g_i g_j)]_{1 \leq i,j \leq v}$ be a Menon Hadamard matrix and A_1 and A_2 be as in Corollary 5.32. Then*

$$a_2(i_1, i_2, \ldots, i_n) = \left(\prod_{j=1}^{n} \varphi(g_{i_j})^{-1} \right) a_1(i_1, i_2, \ldots, i_n).$$

It is easy to see that A_2 is equivalent to A_1 under repeated application of the second operation of Definition 5.18 to A_1. Thus, when $g_1 = 1 \in G$ and $\varphi(1) = 1$, A_2 can be thought of as a *normalised* version of A_1, just as $[\partial\varphi]$ is the normalised version of $[\varphi]$ in the 2-D case.

Similarly, alternative constructions can be applied for any Hadamard matrix $[\psi(g_i, g_j)\phi(g_i g_j)]_{1 \leq i,j \leq v}$, since it will be Hadamard equivalent to the Hadamard matrix $[(\psi \, \partial\phi)(g_i, g_j)]_{1 \leq i,j \leq v}$. In order to prevent any confusion arising in application of Theorem 5.31, it is preferable to isolate an unambiguous version of this construction.

DEFINITION 5.34 *Let $G = \{g_1 = 1, g_2, \ldots, g_v\}$ be a finite group of order v. Let $\psi : G \times G \rightarrow \{\pm 1\}$ with $\psi(1, 1) = 1$ satisfy (5.15) and suppose that $[\psi(g_i, g_j)]_{1 \leq i,j \leq v}$ is a Hadamard matrix. The construction*

$$a(i_1, i_2, \ldots, i_n) = \prod_{k=2}^{n} \psi\left(\prod_{j=1}^{k-1} g_{i_j}, g_{i_k} \right) \tag{5.19}$$

of Theorem 5.31 with $\phi \equiv 1$ is called the cocyclic *or* relative difference set *construction for proper n-dimensional Hadamard matrices, and a proper n-dimensional Hadamard matrix A of order v so constructed is denoted $\mathcal{R}(n, v, G, \psi)$.*

These proper n-dimensional Hadamard matrices have the form

$$\mathcal{R}(n, v, G, \psi) = [\Phi(e_1 e_2 \ldots e_n)]_{e_i \in R},$$

where R is a subset of size v in an extension E_ψ of $\{\pm 1\}$ by G (see (4.14))

$$1 \rightarrow \{\pm 1\} \rightarrow E_\psi \rightarrow G \rightarrow 1 \ .$$

As will be seen in Lemma 7.2, E_ψ has order $2v$, elements $\{\pm 1\} \times G$ and multiplication $(a, g_i)(b, g_j) = (ab\,\psi(g_i, g_j), g_i g_j)$, for all $a, b \in \{\pm 1\}$ and $g_i, g_j \in G$. Define $\Phi : E_\psi \to \{\pm 1\}$ by $\Phi(a, g) = a$. In E_ψ,

$$\prod_{i=1}^{n}(1, g_i) = \left(\prod_{i=2}^{n}\psi\left(\prod_{j=1}^{i-1} g_j, g_i\right), g_1 g_2 \cdots g_n\right)$$

so the right-hand term in (5.19) is the image under Φ of this product of elements from the subset $R = \{(1, g) : g \in G\} \subset E_\psi$. In Chapter 7 (Corollary 7.32), R will be identified as a $(v, 2, v, v/2)$-relative difference set in E_ψ. It was this derivation in terms of the set R which prompted the earlier 'relative difference set' terminology.

As a consequence, calculating a single entry of these proper n-dimensional Hadamard matrices requires $n - 1$ multiplications in a group and one lookup of a table of length $2v$, that is, $O(n)$ operations.

Lemma 5.33 shows that any $\mathcal{G}(n, v, G, \phi)$ is equivalent to a $\mathcal{R}(n, v, G, \partial\phi)$. Ma [229] shows that the number of equivalence classes of $\mathcal{R}(n, v, G, \psi)$ is at least equal to the number of equivalence classes of $\mathcal{R}(2, v, G, \psi)$.

The final results of this section bring us full circle, to link $\mathcal{R}(n, v, G, \psi)$ with the original product construction of n-dimensional Hadamard matrices. They show that for one important class of cocycles, the two constructions are identical. This class consists of those cocycles which are homomorphic in either coordinate.

LEMMA 5.35 *If $\psi : G \times G \to \{\pm 1\}$ satisfies (5.15) and is a homomorphism in one coordinate, then it is a homomorphism in the other coordinate.*

Proof. For all $g, h, k \in G$,

$$\psi(gh, k) = \psi(g, k)\psi(h, k)$$
$$\Leftrightarrow \psi(g, h)\psi(gh, k) = \psi(g, h)\psi(g, k)\psi(h, k)$$
$$\Leftrightarrow \psi(h, k)\psi(g, hk) = \psi(g, h)\psi(g, k)\psi(h, k)$$
$$\Leftrightarrow \psi(g, hk) = \psi(g, h)\psi(g, k).$$

$\square$

THEOREM 5.36 [229, Theorem 4.16] *If $\psi : G \times G \to \{\pm 1\}$ satisfies (5.15) and is a homomorphism in either coordinate and $H = [\psi(g, h)]_{g,h \in G} = \mathcal{R}(2, v, G, \psi)$ is a Hadamard matrix, then for any $n \geq 2$, $\mathcal{R}(n, v, G, \psi) = \mathcal{P}(n, v, H)$.*

Proof. For $2 \leq k \leq n$, $\psi(\prod_{j=1}^{k-1} g_{i_j}, g_{i_k}) = \prod_{j=1}^{k-1} \psi(g_{i_j}, g_{i_k})$, so compare (5.19) with (5.10). $\square$

The cocyclic construction of n-dimensional Hadamard matrices therefore generalises the group-developed construction (5.11) and coincides with the product construction (5.10) on a major subclass. But it depends explicitly on the existence of a Hadamard matrix $[\psi(g_i, g_j)]$, where ψ satisfies (5.15). This is the genesis of the unified theory of cocyclic development of generalised Hadamard matrices which is presented in Part 2.

So first, we must study cocycles.

PART 2
Cocyclic Hadamard Matrices

Chapter Six

Cocycles and Cocyclic Hadamard Matrices

Cocycles occur naturally in many areas. The underlying ideas have been investigated for well over a century. Möbius introduced the concept of 2-complexes into surface topology in the mid-1860s, and Mayer introduced the corresponding algebraic chain complexes, in which n-cocycles map n-simplexes into abelian groups, in the 1920s and 1930s [236, II.9]. During the latter period Schreier [282, 283], Baer [15], Hall and Fitting also considered 2-cocycles (known as 'factor sets') in the study of group extensions, following their introduction by Schur [284, 285, 286] in projective representation theory at the beginning of the twentieth century [236, IV.11]. In the 1960s Mackey [235] analysed the quantum system of a free particle in space, and found that the unitary operator representation of its symmetry group defines a 2-cocycle, mapping to the unit circle in $\mathbb{C}$.

As we have seen, in the mid-1980s de Launey [79] derived the equation (5.15) as a functional constraint on the entries of a 2-D combinatorial design which, if satisfied, enabled the 2-D design to be extended to form a proper n-dimensional design, for any $n \geq 2$. That paper, together with [90], also exhibited large families of Hadamard matrices, weighing matrices, orthogonal designs, generalised Hadamard matrices and generalised weighing matrices whose entries satisfied (5.15), by then recognised as the 2-cocycle equation. In [162] de Launey and the author rederived (5.15) by considering the development of an abstract (2-D) combinatorial design from an initial row. The presence of the cocycle results from there being a group of row and a group of column operations each of which preserves the defining properties of the combinatorial design. When these two groups coincide, they become the target group for the cocycle. In the case of weighing matrices and Hadamard matrices, that group is $\langle -1 \rangle \cong \mathbb{Z}_2$: a row or column can be negated without destroying orthogonality. Hence 2-cocycles also arise naturally in combinatorial design theory.

The inspiration for the work in Part 2 was therefore two-fold. First came the recognition, described in Chapter 5.4, that 2-cocycles may be used to generate n-dimensional combinatorial designs from 2-D ones. Second came the realisation that the cocyclic 2-D designs used in that construction themselves give a rich and unexpected perspective from which to investigate square designs, particularly generalised Hadamard matrices (cf. [86, Section 4]), and their applications. It was soon appreciated that the cocyclic approach would also illuminate the study of semiregular RDS in nonabelian groups (Beth, Jungnickel and Lenz [24, Remarks VIII.3.23]), but Part 2 will prove it can do much, much more.

The first three Sections of this Chapter are devoted to the study of cocycles — examples, properties, constructions and numerical computations. Especially important are the *orthogonal* cocycles — those whose matrices are generalised

Hadamard. Scattered results appear in the literature, principally in [155, 260]. Since individual cocycles are rarely the objects of scrutiny in cohomology theories (their cohomology class groups are the usual focus) these results are collected here for the first time.

The second two Sections of the Chapter deal with those orthogonal cocycles for which the target group is $\{\pm 1\} \cong \mathbb{Z}_2$, or equally, with the *cocyclic* Hadamard matrices. We show that all the constructions of Hadamard matrices in Chapter 2, and more, are cocyclic, except for a single unknown case. **To date, cocyclic construction of Hadamard matrices is the most successful uniform technique known.** Evidence supporting the Cocyclic Hadamard Conjecture, that for every $t \geq 1$ there exists a cocyclic Hadamard matrix of order $4t$, appears in the final Section.

6.1 COCYCLES AND GROUP COHOMOLOGY

We begin with the description of cocycles as they arise within the cohomology theory of groups, in terms of the normalised inhomogeneous bar resolution for the group [38]. The reader is advised that use of left group actions is standard in this context. Because multiplication in symmetric groups is usually defined for consistency with right group actions, the switch is signalled by using the 'opposite' notation ^{op} for multiplication. For instance, multiplication in the group $\mathrm{Aut}(N)^{\mathrm{op}}$ of automorphisms of a group N is given by $\sigma_1 \sigma_2 = \sigma_1 \circ \sigma_2$.

If G is a group, a (left) G-*module* is defined to be a pair (N, ε) where N is an abelian group, written additively, and $\varepsilon : G \to \mathrm{Aut}(N)^{\mathrm{op}}$ is a homomorphism. Equivalently, a (left) G-module structure on N is given by a map $G \times N \to N$ satisfying

$$x(a + b) = xa + xb, \quad (xy)a = x(ya), \quad 1a = a, \quad \forall \, x, y \in G, \; a, b \in N. \quad (6.1)$$

The (left) action of G on N is *trivial* if $\varepsilon \equiv 1$, that is, if, in (6.1), $xa = a$ for all $x \in G$, $a \in N$. An abelian group N carries a trivial G-module structure for every group G.

A G-module and a module over the integral group ring $\mathbb{Z}G$ are equivalent objects. The restriction of scalar multiplication for a $\mathbb{Z}G$-module to the elements of G satisfies (6.1). Conversely, any action of G on N by automorphisms extends uniquely to a $\mathbb{Z}G$-module structure on N by setting $(\sum_{x \in G} k_x x)a = \sum_{x \in G} k_x xa$.

DEFINITION 6.1 *Let* (N, ε) *be a G-module. For $n \geq 0$, the abelian group of n-cochains is the set*

$$C^n(G, N) = \{f : G^n \to N \mid f(x_1, \ldots, x_n) = 0 \text{ whenever } x_i = 1 \text{ for some } i\}$$

with group operation given by pointwise addition. By definition, $|G^0| = 1$, so $C^0(G, N) \cong \{0\}$. The degree n differential is the homomorphism $\partial_n : C^n(G, N) \to C^{n+1}(G, N)$ defined by

$$\partial_n(f)(x_1, \ldots, x_{n+1}) = f(x_2, \ldots, x_{n+1})^{\varepsilon(x_1)}$$

$$+ \sum_{i=1}^{n} (-1)^i f(x_1, \ldots, x_{i-1}, x_i x_{i+1}, \ldots, x_{n+1})$$
$$+ (-1)^{n+1} f(x_1, \ldots, x_n), \qquad (6.2)$$

which satisfies $\partial_n \circ \partial_{n-1} = 0$. *The kernel* $Z_\varepsilon^n(G, N) = \mathrm{Ker}(\partial_n)$ *is called the group of* n*-cocycles, and the image* $B_\varepsilon^n(G, N) = \mathrm{Im}(\partial_{n-1})$, *the group of* n*-coboundaries. The quotient group* $H_\varepsilon^n(G, N) = Z_\varepsilon^n(G, N)/B_\varepsilon^n(G, N)$ *is called the* n^{th}*-cohomology group of* G *with coefficients in* N. *Each coset* $f + B_\varepsilon^n(G, N)$ *is called a* cohomology class *and is denoted* $[f]$. *Two* n*-cocycles* f_1, f_2 *are* cohomologous *if they lie in the same cohomology class.*

Since we will only ever be interested in 1- and 2-cochains, we shall call 2-cocycles and 2-coboundaries *(normalised) cocycles* and *coboundaries*, respectively, and will denote ∂_1 by ∂. Observe from (6.2) that a cocycle is a map $f : G \times G \to N$ satisfying $0 = f(y, z)^{\varepsilon(x)} - f(xy, z) + f(x, yz) - f(x, y)$ and $f(x, 1) = 0 = f(1, x)$ for all $x, y, z \in G$. A coboundary is a cocycle $\partial\phi$ of the form $\partial\phi(x, y) = \phi(x) + \phi(y)^{\varepsilon(x)} - \phi(xy)$ for some 1-cochain $\phi : G \to N$.

At this point, it is necessary to introduce a variation of the above definition of a coboundary, in order to conform with the natural design theoretic interpretation of a coboundary matrix as a normalised group developed matrix (seen in (5.18) in Chapter 5.4).

It is straightforward to check, in Definition 6.1, that f is a cochain if and only if its negation $-f$ is a cochain, and if the degree n differential $\partial_n(f)$ is replaced by its negation $-\partial_n(f) = \partial_n(-f)$, none of the other terms defined is affected.

Henceforward, we make this change without loss of generality, and we also switch from additive to multiplicative notation for the group operation in N, unless indicated otherwise.

DEFINITION 6.2 *Let* (N, ε) *be a* G*-module. A* normalised *cocycle is a map* $\psi : G \times G \to N$ *satisfying*

$$\psi(g, h)\psi(gh, k) = \psi(h, k)^{\varepsilon(g)}\psi(g, hk), \qquad (6.3)$$
$$\psi(g, 1) = 1 = \psi(1, g), \qquad (6.4)$$

for all $g, h, k \in G$. *A* coboundary $\partial\phi$ *is a cocycle of the form*

$$\partial\phi(g, h) = \phi(g)^{-1}(\phi(h)^{\varepsilon(g)})^{-1}\phi(gh) \qquad (6.5)$$

for some 1-cochain $\phi : G \to N$. *The* identity cocycle *is* $1(g, h) = 1$, $g, h \in G$.

For the rest of this Chapter, the G-module N will have trivial G-action $\varepsilon \equiv 1$, and will be denoted C, both because it is known as the *coefficient group* for the cohomology of G and because it is *central* in an extension of itself by G. The groups of cocycles, coboundaries and the second cohomology group will be denoted $Z^2(G, C)$, $B^2(G, C)$ and $H^2(G, C)$, respectively. The cocycle equation (6.3) then equals (5.15), that is,

$$\psi(g, h)\psi(gh, k) = \psi(h, k)\psi(g, hk), \; g, h, k \in G, \qquad (6.6)$$

with $\psi(1,1) = 1$, and the coboundary equation (6.5) becomes

$$\partial\phi(g,h) = \phi(g)^{-1}\phi(h)^{-1}\phi(gh), \quad g,h \in G. \tag{6.7}$$

When G is *finite*, the matrix representation of a cocycle ψ forms a critical link between finite group cohomology and combinatorial design theory. The most general notion of a cocyclic matrix, extending that given now, will however not appear until Chapter 7 (Definition 7.18).

DEFINITION 6.3 *Let G be a finite group of order v, let C be an abelian group and let M be a $v \times v$ matrix with entries in C. Then M is a G-cocyclic matrix over C if there are a cocycle $\psi \in Z^2(G,C)$ and an ordering $G = \{1 = g_1, g_2, \ldots, g_v\}$ such that M is Hadamard equivalent (Definition 4.12) to the (normalised) matrix*

$$M_\psi = [\psi(g_i, g_j)]_{1 \leq i,j \leq v}. \tag{6.8}$$

Plainly, if $\phi : G \to C$ is a 1-cochain, the matrix $[\phi(g_i g_j)]_{1 \leq i,j \leq v}$ is G-developed over C (see Definition 2.17). Its normalisation has entry $\phi(g_i)^{-1}\phi(g_j)^{-1}\phi(g_i g_j)$ in position (i,j), that is, it takes the value $\partial\phi(g_i, g_j)$ of the coboundary $\partial\phi$. Thus the coboundary matrix $M_{\partial\phi}$ is the normalised version of the group developed matrix $[\phi]$. This point has already been made in Chapter 5.4 for $C = \{\pm 1\}$.

6.2 COCYCLES ARE EVERYWHERE!

Many cocycles in $Z^2(G,C)$, or their matrices, will already be familiar to the reader from other areas of mathematics. The following list demonstrates this, and both illustrates some characteristic properties (which are then isolated) and provides some of the fundamental examples needed later.

6.2.1 Examples of cocycles

Example 6.2.1 *If $\phi : G \to C$ is a set map for which $\phi(1) = 1$, and $\partial\phi(g,h) = \phi(g)^{-1}\phi(h)^{-1}\phi(gh)$, then the coboundary $\partial\phi$ is a cocycle. If G is finite, the cocyclic matrix $M_{\partial\phi}$ is the normalised version of the group developed matrix $[\phi]$.*

Example 6.2.2 *If $G = \langle a : a^v = 1 \rangle \cong \mathbb{Z}_v$ with indexing $\{1, a, a^2, \ldots, a^{v-1}\}$, if $\omega \in C$ and if $\psi_\omega(a^i, a^j) = \omega^{\lfloor (i+j)/v \rfloor}$, $0 \leq i, j \leq v - 1$, then ψ_ω is a cocycle and*

$$M_{\psi_\omega} = \begin{bmatrix} 1 & 1 & \cdots & 1 & 1 \\ 1 & 1 & \cdots & 1 & \omega \\ \vdots & \vdots & & \vdots & \vdots \\ 1 & 1 & \cdots & \omega & \omega \\ 1 & \omega & \cdots & \omega & \omega \end{bmatrix}. \tag{6.9}$$

Matrices which are Hadamard (that is, entrywise) products of a back-circulant matrix and (6.9) are termed (back) *ω-cyclic* or *constacyclic* in the literature of combinatorial designs. In particular, if $\omega = -1$, then the matrix is back *negacyclic*.

Example 6.2.3   *If $G = \langle a : a^v = 1 \rangle \cong \mathbb{Z}_v$ with indexing $\{1, a, a^2, \dots, a^{v-1}\}$ and ω is in C, then $\psi(a^k, a^j) = \omega^{kj}$, $0 \leq k, j \leq v - 1$, is a cocycle and M_ψ is a Vandermonde matrix.*

In particular, when $C = \mathbb{C}$ and $\omega = \exp(-2\pi i/v)$, then $M_\psi = \mathcal{F}_v$ is the matrix of the Discrete Fourier Transform (3.6).

Example 6.2.4   *Let $G = \mathbb{Z}_2^n$ and $C = \{\pm 1\} \cong \mathbb{Z}_2$. The vector inner product $\langle \mathbf{u}, \mathbf{v} \rangle$ determines a cocycle ψ, where $\psi(\mathbf{u}, \mathbf{v}) = (-1)^{\langle \mathbf{u}, \mathbf{v} \rangle}$ for all $\mathbf{u}, \mathbf{v} \in G$, and M_ψ is the Sylvester Hadamard matrix (2.4) of order 2^n.*

Example 6.2.5   *Let G be abelian of order v and exponent m and $C = \langle e^{2i\pi/m} \rangle \subset \mathbb{C}$. The character group $\widehat{G}$ of G determines a cocycle χ_G defined by $\chi_G(g, h) = \chi_g(h)$ for all $g, h \in G$ and $M_{\chi_G} = \mathcal{F}_G$ is the FT matrix (Example 4.1.2) for G.*

Example 6.2.3 (with $M_\psi = \mathcal{F}_v$) and Example 6.2.4 are the cyclic and elementary abelian 2-group cases of Example 6.2.5, respectively.

Example 6.2.6   *Let V be a vector space over a field $\mathbb{F}$. Any bilinear form $\psi :$ $(V, +) \times (V, +) \to \mathbb{F}$ is a cocycle. Its diagonal $D\psi(\mathbf{u}) = \psi(\mathbf{u}, \mathbf{u})$ is a quadratic form. Furthermore, if $\phi : V \to \mathbb{F}$ is a quadratic form and char $\mathbb{F} \neq 2$, the unique symmetric bilinear form associated with ϕ (the* polar *form of ϕ) is the coboundary $\partial(2^{-1}\phi)$.*

Example 6.2.7   *Let R be a ring with additive group $G = C = (R, +)$. Multiplication in R is a cocycle μ, with $\mu(g, h) = gh$, $\forall\, g, h \in G$, termed the* multiplication *cocycle. When R is finite, M_μ is the multiplication table of R.*

In Example 6.2.7, R may be a field, as already seen in Example 4.3.2 for $R = GF(p^n)$ and illustrated for $GF(4)$ in Example 4.3.1. This type of cocycle is not confined to rings. In Chapter 9 the multiplication of a finite presemifield is shown to be a cocycle (Theorem 9.32), and in Chapter 9.3.2.1 two concrete examples — multiplication tables for nonisotopic presemifields of order 16 — are displayed.

6.2.2 New from old

Once a cocycle is identified, it is possible to construct others from it. Some basic constructions follow; verification that they do give cocycles, together with their specialisation to the coboundary case, is left to the reader.

Example 6.2.8   *If $\psi \in Z^2(G, C)$ and $n \in \mathbb{Z}$, then the nth* power *of ψ is $\psi^n \in Z^2(G, C)$, where $\psi^n(g, h) = \psi(g, h)^n$. In particular, the* inverse *ψ^{-1} is a cocycle, and $M_{\psi^{-1}} = M_\psi^{(-1)}$ is the matrix of inverses (cf. Lemma 4.10.1).*

When the abelian group C is written additively, Example 6.2.8 simply tells us that the natural $\mathbb{Z}$-module structure of the abelian group $Z^2(G, C)$ is inherited from the natural $\mathbb{Z}$-module structure of C. The same is true if C is an R-module for some ring R.

Example 6.2.9   *Write C and $Z^2(G,C)$ additively. If C is a (left) R-module for some ring R, $r \in R$ and $\psi \in Z^2(G,C)$, then the* scalar multiple *of ψ by r is $r\psi \in Z^2(G,C)$, where $(r\psi)(g,h) = r\psi(g,h)$. Under scalar multiplication, $Z^2(G,C)$ is a (left) R-module.*

Example 6.2.10   *If K is a subgroup of G and $\psi \in Z^2(G,C)$, then the* restriction *of ψ to K is $\operatorname{res}\psi \in Z^2(K,C)$, where*

$$\operatorname{res}\psi(g,h) = \psi(g,h), \quad g,h \in K.$$

Example 6.2.11   *If K is a normal subgroup of G and $\psi \in Z^2(G/K,C)$, then the* inflation *of ψ to G is $\inf\psi \in Z^2(G,C)$, where*

$$\inf\psi(g,h) = \psi(gK,hK),$$

and $M_{\inf\psi} = M_\psi \otimes J_{|K|}$. Informally (cf. [111, p. 182]), inflation is just 'tensoring up'.

Example 6.2.12   *If $\psi \in Z^2(G,C)$ and $\theta : G' \to G$ and $\gamma : C \to C'$ are group homomorphisms, then the composition $\gamma \circ \psi \circ (\theta \times \theta) \in Z^2(G',C')$, where*

$$\gamma \circ \psi \circ (\theta \times \theta)(g',h') = \gamma(\psi(\theta(g'),\theta(h'))).$$

Example 6.2.13   *If $\psi_1 \in Z^2(G,C_1)$ and $\psi_2 \in Z^2(G,C_2)$, then their* direct sum[1] *$(\psi_1,\psi_2) \in Z^2(G,C_1 \times C_2)$, where*

$$(\psi_1,\psi_2)(g,h) = (\psi_1(g,h),\psi_2(g,h)).$$

Conversely, if $\psi \in Z^2(G,C)$ and C decomposes as an internal direct sum of subgroups C_1 and C_2, that is, $C = C_1 \oplus C_2$, then there is an isomorphism $\beta : C \cong C_1 \times C_2$ and $\beta \circ \psi = (\psi_1,\psi_2)$ is the direct sum of $\psi_i = \pi_i \circ \beta \circ \psi \in Z^2(G,C_i)$, where $\pi_i : C_1 \times C_2 \to C_i$ is the projection epimorphism, $i = 1,2$.

Example 6.2.14   *If $\psi \in Z^2(G,C)$ and $\psi' \in Z^2(G',C)$, then their* tensor product *$\psi \otimes \psi' \in Z^2(G \times G',C)$, where*

$$(\psi \otimes \psi')((g,g'),(h,h')) = \psi(g,h)\psi'(g',h'),$$

and $M_{\psi \otimes \psi'} = M_\psi \otimes M_{\psi'}$.

Example 6.2.15   *If $\psi \in Z^2(G,C)$ and $\psi' \in Z^2(G',C')$, then their* direct product *$\psi \times \psi' \in Z^2(G \times G', C \times C')$, where*

$$(\psi \times \psi')((g,g'),(h,h')) = (\psi(g,h),\psi'(g',h')).$$

---

[1]The direct sum is so-called to avoid confusion with the direct product of Example 6.2.15. Alternatively, if $C = C_1 \oplus C_2$, use of the cocycle $\psi = \psi_1 \oplus \psi_2 = \beta^{-1} \circ (\psi_1,\psi_2) \in Z^2(G,C)$ may be preferred.

If $\psi \in Z^2(G,C)$, the *transpose* $\psi^\top \in C^2(G,C)$ of ψ is the 2-cochain

$$\psi^\top(g,h) = \psi(h,g). \tag{6.10}$$

Unless G is abelian, the transpose of a cocycle is not necessarily a cocycle. In view of this restriction, the transpose is probably not the natural 2-cochain with which to work, despite its utility when G is abelian. Instead, by the insights gained in Chapter 7, the rôle of the transpose should really be performed by a newly identified cocycle, the *dual*, an 'inverse transpose of inverses'.

Example 6.2.16 *If $\psi \in Z^2(G,C)$, its dual $\psi^* \in Z^2(G,C)$ is defined (see Theorem 7.25) to be*

$$\psi^*(g,h) = \psi^{-1}(h^{-1},g^{-1}).$$

Since inversion is an automorphism of C and a permutation of G, $M_{\psi^*} \sim [\psi^\top]$ (see Definition 4.12). If G is abelian, then $\psi^* = \gamma \circ \psi^\top \circ (\theta \times \theta)$, where γ, θ are the inversion automorphisms on C, G, respectively, so that $M_{\psi^*} \sim M_{\psi^\top}$.

6.2.3 Characteristic properties

First note that the order of any cohomology class $[\psi]$ in $H^2(G,C)$ divides $|G|$.

LEMMA 6.4 *If G is finite of order v and $\psi \in Z^2(G,C)$, then $\psi^v \in B^2(G,C)$. Hence the exponent of $H^2(G,C)$ divides v.*

Proof. Given ψ, define $\phi : G \to C$ to be $\phi(g) = \prod_{h \in G} \psi(g,h)$. Then

$$(\partial\phi)^{-1}(g,k)$$
$$= \prod_{h \in G} \psi(gk,h)^{-1} \prod_{h \in G} \psi(g,h) \prod_{h \in G} \psi(k,h)$$
$$= \prod_{h \in G} \{\psi(gk,h)^{-1}\psi(g,kh)\psi(k,h)\}$$
$$= \prod_{h \in G} \psi(g,k) = \psi^v(g,k).$$

$\square$

Three important properties of cocycles and their matrices determine corresponding subgroups of $Z^2(G,C)$.

DEFINITION 6.5 *Let $\psi \in Z^2(G,C)$.*

1. *ψ is symmetric if $\psi = \psi^\top$. Equivalently, M_ψ is a symmetric matrix. The set of symmetric cocycles forms a subgroup of $Z^2(G,C)$ which we denote $S^2_+(G,C)$.*

2. *ψ is skew-symmetric if $\psi(g,h) = \psi(h,g)^{-1}$ and $\psi(g,g) = 1$ for all $g,h \in G$. Equivalently, M_ψ is a skew-symmetric matrix (provided C is written additively). Note that when $|C|$ is odd, condition $\psi\psi^\top = 1$ implies $\psi(g,g) = 1$ for all $g \in G$. The set of skew-symmetric cocycles forms a subgroup of $Z^2(G,C)$ which we denote $S^2_-(G,C)$.*

3. ψ is multiplicative *if it is a homomorphism of groups in either coordinate (and hence in both — the proof of Lemma 5.35 holds when $\{\pm 1\}$ is replaced by any abelian group C). The set of multiplicative cocycles forms a subgroup of $Z^2(G,C)$ which we denote $M^2(G,C)$.*

If ψ is a coboundary, the symmetry property will hold for any commuting pairs $g, h \in G : gh = hg$, and cocycles which are symmetric on all commuting pairs of elements in G are called *almost symmetric*. The set of almost symmetric cocycles forms a subgroup $A^2(G,C) \leq Z^2(G,C)$, containing $B^2(G,C)$. It is an open question to identify $A^2(G,C)$. See the work of Flannery [111, §4] for more on this problem.

Research Problem 32 *Given G, for each C, what is the subgroup $A^2(G,C) \leq Z^2(G,C)$ of almost symmetric cocycles?*

Some further cochain constructions, like the transpose, are not always cocycles.

Example 6.2.17 *If $\psi \in Z^2(G,C)$, its symmetrisation $\psi^+ \in C^2(G,C)$ is*

$$\psi^+(g,h) = \psi(g,h)\psi(h,g),$$

and its skew-symmetrisation $\psi^- \in C^2(G,C)$ is

$$\psi^-(g,h) = \psi(g,h)\psi(h,g)^{-1}.$$

These are not necessarily cocycles, but if G is abelian, $\psi^+ = \psi\,\psi^\top \in S^2_+(G,C)$ and $\psi^- = \psi\,(\psi^\top)^{-1} \in S^2_-(G,C)$. Then the factorisation $\psi = \psi^\top\psi^-$ is unique, and ψ^- is multiplicative and is known as the commutator pairing *(cf. [38, Exercises IV.3.8 and V.6.5] and (6.14)).*

If $\psi = \psi^\top$, then $\psi^+ = \psi^2$, which is a cocycle (Example 6.2.8), though if C is an elementary abelian 2-group, this symmetrisation is the trivial cocycle 1.

The property embodied in the next result is fundamental to the study of generalised Hadamard matrices through cocycles.

LEMMA 6.6 [260, Lemmas 2.1, 2.2] *If G is finite of order v and $\psi \in Z^2(G,C)$, then in $\mathbb{Z}C$, for each pair of elements $h, k \in G$,*

$$\sum_{g\in G} \psi(h,g)\psi(k,g)^{-1} = \psi(hk^{-1},k)^{-1} \sum_{g\in G} \psi(hk^{-1},g) . \qquad (6.11)$$

Consequently M_ψ is row pairwise balanced (Definition 4.9) if and only if M_ψ is row balanced (Definition 4.6).

Proof. Put $d = hk^{-1}$. Then in (6.11)

$$\text{LHS} = \sum_{g\in G} \psi(dk,g)\psi(k,g)^{-1} = \sum_{g\in G} (\psi(d,k)^{-1}\psi(d,kg)\psi(k,g))\psi(k,g)^{-1}$$

$$= \psi(d,k)^{-1} \sum_{g\in G} \psi(d,kg) = \text{RHS}.$$

For the second part, we may suppose C is finite of order w. If M_ψ is row balanced, then by Definition 4.6, for any $d \neq 1 \in G$, $\sum_{g\in G} \psi(d,g) = (v/w)\sum_{a\in C} a$. Thus

(6.11) equals $(v/w) \sum_{a \in C} a$, and by Definition 4.9, M_ψ is row pairwise balanced, and conversely. □

The significance of Lemma 6.6 is that a G-cocyclic matrix M_ψ over C is a normalised generalised Hadamard matrix $GH(v, v/w)$ if and only if it is row balanced. In other words, the additional internal structure in a matrix which represents a cocycle is sufficient to provide a substantial cut-down in computational complexity of the problem of testing if it is generalised Hadamard.

6.2.4 Orthogonality and its inheritance

The term (coined in 1996 [152]) used to describe a cocycle whose matrix is row balanced is *orthogonal*. (When $C = \{\pm 1\}$, de Launey and others call this property *pure Hadamard* [91].)

DEFINITION 6.7 *Suppose G has order v and C has order w. A cocycle $\psi \in Z^2(G, C)$ is orthogonal if M_ψ is row balanced. That is, ψ is orthogonal if and only if*

1. M_ψ is a normalised $GH(w, v/w)$ over C, or

2. in $\mathbb{Z}C$, for each $g \neq 1 \in G$, $\sum_{h \in G} \psi(g, h) = v/w \left(\sum_{a \in C} a \right)$, or

3. for each $g \neq 1 \in G$ and each $c \in C$, $|\{h \in G : \psi(g, h) = c\}| = v/w$.

Orthogonality is inherited by some of the constructions in Section 6.2.2. Obviously the inverse ψ^{-1} and dual ψ^* of an orthogonal cocycle ψ are orthogonal, since both matrices $M_{\psi^{-1}}$ and $M_{\psi^*} \sim [\psi^\top]$ are $GH(w, v/w)$, by Lemma 4.10.

A scalar multiple $r\psi$ (Example 6.2.9) of an orthogonal cocycle ψ is itself orthogonal if and only if the homomorphism $r : C \to C$ defined by $c \mapsto rc$ is an automorphism. This is a special case of the next result.

COROLLARY 6.8 *If $\psi \in Z^2(G, C)$ is orthogonal, $\theta : G' \to G$ is an isomorphism and $\gamma : C \to C'$ is a homomorphism, then the composition $\gamma \circ \psi \circ (\theta \times \theta) \in Z^2(G', C')$ is orthogonal if and only if γ is an epimorphism.*

Proof. Lemma 4.13 and permutation equivalence (Definition 4.12) suffice. □

Whilst not every cocycle in $Z^2(G_1 \times G_2, C)$ is a tensor product of cocycles in $Z^2(G_i, C)$, $i = 1, 2$, for those that are, orthogonality is inherited from the factors, and vice versa.

THEOREM 6.9 *Let $\psi_i \in Z^2(G_i, C)$, $1 \leq i \leq n$ and $\psi = \psi_1 \otimes \cdots \otimes \psi_n \in Z^2(G_1 \times \cdots \times G_n, C)$. Then ψ is orthogonal if and only if ψ_i is orthogonal, $1 \leq i \leq n$.*

Proof. [173, Theorem 4.iii] Suppose $n = 2$ and $|G_i| = v_i$. If $\psi = \psi_1 \otimes \psi_2$ is orthogonal and $g \neq 1 \in G_1$, then in $\mathbb{Z}C$, $\sum_{(h_1, h_2) \in G_1 \times G_2} \psi((g, 1), (h_1, h_2)) = (v_1 v_2/w) \left(\sum_{a \in C} a \right) = \sum_{(h_1, h_2) \in G_1 \times G_2} \psi_1(g, h_1) = v_2 \sum_{h_1 \in G_1} \psi_1(g, h_1)$, so ψ_1

is orthogonal, and similarly for ψ_2. The converse is [260, Theorem 5.1] and the general case follows by induction.   □

Every cocycle in $Z^2(G, C_1 \times C_2)$ is a direct sum of cocycles in $Z^2(G, C_i)$, $i = 1, 2$ (Example 6.2.13). It is tempting to hope that an analogue of Theorem 6.9 exists for the direct sum of orthogonal cocycles. Certainly, by Corollary 6.8, if (ψ_1, ψ_2) is orthogonal, then each of the cocycles ψ_i is orthogonal. However, the converse does not hold, even for multiplicative cocycles. The characterisation of orthogonality of a direct sum in terms of orthogonality of its direct summands is a subtle problem, and further discussion is postponed until Chapter 9.3.1.

The best-understood orthogonal cocycles are the multiplicative ones, which exist only for elementary abelian p-groups G and C and are easily identified and enumerated when $C = \mathbb{Z}_p$.

THEOREM 6.10 *Suppose $Z^2(G, C)$ contains a multiplicative orthogonal cocycle.*

1. (Chen [168, Lemma 2.11]) *There is a prime p such that G and C are both elementary abelian p-groups.*

2. *If $G = \mathbb{Z}_p^m$ and $C = \mathbb{Z}_p$, then the number of orthogonal cocycles in $M^2(\mathbb{Z}_p^m, \mathbb{Z}_p)$ is $|GL(m, p)| = \prod_{j=0}^{m-1}(p^m - p^j)$.*

Proof. 1. Let $\psi \in M^2(G, C)$ be orthogonal. First, G must be abelian, since $1 = \psi(1, h) = \psi(g, h)\psi(g^{-1}, h)$, so $\psi(g^{-1}, h) = \psi(g, h)^{-1}$, hence for each fixed pair $g, k \in G$, $\psi([g, k], h) = 1$ for every $h \in G$. Since ψ is orthogonal, $[g, k] = 1$ for all $g, k \in G$. Next, suppose p is a prime dividing $|C|$, and suppose $g \neq 1 \in G$ exists with $g^p \neq 1$. Then $C = \{\psi(g^p, h) : h \in G\} = \{\psi(g, h)^p : h \in G\} = \{c^p : c \in C\}$. But the Sylow p-subgroup of C drops its exponent by 1 in $\{c^p : c \in C\}$ so these sets cannot be equal. Thus p is unique, G must be an elementary abelian p-group and C must be an abelian p-group. Finally, for any $c \neq 1 \in C$, there exist $g \neq 1 \in G$ and $h \in G$ such that $\psi(g, h) = c$. Then $1 = \psi(g^p, h) = \psi(g, h)^p = c^p$, so C is elementary abelian.

2. Each $\psi \in M^2(\mathbb{Z}_p^m, \mathbb{Z}_p)$ is a bilinear form over $GF(p)$, so $\psi(\mathbf{x}, \mathbf{y}) = \mathbf{x}M\mathbf{y}^\top$ for a uniquely defined square matrix M, and ψ is orthogonal if and only if M is nonsingular, that is, $M \in GL(m, p)$.   □

6.3 COMPUTATION OF COCYCLES

Research in group cohomology usually focusses on global properties of the cohomology groups $H^n(G, C)$ such as their spectral sequences and the graded rings they form.

Consequently, most attempts to exploit newly available computational algebra software systems, such as GAP, MAGMA and AXIOM, concentrate on algorithms to compute free resolutions and cohomology groups. For example, Grabmeier and Lambe [130] have an implementation in AXIOM of their algorithm for computation of resolutions and (co)homology for any finite p-group. Some facilities for cohomology computations are available in GAP 4.

The reason for computing resolutions is that the homology and cohomology groups of G are calculated from a free $\mathbb{Z}G$-resolution, and are independent of the resolution chosen. The derivation of cohomology of groups given in Definition 6.1 uses a particular free $\mathbb{Z}G$-resolution, the *unnormalised standard* or *bar* resolution, for G. This is relatively expensive computationally: in dimension n, the unnormalised bar resolution of G has $|G|^n$ free generators. Lambe has a substantial body of work devoted to finding smaller models for homology than the bar resolution.

The emphasis when computing cohomology groups is on identifying their isotypes and performing large-scale operations with them, rather than in finding representatives of each cohomology class.

So, even though we know that each cocycle $\psi \in Z^2(G, C)$ may be written as a product $\psi = \varphi\, \partial\phi$, where $\psi \in [\varphi]$ and $\partial\phi \in B^2(G, C)$, before 1990 very little was known about the computation of individual cocycles, or how to list a set of cohomology class representatives, or how to list all the cocycles (or even all the coboundaries) for a given group. Holt [149] wrote procedures for computing the second integral homology group $H_2(G)$ of a finite permutation group G — cf. (6.14) below — which are distributed as part of MAGMA. Only for the cyclic group $G = \mathbb{Z}_v$ was $Z^2(G, C)$ completely understood.

It has been possible to apply results from cohomology theory to derive three algorithms for the computation of all cocyclic matrices for a given finite group G. The first applies to any finite abelian group. The second and third apply without restriction, with the third aiming for more efficient computation by using a smaller homological model.

The key to the first two algorithms has been to break the computation into two parts: the first removes from consideration the actual target or coefficient group C by deriving a minimal generating set of cocycles in $Z^2(G, U(G))$ for a "universal" coefficient group $U(G)$, and the second then maps these generators to whichever coefficient group C is presently of interest.

DEFINITION 6.11 *Let $F_G = (\mathbb{Z}(G \times G), +)$ be the additive group of the integral group ring of $G \times G$ and let R_G be the subgroup of F_G generated by*

$$\{(1,1); \ (g,h) + (gh,k) - (g,hk) - (h,k), \ g,h,k \in G\}. \tag{6.12}$$

The universal coefficient group $U(G)$ *for G is defined to be the quotient group* $U(G) = F_G/R_G$. *Denote the coset $(g,h) + R_G$ by $[[g,h]]$. The* universal cocycle $\Gamma_G : G \times G \to U(G)$ *for G is defined by $\Gamma_G(g,h) = [[g,h]], \ g,h \in G$.*

The universality of this abelian coefficient group and cocycle derives from the fact that for each coefficient group C, any cocycle $\psi \in Z^2(G, C)$ must factor through the universal cocycle Γ_G. That is, if the map $\psi_c : U(G) \to C$ is given by linear extension of $\psi_c([[g,h]]) = \psi(g,h)$ to all of $U(G)$, then ψ_c is a well-defined homomorphism of abelian groups and $\psi_c \circ \Gamma_G = \psi$. Conversely, if $\psi_c : U(G) \to C$ is an abelian group homomorphism, then $\psi : G \times G \to C$ defined by $\psi(g,h) = \psi_c([[g,h]])$ is a cocycle, by Example 6.2.12.

COROLLARY 6.12 *Let $\psi : G \times G \to C$ and $\psi_c : U(G) \to C$ satisfy $\psi = \psi_c \circ \Gamma_G$, where $U(G)$ is the universal coefficient group and Γ_G is the universal cocycle for G. Then ψ is a cocycle if and only if ψ_c is a homomorphism.*

The next step is to identify a minimal set of generators for the finitely generated abelian group $U(G)$, since by Examples 6.2.9 and 6.2.13, every cocycle in $Z^2(G, U(G))$, including Γ_G, is a unique $\mathbb{Z}$-linear combination of generator cocycles.

THEOREM 6.13 [162, Theorem 11.1] *Let G be finite of order v. The universal coefficient group has torsion-free and torsion components whose isotypes are given by the isomorphism*

$$U(G) \cong \mathbb{Z}^{v-1} \oplus H_2(G), \tag{6.13}$$

where the finite abelian group $H_2(G)$ is the second integral homology group, *or* Schur multiplier, *of G. Consequently,*

$$Z^2(G, C) \cong C^{v-1} \oplus \mathrm{Hom}(H_2(G), C). \tag{6.14}$$

When (6.14) is factored out by the group of coboundaries $B^2(G, C)$, we obtain the Universal Coefficient Theorem, a standard cohomological result:

$$H^2(G, C) \cong \mathrm{Ext}_{\mathbb{Z}}(G/G', C) \oplus \mathrm{Hom}(H_2(G), C), \tag{6.15}$$

where $G' = [G, G]$ is the commutator subgroup of G.

6.3.1 Algorithm 1 — abelian groups

For the first algorithm, where G is abelian, we use (6.12) to derive a standard minimal set of generators for $U(G)$, which is then mapped to C to provide a standard minimal set of generators for $Z^2(G, C)$. The characterisation appearing next is as given in [217]; the original version [90, Lemma 4.5.i] misstates the form of the second sum $\sum_{j=2}^m \partial \theta_j(a, b)$ in (6.16).

LEMMA 6.14 *Let G be a finite abelian group of order v, with torsion invariant form*

$$G \cong \mathbb{Z}_{n_1} \times \mathbb{Z}_{n_2} \times \cdots \times \mathbb{Z}_{n_m}, \ \ n_i | \, n_{i+1}, \ \ 1 \le i < m,$$

where $\mathbb{Z}_{n_i} \cong \langle x_i : x_i^{n_i} = 1 \rangle$, so that each $a \in G$ has a unique representation $a = \prod_{i=1}^m x_i^{a_i}$, where $0 \le a_i < n_i$.

For $2 \le j \le m$, let $\theta_j : G \to U(G)$ be the set map $\theta_j(a) = [[\prod_{i=1}^{j-1} x_i^{a_i}, x_j^{a_j}]]$.

For $1 \le i < j \le m$, define $c_{ij} \in U(G)$, of order $o(c_{ij}) = n_i$, to be $c_{ij} = [[x_i, x_j]] - [[x_j, x_i]]$.

Then, for all $a, b \in G$, the coset $[[a, b]] \in U(G)$ may be expressed in the following normal form:

$$[[a, b]] = \sum_{i=1}^m \sum_{j=0}^{b_i-1} ([[x_i, x_i^{(a_i+j)}]] - [[x_i, x_i^j]]) + \sum_{j=2}^m \partial \theta_j(a, b) - \sum_{i=1}^{m-1} \sum_{j=i+1}^m b_i a_j c_{ij}, \tag{6.16}$$

where the coefficient $b_i a_j$ is reduced $\mod n_i$.

THEOREM 6.15 *Let G be a finite abelian group. With the notation of Lemma 6.14, the normal form (6.16) is unique.*

Proof. (Sketch) If $b > a$ under lexicographic order on G, then the normal form for $[[b, a]]$ is unique if and only if the normal form for $[[a, b]]$ is, so assume $a \leq b$. Partition the set of relators $\neq (1, 1)$ in (6.12) into those for which one term equals another (c, d) or its transpose (d, c) and those where all four terms are distinct and nontransposes. Partition the latter set of relators according to the term which is largest in lexicographic order on $G \times G$. Then the normal form of $[[a, b]]$ is invariant under the application of any such relator with largest term less than or equal to (a, b) and, conversely, no relator with greater largest term can transform a normal form to another normal form. $\square$

COROLLARY 6.16 *Let G be a finite abelian group of order v as in Lemma 6.14.*

1. *With the notation of Lemma 6.14, $U(G) = S(G) \oplus B(G) \oplus T(G)$, where $S(G)$ is freely generated by the $(\sum_{i=1}^{m} n_i - m)$ cosets*

$$[[x_k, x_k^{a_k}]], \ \ 0 < a_k < n_k, \ 1 \leq k \leq m;$$

   $B(G)$ is freely generated by the $(v - 1 + m - \sum_{i=1}^{m} n_i)$ cosets

$$\left[\left[\prod_{i=1}^{k-1} x_i^{a_i}, \ x_k^{a_k}\right]\right], \ \ a_k \neq 0, \ k \geq 2, \ \prod_{i=1}^{k-1} x_i^{a_i} \neq 1;$$

   and $T(G) \cong H_2(G)$, the torsion subgroup, is generated by the $m(m-1)/2$ cosets

$$c_{ij} = [[x_i, x_j]] - [[x_j, x_i]], \ \ 1 \leq i < j \leq m$$

   of finite order $o(c_{ij}) = n_i$.

2. *Let C be a finite abelian group of order w and let w_i be the number of elements of C with order dividing n_i, $1 \leq i \leq m$. Then*

$$|Z^2(G, C)| = w^{v-1} \prod_{i=1}^{m-1} w_i^{m-i}.$$

As a consequence, we can decompose Γ_G as a direct sum $\Gamma_G = \Gamma_S \oplus \partial\gamma_B \oplus \Gamma_T$, where Γ_S is the composition of Γ with the projection $U(G) \to S(G)$, and so on, and $\Gamma_S, \partial\gamma_B$ and Γ_T are cocycles with images in $S(G), B(G)$ and $T(G)$, respectively — see the footnote to Example 6.2.13. Even though $B(G) \subset B^2(G, U(G))$, in general $B(G) \neq B^2(G, U(G))$. For instance, $v\Gamma_S \in S(G) \cap B^2(G, U(G))$ by Lemma 6.4.

Therefore, each cocycle $\psi = \psi^* \circ \Gamma_G \in Z^2(G, C)$ factors *uniquely* as a triple product

$$\psi = \psi_S \, \partial\phi_B \, \psi_T, \tag{6.17}$$

where $\psi_S = \psi^* \circ \Gamma_S$, $\partial\phi_B = \psi^* \circ \partial\gamma_B$, $\psi_T = \psi^* \circ \Gamma_T$ and ψ_T is multiplicative.

Example 6.3.1 *(Example 6.2.2 continued)* [79, 90, 163] *If $G = \langle a : a^v = 1 \rangle \cong \mathbb{Z}_v$, then $B(G) = 0$ and $H_2(\mathbb{Z}_v) = 0$, so $\partial\gamma_B = \Gamma_T = 0$. Each cocycle $\psi = \psi_S$ in $Z^2(\mathbb{Z}_v, C)$ is uniquely determined by the values in C it takes on the $v - 1$ elements $(a, a), (a, a^2), \ldots, (a, a^{v-1})$. In particular, for the smallest cases $v = 2, 3$,*

$$\psi \in Z^2(\mathbb{Z}_2, C) \Leftrightarrow M_\psi = \begin{bmatrix} 1 & 1 \\ 1 & \alpha \end{bmatrix}, \psi(a, a) = \alpha \in C;$$

$$\psi \in Z^2(\mathbb{Z}_3, C) \Leftrightarrow M_\psi = \begin{bmatrix} 1 & 1 & 1 \\ 1 & \alpha & \gamma \\ 1 & \gamma & \alpha^{-1}\gamma \end{bmatrix}, \psi(a, a) = \alpha, \psi(a, a^2) = \gamma \in C.$$

Example 6.3.2 [79, 3.10] *If* $G = \langle a, b : a^2 = b^2 = (ab)^2 = 1 \rangle \cong \mathbb{Z}_2^2$ *with indexing* $\{1, a, b, ab\}$, *then* ψ *is in* $Z^2(\mathbb{Z}_2^2, C)$ *if and only if* M_ψ *is of the form*

$$M_\psi = \begin{bmatrix} 1 & 1 & 1 & 1 \\ 1 & \alpha & \gamma^{-1} & \alpha\gamma \\ 1 & \gamma^{-1}\kappa & \beta & \beta\gamma\kappa \\ 1 & \alpha\gamma\kappa & \beta\gamma & \alpha\beta\gamma^2\kappa \end{bmatrix},$$

where $\psi(a, a) = \alpha$, $\psi(b, b) = \beta$, $\psi(a, b) = \gamma^{-1}$, $\psi(a, b)\psi(b, a)^{-1} = \kappa \in C$, *and* κ *is an element of order dividing* 2. *(If* C *has no elements of order* 2, *then* $\kappa = 1$.*)*

6.3.2 Algorithm 2 — MAGMA implementation

For the second algorithm, we use the Universal Coefficient Theorem (6.15), a presentation $G = F/R$ of G, a short exact sequence of groups $1 \to R/S \to F/S \to F/R \to 1$ and [111, Theorem 3.6] to identify $H^2(G, -)$ as an internal direct sum

$$H^2(G, -) = \text{im inf} \oplus \text{im } \tau_S.$$

Here inf $: \text{Ext}_{\mathbb{Z}}(G/G', -) \to H^2(G, -)$ is the restriction to $\text{Ext}_{\mathbb{Z}}(G/G', -)$ of the inflation map $H^2(G/G', -) \to H^2(G, -)$ and $\tau_S : \text{Hom}(R/S, -) \to H^2(F/R, -)$ is the transgression map. When this second algorithm is applied to an abelian group, it may output a different minimal set of generators for $Z^2(G, -)$ from the first algorithm (see [112, p. 769]).

Thus each cocycle $\psi \in Z^2(G, -)$ may be written as a triple product

$$\psi = \psi_I \, \partial\varphi_B \, \psi_T \tag{6.18}$$

of an *inflation* cocycle ψ_I, a coboundary and a *transgression* cocycle ψ_T, though the factorisation is *not* unique.

The inflation cocycle is the image of a nontrivial coset representative in $\text{Ext}_{\mathbb{Z}}(G/G', -)$ and may be selected to have a standard form (cf. [162, Definition 13.2]). If the primary invariant decomposition of G/G' is

$$G/G' = \bigoplus_{j=1}^{k} \mathbb{Z}_{q_j}, \ q_j = p_j^{t_j}, \ p_j \text{ prime}, \ 1 \leq j \leq k,$$

the corresponding cocyclic matrix is a tensor product of k back ω-cyclic matrices — one $q_j \times q_j$ matrix $M_{\psi_{\omega_j}}$ for each cyclic component $\mathbb{Z}_{q_j}$ (Example 6.2.2) — and the $|G'| \times |G'|$ all-1s matrix (cf. Example 6.2.11)

$$M_{\psi_I} = M_{\psi_{\omega_1}} \otimes \cdots \otimes M_{\psi_{\omega_k}} \otimes J_{|G'|}.$$

The coboundary matrix may be derived from the multiplication table of G (cf. Example 6.2.1).

The most difficult component to compute is the transgression cocycle — the representative of $\mathrm{Hom}(H_2(G), -)$. This depends on the presentation $G = F/R$ chosen for G, and in particular on the choice of Schur complement $S/[R, F] \cong F/F'$ of the Schur multiplier $(R \cap F')/[R, F] \cong R/S \cong H_2(G)$ in $R/[R, F]$ and on the choice of transversal map $F/R \to F/S$ from G to the covering group F/S.

Flannery provides the theoretical basis for this computation in [110, 111]. Ellis and Kholodna [107] also use the Universal Coefficient Theorem to describe cocycles in $H^2(G, C)$, and detail an implementation in MAGMA which uses the LLL algorithm to construct covering groups F/S. Refer to these and to [162] for further details.

Flannery's method [111] as implemented by Flannery and O'Brien [114], is distributed as a module in MAGMA. This module explicitly outputs a full set of representative cocycles for the elements of $H^2(G, C)$. An outline of the module appears in the MAGMA online help manual [238], in the Central Extensions subsection of the section Finite Soluble Groups within Finite Groups.

We illustrate with four abelian examples.

Example 6.3.3 *(i)* $G = \mathbb{Z}_2^n$. *The Sylvester Hadamard matrix of Example 6.2.4 represents an inflation cocycle ψ_I only, since it equals* $\displaystyle\bigotimes_{i=1}^{n} \begin{bmatrix} 1 & 1 \\ 1 & -1 \end{bmatrix}$.

(ii) $G = \mathbb{Z}_p^n$, p *odd. If* $R = GF(p^n)$, *the multiplication cocycle of Example 6.2.7 is a coboundary only (cf. [168, Corollary 4.1]).*

(iii) $G = \mathbb{Z}_v$. *The cocycle ψ_S of Example 6.3.1 is a product of one inflation cocycle ψ_I (having a $v \times v$ back ω-cyclic matrix as in Example 6.2.2) and $v - 2$ generator coboundaries $\partial\phi_i$, one for each of the elements (a, a^i), $1 \leq i \leq v - 2$, with*

$$\psi_S(a, a^i) = \partial\phi_i(a, a^i),\ 1 \leq i \leq v - 2,$$

$$\psi_S(a, a^{v-1}) = \left(\prod_{i=1}^{v-2} \psi_S(a, a^i) \right)^{-1} \psi_I(a, a^{v-1}).$$

Its value elsewhere is uniquely determined by the cocycle equation (6.3).

(iv) $G = \mathbb{Z}_2^2$. *The matrix of Example 6.3.2 is a Hadamard (entrywise) product*

$$\left(\begin{bmatrix} 1 & 1 \\ 1 & \beta \end{bmatrix} \otimes \begin{bmatrix} 1 & 1 \\ 1 & \alpha \end{bmatrix} \right) \bullet \begin{bmatrix} 1 & 1 & 1 & 1 \\ 1 & 1 & \gamma^{-1} & \gamma \\ 1 & \gamma^{-1} & 1 & \gamma \\ 1 & \gamma & \gamma & \gamma^2 \end{bmatrix} \bullet \begin{bmatrix} 1 & 1 & 1 & 1 \\ 1 & 1 & 1 & 1 \\ 1 & \kappa & 1 & \kappa \\ 1 & \kappa & 1 & \kappa \end{bmatrix},$$

in which α and β determine an inflation cocycle (unless either one is a square in C, in which case it determines a coboundary by Lemma 6.4), γ determines a coboundary and κ determines a trangression cocycle.

6.3.3 Algorithm 3 — Homological perturbation

As mentioned earlier, working with the unnormalised bar resolution to compute cocycles is computationally expensive. The Sevilla group (Alvarez, Armario, Frau and Real) uses a mix of Flannery's approach in the second algorithm [111] and Lambe's homological perturbation methods [130]. The basic idea is to determine a contraction (that is, a strong deformation retraction [130]) — a special form of homotopy equivalence — between the unnormalised bar resolution for G and a 'smaller' differentially graded module, and then to perturb the differential ∂ of the bar resolution so as to obtain a new contraction. The smaller differentially graded module hG, or *homological model*, will have the same homology groups but be faster to compute. Representative cycles computed for the first homology group $H_1(G) \cong H_1(hG) \cong G/G'$ and second homology group $H_2(G) \cong H_2(hG)$ of the homological model are then mapped back to $U(G)$ via the contraction. See [4] for the theoretical basis for this algorithm whenever G is a semidirect product of cyclic groups, that is, $G = \langle a, b : a^r = b^s = 1, b^{-1}ab = a^{\varrho(b)} \rangle \cong \mathbb{Z}_r \rtimes_\varrho \mathbb{Z}_s$ for some right action $\varrho : \langle b \rangle \rightarrow \mathrm{Aut}(\langle a \rangle)$, illustrated for the dihedral groups $D_{4t} \cong \mathbb{Z}_{2t} \rtimes \mathbb{Z}_2$. The cutdown is quite striking: their homological model has only 3 free abelian generators of degree 2, whereas there are $r^2 s^2$ in dimension 2 of the bar resolution. However more preprocessing is required.

In [116] (and the references cited in [5]), these techniques are extended to provide homological models for iterated central extensions and semidirect products of finite abelian groups.

Research Problem 33 *Determine more computationally efficient algorithms for listing a minimal set of generators (6.13) for the universal coefficient group $U(G)$.*

6.4 COCYCLIC HADAMARD MATRICES

In this Section we demonstrate that many Hadamard matrices H are indeed cocyclic, that is, that there is a finite group G and a cocycle $\psi : G \times G \rightarrow \{\pm 1\}$ such that $H \sim M_\psi$. When $N = \{\pm 1\}$, G can *only* act trivially on N, since $\mathrm{Aut}(N)^{\mathrm{op}} = \{1\}$. The simpler equation (6.6) is sufficient to define cocycles mapping to $\{\pm 1\}$.

First, using the results of Sections 6.2.2 and 6.2.4, we record the cocyclic version of Lemma 2.2 for elementary constructions of Hadamard matrices.

LEMMA 6.17 *Suppose M_ψ is a cocyclic Hadamard matrix of order n. Then*

1. *the negation $-M_\psi$ is a cocyclic Hadamard matrix (since $-M_\psi \sim M_\psi$);*

2. *the transpose $M_\psi^\top$ is a cocyclic Hadamard matrix (since $M_\psi^\top \sim M_{\psi^*}$);*

3. *if $M_{\psi'}$ is a cocyclic Hadamard matrix, then the tensor product $M_\psi \otimes M_{\psi'} = M_{\psi \otimes \psi'}$ is a cocyclic Hadamard matrix (by Theorem 6.9);*

4. *for $t \geq 1$, $(\otimes^t \mathcal{S}_1) \otimes M_\psi$ is a cocyclic Hadamard matrix of order $2^t n$.*

6.4.1 Sylvester Hadamard matrices

From Example 6.2.4, the Sylvester Hadamard matrix $\mathcal{S}_n$ is $\mathbb{Z}_2^n$-cocyclic. (This also follows from Lemma 6.17.)

6.4.2 Menon Hadamard matrices

Recall that a Menon Hadamard matrix is the same as a group developed Hadamard matrix (Definition 2.20). If $\phi : G \to \{\pm 1\}$ is a set map and $\phi(1) = -1$, then $[\phi]$ is Hadamard if and only if $[-\phi]$ is Hadamard, so without loss of generality we may assume that $\phi(1) = 1$ and $\partial \phi$ is a coboundary. By Example 6.2.1, a G-developed Menon Hadamard matrix is G-cocyclic.

6.4.3 Williamson Hadamard matrices

Suppose the matrix of (2.14)

$$H_1 = \begin{bmatrix} A & B & C & D \\ B & -A & D & -C \\ C & -D & -A & B \\ D & C & -B & -A \end{bmatrix}$$

has (back) circulant components A, B, C and D of order w. Then it is $(\mathbb{Z}_2^2 \times \mathbb{Z}_w)$-cocyclic.

The argument appears in [79, 90]; see also [18]. First, if the signs are ignored, H_1 is seen to be $(\mathbb{Z}_2^2 \times \mathbb{Z}_w)$-developed, and on normalising, we obtain a $(\mathbb{Z}_2^2 \times \mathbb{Z}_w)$-coboundary matrix $[\partial \phi]$, say, where we may assume $\phi(1) = a_{11} = 1$. Second, if the letters are ignored, on setting $\alpha = \beta = \kappa = -1, \gamma = 1$ in Example 6.3.2, we obtain a $\mathbb{Z}_2^2$-cocyclic matrix over $\{\pm 1\}$. The $w \times w$ all-1s matrix J_w is the $\mathbb{Z}_w$-cocyclic matrix corresponding to the identity cocycle 1. The tensor product of the 4×4 matrix with J_w is the matrix for a (noncoboundary) cocycle ψ over $\mathbb{Z}_2^2 \times \mathbb{Z}_w$. Thus the Hadamard product $M_{\psi \, \partial \phi} = M_\psi \bullet M_{\partial \phi}$ is the normalised version of H_1.

Consequently any Williamson Hadamard matrix of order $4w$, with *circulant* components, is $(\mathbb{Z}_2^2 \times \mathbb{Z}_w)$-cocyclic.

6.4.4 Ito Hadamard matrices

As indicated in Chapter 2.3.2, and first noted in [82], Ito Hadamard matrices are cocyclic over the dihedral group D_{4w}. Let w be odd, let H be the matrix (2.17), and let R be the $(0, 1)$ matrix of order w with all 1s on the back diagonal, and 0s elsewhere. Let $H_2 = R^* H R^*$, where

$$R^* = \begin{bmatrix} I & 0 & & \\ 0 & I & & \\ & & 0 & R \\ & & R & 0 \end{bmatrix}, \quad \text{that is,} \quad H_2 = \begin{bmatrix} A & B & DR & CR \\ B & -A & -CR & DR \\ DR & CR & -A & -B \\ CR & -DR & B & -A \end{bmatrix}.$$

Since $H_2 \sim H$, it is sufficient to show that H_2 is cocyclic over D_{4w}. The argument is similar to that applied to H_1 in Section 6.4.3 above.

Since w is odd, the isomorphism $\mathbb{Z}_{2w} \cong \mathbb{Z}_2 \times \mathbb{Z}_w$ is invoked, giving $D_{4w} \cong (\mathbb{Z}_2 \times \mathbb{Z}_w) \rtimes \mathbb{Z}_2$. Suppose $G = (\mathbb{Z}_2 \times \mathbb{Z}_w) \rtimes \mathbb{Z}_2$ has the presentation (cf. (2.16))

$$G = \langle a, x, b \mid a^2 = x^w = b^2 = 1, x^a = x, a^b = a, x^b = x^{-1} \rangle.$$

By equivalence, an indexing of the rows of H_2 by the elements of D_{4w} which differs from that of the columns may be used. Choose row indexing

$$1, x^{w-1}, x^{w-2}, \dots, x, a, ax^{w-1}, \dots, ax, b, bx^{w-1}, \dots, bx, ab, abx^{w-1}, \dots, abx,$$

and column indexing

$$1, x, x^2, \ldots, x^{w-1}, a, ax, \ldots, ax^{w-1}, b, bx, \ldots, bx^{w-1}, ab, abx, \ldots, abx^{w-1}.$$

With this indexing, the unsigned matrix is G-developed by ϕ, say, and normalising gives a coboundary matrix $[\partial\phi]$.

Let φ be the cocycle on $\mathbb{Z}_2^2$ defined by the case $\alpha = \beta = \gamma = \kappa = -1$ of Example 6.3.2. Since $\mathbb{Z}_w$ is normal in G and $G/\mathbb{Z}_w \cong \mathbb{Z}_2^2$, define $\psi = \inf\varphi$ (see Example 6.2.11, that is, $\psi(a^i b^j x^k, a^l b^m x^n) = \varphi(a^i b^j, a^l b^m)$ always) and then $H_2 \sim M_{\psi\, \partial\phi}$.

Consequently any Ito Hadamard matrix of order $4w$ for w odd, is D_{4w}-cocyclic. Note that the cocycle $\psi\, \partial\phi$ is not a coboundary, but this to be expected, since existence of a D_{4w}-coboundary Hadamard matrix is equivalent to existence of a D_{4w}-developed Menon Hadamard matrix (Section 6.4.2) and would overturn the Circulant Hadamard Conjecture by Theorem 2.23.

We can be more specific. We know $H^2(D_{4w}, \{\pm1\}) \cong \mathbb{Z}_2^3$. In [112, Proposition 6.5.ii] Flannery shows that, if $(1, -1, -1) \in Z^2(D_{4w}, \{\pm1\})$ denotes the noncoboundary cocycle which maps the two inflation generators to 1 and -1, and the single transgression generator to -1, and $\psi \in [(1, -1, -1)]$, then M_ψ is a D_{4w}-cocyclic Hadamard matrix if and only if there exists a pair of $(1, -1)$ matrices M, N of order $2w$, each the entrywise product of a back circulant and negacyclic matrix, such that

$$MM^\top + NN^\top = 4wI_{2w}. \tag{6.19}$$

It may be shown that M and N are each equivalent to 2×2 block negacyclic matrices where each $w \times w$ block is circulant. If M and N are respectively equivalent to

$$\begin{bmatrix} A & B \\ B & -A \end{bmatrix} \quad \text{and} \quad \begin{bmatrix} C & D \\ D & -C \end{bmatrix}, \tag{6.20}$$

where A, B, C, D are circulant, then it is easy to see that (2.13) and (2.18) together are equivalent to (6.19).

THEOREM 6.18 *For odd w, there is a D_{4w}-cocyclic Hadamard matrix M_ψ with $\psi \in [(1, -1, -1)] \in H^2(D_{4w}, \{\pm1\})$ if and only if there is an Ito Hadamard matrix of order $4w$.*

6.4.5 Generalisations of Ito Hadamard matrices

The type Q template (2.17) of Ito has been generalised in two ways, both of which give cocyclic Hadamard matrices.

Recall that Schmidt [280] considered Hadamard matrices of the forms H_1 in Section 6.4.3 and H in (2.17) in which the components A, B, C, D are all group developed over an arbitrary abelian group K rather than over $\mathbb{Z}_w$. He noted that they are cocyclic. It may be readily checked that these Hadamard matrices are cocyclic over $\mathbb{Z}_2^2 \times K$ and $\mathbb{Z}_2 \ltimes (\mathbb{Z}_2 \times K)$, respectively.

An earlier generalisation of Ito Hadamard matrices is the *generalised quaternion* Hadamard matrices, introduced by Yamada [328]. In her construction, 2×2 block negacyclic matrices with circulant components of order w, equivalent to those in (6.20), are generalised to $2^s \times 2^s$ block negacyclic matrices with circulant components of order w. She notes that her definition is the same as Ito's when $s = 1$. Yamada describes several infinite classes of generalised quaternion Hadamard matrices, including the Paley Type I Hadamard matrices.

The argument given for Ito Hadamard matrices extends without difficulty to generalised quaternion Hadamard matrices. These matrices are cocyclic over $D_{2^{s+1}w}$.

6.4.6 Numerical results

The theoretical results above, which prove that many known constructions of Hadamard matrices are, in fact, cocyclic, have been informed and supplemented by computational results using the algorithms of Section 6.3.

Several authors have demonstrated the existence of G-cocyclic Hadamard matrices for particular classes of groups and each order $|G| = 4t$, for restricted values of t.

Of course, there is a $\mathbb{Z}_2^t$-cocyclic Sylvester Hadamard matrix for every t. Rao (= Baliga) and Horadam [18] list instances of G-cocyclic Hadamard matrices for $G = \mathbb{Z}_2^2 \times \mathbb{Z}_t$, $t \leq 25$ odd and Horton et al. [170] the same (by Section 6.4.3) for odd $t \neq 35 \leq 45$. Flannery [112, Table 4] lists instances of orthogonal cocycles for $G = D_{4t}$, $t \leq 11$ and Alvarez et al. [5] the same for $t \leq 13$, using a genetic algorithm to search, while by the remarks at the end of Chapter 2.3.2, there are instances for all odd $t \leq 45$.

In parallel, researchers have searched the set of cocyclic matrices M_ψ, with a fixed ordering of G, using either the row balance property (orthogonality — Definition 6.7) or the MAGMA Hadamard matrix testing module, to produce exact or estimated total counts of the number of distinct cocyclic Hadamard matrices of the form M_ψ. In all cases, these lists are bounded by the algorithmic complexity and computational power available. This underscores both the importance of Research Problem 33 and the need for effective search procedures for orthogonal cocycles among all cocycles. Chua and Rao [56] have used image restoration techniques to plan the search and Alvarez et al. [5] have developed a genetic algorithm for this purpose.

Table 6.1 is due to LeBel [217, 7.2.1]. For $G = \mathbb{Z}_2^t$, $t \leq 4$, it lists the number x of multiplicative orthogonal cocycles and the total number o of orthogonal cocycles he found by exhaustive checking using MAGMA. For $t = 5$, x is calculated from Theorem 6.10 with $p = 2$, and o is estimated by Monte Carlo sampling. The total number $z = 2^{2^t-1+t(t-1)/2} = |Z^2(\mathbb{Z}_2^t, \mathbb{Z}_2)|$ of cocycles, found from Corollary 6.16.2, is included for comparison purposes. When t is even, some of the orthogonal cocycles will be orthogonal coboundaries, corresponding to the bent functions — $\mathbb{Z}_2^t$-developed Menon Hadamard matrices — by Corollary 3.30.

Several authors [18, 152, 16, 116, 56] have published counts of the total numbers of orthogonal cocycles for $G = \mathbb{Z}_2^2 \times \mathbb{Z}_t$ and $G = D_{4t}$. Since $|Z^2(G, \mathbb{Z}_2)| = 2^{4t}$ for both these groups, they provide good testbeds for comparison of the three algorithms in Section 6.3. To simplify the search through all cocycles, some of these researchers check all unnormalised functions $\phi : G \to \mathbb{Z}_2$, with G-developed

t	1	2	3	4	5
x	1	6	168	$20,160$	$9,999,360$
o	1	6	168	$26,880$	$\approx 7.34 \times 10^7$
x/o	1	1	1	0.75	≈ 0.136
z	2	16	$1,024$	$2^{21} \approx 2.1 \times 10^6$	$2^{41} \approx 2.2 \times 10^{12}$
o/z	0.5	0.375	≈ 0.164	$\approx 1.28 \times 10^{-2}$	$\approx 3.34 \times 10^{-5}$

Table 6.1 (LeBel [217]) Number x of orthogonal cocycles in $M^2(\mathbb{Z}_2^t, \mathbb{Z}_2)$ versus number o in $Z^2(\mathbb{Z}_2^t, \mathbb{Z}_2)$ and total number $z = |Z^2(\mathbb{Z}_2^t, \mathbb{Z}_2)|$ of cocycles

t	$\mathbb{Z}_{4t}$	Q_{4t}	$\mathbb{Z}_2^2 \times \mathbb{Z}_t$	D_{4t}
1	2	2	6	6
2	0	0	168	32
3	0	0	24	72
4	0		$1,984$	768
5	0	0	120	$2,200$
7	0	0	840	$11,368^\dagger$
9	0		$3,240$	$130,248^\dagger$

Table 6.2 Total numbers of orthogonal cocycles in $Z^2(G, \mathbb{Z}_2)$ for various G; †numbers in cohomology class $[(1, -1, -1)]$ only

matrix $[\phi]$, instead of the coboundary component $\partial\phi$ of a cocycle. Their totals must be reduced to take into account the fact that $\phi(1) = 0$ and that $|\text{Hom}(G, \mathbb{Z}_2)|$ different functions ϕ all give exactly the same normalised cocyclic matrix $M_{\partial\phi}$.

Table 6.2 indicates the relative productivity in small orders of several families of groups, as sources of cocyclic Hadamard matrices. Columns 2, 4 and 3 are from [152, 116] and the remarks below: that if odd $t \neq u^2$, there can be no Q_{4t}-cocyclic Hadamard matrices. Column 5 is from [6, 56, 116, 152]. The total count given for D_{20} in [152] is incorrect. In the prolific 'Ito' cohomology class $[(1, -1, -1)]$ for D_{20}, there are 1,400 orthogonal cocycles [56]. Totals for other groups of these orders $4t$ are given in [6, 116].

Research Problem 34 *Complete Table 6.2. Extend it to all isotypes of groups of order 4t, and to odd t > 9.*

Some features are obvious on inspection of Table 6.2: the scarcity of solutions in the leftmost two groups and the relatively high numbers for the rightmost two groups, with the dihedral groups appearing to support the most solutions. These experimental observations are consistent with theory.

Flannery has proved [112, Lemma 5.2] that if G has a cyclic Sylow 2-subgroup H then $H^2(G, \mathbb{Z}_2) \cong H^2(H, \mathbb{Z}_2) \cong \mathbb{Z}_2$ (so $H_2(G, \mathbb{Z}_2) = 0$) and any G-cocyclic

Hadamard matrix M_ψ must have ψ a coboundary, that is, have trivial inflation and transgression components. Hence for odd $t > 1$, if $G = \mathbb{Z}_{4t}$ or $G = Q_{4t} \cong \mathbb{Z}_t \rtimes \mathbb{Z}_4$, any G-cocyclic Hadamard matrix M_ψ must be equivalent to a G-developed matrix, so $t = u^2$ for some odd $u > 1$. We believe there are no G-developed Menon Hadamard matrices in either of these cases (see Research Problem 5).

Research Problem 35 *For odd $t = u^2 > 1$, prove that no Q_{4t}-developed Menon Hadamard matrix can exist.*

Research Problem 36 [152, Problem 3] *If $H_2(G) = 0$ and M_ψ is a G-cocyclic Hadamard matrix, is $\psi \in B^2(G, \mathbb{Z}_2)$? That is, is the inflation component ψ_I in (6.18) trivial?*

For odd t, $H^2(\mathbb{Z}_2^2 \times \mathbb{Z}_t, \mathbb{Z}_2) \cong H^2(D_{4t}, \mathbb{Z}_2) \cong \mathbb{Z}_2^3$, with two inflation generators and one transgression generator. For odd $t > 1$, every example of a $\mathbb{Z}_2^2 \times \mathbb{Z}_t$-cocyclic Hadamard matrix found to date has nontrivial transgression generator ($\kappa = -1$ in the terminology of Section 6.4.3), from which it follows by [152, Lemma 3.6] that both inflation generators must be nontrivial ($\alpha = \beta = -1$). It remains an open question as to whether this must be the case. (It is not the case for D_{4t}-cocyclic Hadamard matrices, by Theorem 6.18.)

Research Problem 37 [152, Problem 2] *Prove that if M_ψ is a $\mathbb{Z}_2^2 \times \mathbb{Z}_t$-cocyclic Hadamard matrix and $t > 1$ is odd, the transgression component ψ_T in (6.18) is nontrivial.*

6.5 THE COCYCLIC HADAMARD CONJECTURE

It is obvious from the results above that many of the classical and more recently discovered constructions of Hadamard matrices are in fact cocyclic. Cocyclic construction is the most uniform construction technique for Hadamard matrices yet known. Every Hadamard matrix of order ≤ 20 is cocyclic (see Example 7.4.1). For orders ≤ 200, only $4t = 188 = 4 \cdot 47$ is not yet known to have a cocyclic construction [144, Table 7.29].

THEOREM 6.19 *For orders $4t \leq 200$, a cocyclic Hadamard matrix is known for every $t \neq 47$.*

Naturally, Menon Hadamard matrices form the basic class of cocyclic Hadamard matrices, since coboundaries form the trivial cohomology class of cocycles. However, they arise in orders $4u^2$ only.

As early as 1993 the author and de Launey [90, Conjecture 3.6] conjectured that cocyclic Hadamard matrices exist for every order $4t$, on the strength of the knowledge that Williamson Hadamard matrices with symmetric circulant components are $(\mathbb{Z}_2^2 \times \mathbb{Z}_t)$-cocyclic. By 1997 the equivalence of Ito Hadamard matrices and D_{4t}-cocyclic matrices was known [112]. By Lemma 6.17.4, to prove existence of a cocyclic Hadamard matrix of every order $4t$ it is sufficient to prove it for odd values of t only.

Research Problem 38 *The Cocyclic Hadamard Conjecture. Show that for each odd t there is a cocyclic Hadamard matrix of order 4t.*

In [84] de Launey proves the existence, first announced in [82], of another infinite family of cocyclic Hadamard matrices. This includes Hadamard matrices with the same orders as all the Paley Hadamard matrices (Definition 2.5) and the additional family of Hadamard matrices of orders $4 \cdot 3^d$, $d \geq 0$, listed in [288, Corollary 9.13].

THEOREM 6.20 (de Launey [84, Theorem 1.1]) *If $q_1, q_2, \ldots, q_r \equiv 1 \bmod 4$ and $p_1, p_2, \ldots, p_s \equiv 3 \bmod 4$ are prime powers, and $k_1, k_2, \ldots, k_r$ and $m_1, m_2, \ldots, m_s$ are non-negative integers, then there exists a cocyclic Hadamard matrix of order*

$$\left\{ \prod_{i=1}^{r} 2(q_i + 1) \right\} \left\{ \prod_{j=1}^{s} (p_j + 1) \right\} \left\{ \prod_{i=1}^{r} q_i^{k_i} \right\} \left\{ \prod_{j=1}^{s} p_j^{m_j} \right\}.$$

De Launey's proof depends on three main techniques: use of special properties of the Paley conference matrix; substitutions in which matrices are plugged into cocyclic orthogonal matrices and the result proven to be a cocyclic orthogonal design; and use of the correspondence between cocyclic Hadamard matrices and semiregular relative difference sets which is to be presented in the next Chapter (see Corollary 7.32).

Finally, by arguments based on some of the key ideas in [313], in which Seberry proves her asymptotic result on existence of Hadamard matrices (Figure 2.1), de Launey and Smith obtain an asymptotic result for the existence of cocyclic Hadamard matrices.

THEOREM 6.21 (de Launey and Smith [91, Theorem 1.1.2]) *For any odd positive integer m there exists a $(\mathbb{Z}_2^t \times K)$-cocyclic Hadamard matrix of order $2^t m$ for any integer $t \geq \lfloor 8 \log_2 m \rfloor$, where K has order m and is a direct product of elementary abelian groups.*

6.5.1 Noncocyclic Hadamard matrix constructions?

The results above tell us that the Sylvester, Williamson (and hence by Lemma 2.13 the Paley Type II), Menon and Ito (and hence the Paley Type I) Hadamard matrix constructions are all cocyclic.

Of the three known families of Hadamard difference sets (Example 2.1.1), the Paley difference sets determine Paley Type I Hadamard matrices. The Singer difference sets determine Hadamard matrices H which are equivalent to Sylvester Hadamard matrices (Lemma 2.14), but they also have a direct cocyclic construction. Let $\psi = \gamma \circ \mathrm{tr} \circ \mu$ where μ is multiplication (Example 6.2.7) in $GF(2^t)$, tr is the trace map and $\gamma : \mathbb{Z}_2 \to \{\pm 1\}$ is the exponential isomorphism of (3.25). By Example 6.2.12, $\psi \in Z^2(\mathbb{Z}_2^t, \{\pm 1\})$ and $H = M_\psi$.

Of all the construction techniques described in Chapter 2, only the twin prime power difference set construction (2.11) gives orders of Hadamard matrices which may prove not to be cocyclic. The smallest example, the order 16 Hadamard matrix derived from the $(15, 7, 3)$ twin prime difference set in $\mathbb{Z}_{15}$, is certainly cocyclic (see Example 7.4.1).

Research Problem 39 *Are the Hadamard matrices of order $\geq$ 36, constructed from twin prime power difference sets (2.11), cocyclic?*

For the aficionado, well-versed in Hadamard matrix theory, there is at least one gaping hole in the list of construction techniques given in Chapter 2. No mention has yet been made of one of the most prolific methods for constructing Hadamard matrices: a plug-in technique due to Goethals and Seidel [124]. They showed (cf. [288, p. 450]) that

$$
\begin{bmatrix}
A & BR & CR & DR \\
BR & -A & RD & -RC \\
CR & -RD & -A & RB \\
DR & RC & -RB & -A
\end{bmatrix}
\tag{6.21}
$$

is Hadamard whenever the matrices A, B, C and D are circulants of odd order w satisfying (2.13), and R is again the $(0, 1)$ matrix of order w with all 1s on the back diagonal and 0s elsewhere. Such matrices can be constructed by the method of [288, Theorem 3.6], from 4 circulant $(0, \pm 1)$ *T-matrices* (named for Turyn — see [288, Definition 3.6] for the definition). For instance, this is the method used in [202] to construct a Hadamard matrix of order 428.

The matrix (6.21) is highly structured, and can be viewed as a variant of both the Williamson Hadamard construction H_1 of Section 6.4.3 and the Ito Hadamard construction H_3 of Section 6.4.4. However, it has not been possible to show it is always cocyclic, although by Example 7.4.1 we may assume $w \geq 7$.

Research Problem 40 *Let $w \geq 7$ be odd and let A, B, C and D be circulants of order w satisfying (2.13). Is the Goethals-Seidel Hadamard matrix (6.21) cocyclic?*

A construction by Kimura of Hadamard matrices of 'dihedral group type' has been successfully employed by Kimura and Niwasaki [203, Theorem 2.2] to produce Hadamard matrices of order $4(2k + 1)$ for all *odd* $3 \leq k \neq 15 \leq 29$. The construction uses the binary version $1 \mapsto 1, -1 \mapsto 0$ of a matrix of the form

$$
\begin{bmatrix}
M & K \\
K^\top & W
\end{bmatrix},
\tag{6.22}
$$

where M is the matrix of Example 6.3.2 with $\alpha = \beta = \kappa = 1, \gamma = -1$, the $4 \times 8k$ matrix K is obtained from M by replacing each entry by its length $2k$ repetition, then negating the 4th row, and W is a matrix of a form equivalent to the Williamson matrix (2.14). The $(0, 1)$ matrices A, B, C, D which comprise the binary version of W are images of elements in D_{2k} under the left regular representation of $\mathbb{Z}D_{2k}$, and must satisfy additional conditions much like (2.12) and (2.13).

Research Problem 41 *Is the Kimura construction (6.22) of Hadamard matrices cocyclic?*

A construction by Fletcher, Gysin and Seberry using 'two circulant cores' has been successfully employed (see [210]) to produce Hadamard matrices of order $4t = 2\ell + 2$, for all odd $3 \le \ell \le 75$, so the first unresolved value is $t = 39$, and there are 5 less than $t = 50$. The construction uses a matrix of the form

$$
\begin{bmatrix}
- & - & \mathbf{1} & \mathbf{1} \\
- & 1 & \mathbf{1} & -\mathbf{1} \\
\mathbf{1}^\top & \mathbf{1}^\top & A & B \\
\mathbf{1}^\top & -\mathbf{1}^\top & B^\top & -A^\top
\end{bmatrix},
\tag{6.23}
$$

where A and B are circulant matrices of order ℓ satisfying $AA^\top + BB^\top = (2\ell + 2)I_\ell - 2J_\ell$, and in [210, Conjecture 1] it is conjectured that Hadamard matrices of this form always exist.

Research Problem 42 *Is the 'two circulant cores' construction (6.23) of Hadamard matrices cocyclic?*

From Section 6.4.6 we know that some groups G (such as the dihedral groups) are prolific producers of G-cocyclic Hadamard matrices, while some (such as the cyclic and quaternion groups) appear quite barren. These differences are partly explained by results from two approaches, discussed in Chapter 7.4.2, to the study of cocyclic Hadamard matrices.

We know the problem of classifying Hadamard matrices into equivalence classes is at least as difficult as the Hadamard Conjecture. For a final degree of difficulty we might ask whether all inequivalent constructions of Hadamard matrices are cocyclic. The answer is yes for Hadamard matrices of orders 2, 4, 8 and 12, since there is only a single equivalence class in each case, and we may quote the cocyclic construction of $\mathcal{S}_1, \mathcal{S}_2, \mathcal{S}_3$ and $P_{11} \sim P_5'$, respectively.

In order 16, we know $\mathcal{S}_4$ is cocyclic; of the other 4 equivalence classes, two contain transposed matrices so by Example 6.2.16 all the Hadamard matrices in both of them are cocyclic or none are. The Hadamard matrices P_{19} and P_9' of order 20 are inequivalent by Lemma 2.16, though we are faced with the perhaps surprising fact that both are cocyclic (by Chapter 2.3.2 and Section 6.4.4), over the *same* group D_{20}. The third equivalence class of Hadamard matrices of order 20 contains a matrix (known as Hall's Type N) which is not known to belong to any general family.

In fact, de Launey has shown that *all* Hadamard matrices of orders 16 and 20 are cocyclic (see Example 7.4.1). Consequently there are cocyclic matrices in every equivalence class of Hadamard matrices of order ≤ 20. However, the sheer number of equivalence classes in higher orders (see Chapter 2.2) makes it unlikely that all will be cocyclic.

Research Problem 43 *For $t > 5$, are there cocyclic matrices in each equivalence class of Hadamard matrices of order $4t$? If not, what proportion of the equivalence classes are cocyclic?*

Since the Ito Hadamard matrices (cocyclic over D_{4t}) include the Williamson Hadamard matrices with circulant components (cocyclic over $\mathbb{Z}_2^2 \times \mathbb{Z}_t$), a Hadamard matrix may be cocyclic over nonisomorphic groups of the same order. To pursue the problem of cocyclic inequivalent Hadamard matrices, it is therefore not sufficient to demonstrate the existence of cocyclic Hadamard matrices over nonisomorphic groups of the same order. By the time we reach Chapter 8 (Section 8.3.2.1) we will have the tools to bring some order to this problem.

6.5.2 Status report — research problems in cocyclic Hadamard matrices

Two early papers of the author's [163, 152] contained problem lists for researchers in cocyclic Hadamard matrices. This Chapter closes with an update.

1. *Find a suitable generating set for $H_2(G)$ when G is nonabelian* [163, Problem 5.1]. This problem is solved (see Section 6.3.2), and the algorithm is distributed as a module in MAGMA [238].

2. *Exhibit the Menon-Hadamard difference set in G corresponding to a G-developed Hadamard matrix directly in terms of the coboundary* [152, Problem 5]. This problem is solved (Lemma 2.19).

3. *What is the relationship between the various difference set constructions for cocyclic Hadamard matrices?* [152, Problem 6]. This problem, which refers to the relative difference sets mentioned in Chapters 2.3.2 and 5.4, is solved [88]; see Chapter 7, Corollaries 7.32 and 7.33.

4. *Is there a $\mathbb{Z}_2^2 \times \mathbb{Z}_{35}$-cocyclic Hadamard matrix?* [152, Problem 1]. There can be none with symmetric circulant components [98]; the question refers to arbitrary circulant components. The problem is open, but a D_{140}-cocyclic (Ito) Hadamard matrix exists.

5. *How do different classes of groups compare as sources of cocyclic Hadamard matrices?* [163, Problem 5.2]. The dihedral groups D_{4t} appear the most uniform source of cocyclic Hadamard matrices (see Section 6.4.6 and Research Problem 6). For nonexistence results, see Chapter 7.4.1.

6. *What is the relationship between the Sylow p-subgroup structure of G and the existence of a G-cocyclic Hadamard matrix?* [152, Problem 4]. This problem is partially solved for those G with a cyclic Sylow 2-subgroup [112], and solved for those $G \cong E/\mathbb{Z}_2$ where E has all Sylow subgroups cyclic (see Chapter 7, Theorem 7.45).

7. (Research Problem 36) *If $H_2(G) = 0$ and M_ψ is a G-cocyclic Hadamard matrix, must ψ be a coboundary?* [152, Problem 3]. Except insofar as Problem 6 above applies trivially for G cyclic or dicyclic, this problem is open.

8. (Research Problem 37) *Prove that if M_ψ is a $\mathbb{Z}_2^2 \times \mathbb{Z}_t$-cocyclic Hadamard matrix and $t > 1$ is odd, the transgression component of ψ is nontrivial.* [18, Conjecture 4.2] and [152, Problem 2]. This problem is open.

9. (Research Problem 38) *Does a cocyclic Hadamard matrix of order v exist for every $v \equiv 0$* mod 4 *?* [90, Conjecture 3.6] and [163, Problem 5.4]. This problem is open (see Problem 5 above and this Section 6.5). A positive answer obviously confirms the Hadamard Conjecture (Research Problem 1).

10. *How do different equivalence relations on binary matrices interact with cocycle equivalence?* [163, Problem 5.3]. In view of the solution of Problem 3 above, this problem has bifurcated.

    (a) *What is the relationship between equivalence of $(4t, 2, 4t, 2t)$-RDS and cocycle equivalence?* This problem is solved (see Chapter 8).

    (b) (Research Problem 43) *Which equivalence classes of Hadamard matrices contain cocyclic Hadamard matrices?* This problem is open for $4t \geq 24$.

11. *What are the applications of n-dimensional cocyclic Hadamard matrices to digital communications?* [163, Problem 5.5]. This problem is open (see Chapter 5.3).

12. *What characteristics do cocyclic Hadamard matrices have for communications security, CDMA or data storage?* [152, Problem 7]. *What distance properties and profiles do codes constructed from cocyclic Hadamard matrices have?* [152, Problem 8]. Insofar as many of the Hadamard matrices applied in Chapter 3 are cocyclic, much is already known. More applications-based properties and more specific problems are the subject of Chapter 9.

The evidence that cocycles unlock a treasury of Hadamard matrices is overwhelming. It is time to apply the same key to generalised Hadamard matrices.

Chapter Seven

The Five-fold Constellation

The full theoretical framework rising in this Chapter is due to Galati [118, 120], building on work of the author, de Launey, Flannery, Perera and Hughes, and the treatment here follows his. However, the material in Section 7.1 is standard, and more details may be found in texts such as [3, 276]. In this Section, the class of cocycles is expanded to its limit, the class of *factor pairs*, within the theory of group extensions, and basic properties of factor pairs are noted. As usual, we are interested in equivalence classes of factor pairs, particularly equivalence classes of the *splitting* factor pairs, which generalise coboundaries.

In the second Section, the class of orthogonal cocycles is expanded to the class of orthogonal factor pairs. In Section 7.3, the dual of a factor pair is introduced. More importantly we describe the maximal class of generalised Hadamard matrices, the *coupled cocyclic* generalised Hadamard matrices, obtained by this unifying group extensions approach.

The significance of this class is demonstrated in Section 7.4 by locating four equivalent classes, known by other names, in different areas of mathematics and engineering. These form the *Five-fold Constellation*: coupled cocyclic generalised Hadamard matrices, orthogonal factor pairs, semiregular relative difference sets, semiregular class regular divisible designs with regular action, and a fifth class of well-correlated arrays, *base sequences*, presented here in full generality for the first time.

The Chapter closes with application of the Five-fold Constellation to derive nonexistence results for generalised Hadamard matrices, and a commentary on the wider implications of the cocyclic approach.

7.1 FACTOR PAIRS AND EXTENSIONS

Whilst cocycles are the principal subject of Chapter 6, this Chapter deals with the more general notion of a *factor pair*, which is now introduced. Factor pairs are the mechanism by which extensions (4.14)

$$1 \to N \xrightarrow{\imath} E \xrightarrow{\pi} G \to 1$$

may be constructed.

DEFINITION 7.1 *Let N and G be groups. For $a \in N$, let $\bar{a}$ denote the inner automorphism $\bar{a}(b) = aba^{-1}$ for all $b \in N$. A (normalised)* factor pair *of N by G is a pair (ψ, ε) of functions $\psi : G \times G \to N$ (the* factor set*) and $\varepsilon : G \to \mathrm{Aut}(N)^{\mathrm{op}}$*

(the coupling*) satisfying, for all* $x, y, z \in G$,

$$\varepsilon(x)\varepsilon(y) = \overline{\psi(x,y)}\varepsilon(xy), \tag{7.1}$$

$$\psi(x,y)\psi(xy,z) = \psi(y,z)^{\varepsilon(x)}\psi(x,yz), \tag{7.2}$$

$$\psi(x,1) = 1 = \psi(1,x). \tag{7.3}$$

The set of all factor pairs of N by G is denoted $F^2(G, N)$. *The* identity *factor pair* $(1, 1)$ *consists of the factor set* 1, *defined by* $1(x, y) = 1 \in N$ *and the coupling* 1, *defined by* $1(x) = \mathrm{id}_N$ *for all* $x, y \in G$.

If G is finite, define the decoupled matrix *for* (ψ, ε) *to be the (normalised) matrix*

$$M_\psi = [\psi(x,y)]_{x,y \in G}. \tag{7.4}$$

Suppose N is abelian. If (N, ε) is a G-module and $\psi \in Z_\varepsilon^2(G, N)$, $\overline{\psi(x,y)} = \mathrm{id}_N$ for all x, y in G, giving $\varepsilon(x)\varepsilon(y) = \varepsilon(xy) = \overline{\psi(x,y)}\varepsilon(xy)$, so $(\psi, \varepsilon) \in F^2(G, N)$. Conversely, for any factor pair (ψ, ε) of N by G, it follows from (7.1) that (N, ε) is a G-module (cf. (6.1)), and from (7.2) and (7.3) that $\psi \in Z_\varepsilon^2(G, N)$. Consequently, when N is abelian, the factor pairs of N by G are precisely the cocycles with coefficients in the various G-modules (N, ε). Then, when G is finite and $\varepsilon \equiv 1$, the decoupled matrix coincides with the G-cocyclic matrix M_ψ of (6.8).

From each factor pair in $F^2(G, N)$, a specific extension of N by G may be constructed.

LEMMA 7.2 *If* $(\psi, \varepsilon) \in F^2(G, N)$, *there is an extension (the* canonical extension*)*

$$N \overset{\iota}{\rightarrowtail} E_{(\psi,\varepsilon)} \overset{\kappa}{\twoheadrightarrow} G \tag{7.5}$$

of N by G. The extension group $E_{(\psi,\varepsilon)}$ *consists of the set* $N \times G$ *with multiplication*

$$(a,x)(b,y) = (ab^{\varepsilon(x)}\psi(x,y), xy), \tag{7.6}$$

for all $a, b \in N$ *and* $x, y \in G$. *The identity of* $E_{(\psi,\varepsilon)}$ *is* $(1, 1)$ *and the maps* ι, κ *are given by* $a \mapsto (a, 1)$ *and* $(a, x) \mapsto x$, *respectively. When* $\varepsilon \equiv 1$, $E_{(\psi,1)}$ *is denoted* E_ψ. *Set* $\overline{N} = \iota(N) = N \times \{1\}$, *and note*

$$(a,x)^{-1} = (\psi^{-1}(x^{-1},x)(a^{-1})^{\varepsilon(x^{-1})}, x^{-1}). \tag{7.7}$$

The following technical lemma collects some useful identities, which are easily verified from the definitions.

LEMMA 7.3 *Let* (ψ, ε) *be a factor pair of N by G. Then, for all* $x, y \in G$,

1. $\varepsilon(1) = \mathrm{id}_N$,

2. $\psi^{-1}(x^{-1}, y)^{\varepsilon(x)} = \psi(x, x^{-1}y)\,\psi^{-1}(x, x^{-1})$,

3. $\psi(x, x^{-1}) = \psi(x^{-1}, x)^{\varepsilon(x)}$,

4. $\varepsilon(x)^{-1} = \varepsilon(x^{-1})\,\overline{\psi^{-1}(x, x^{-1})} = \overline{\psi^{-1}(x^{-1}, x)}\varepsilon(x^{-1})$,

5. $(1, x)(1, y)^{-1} = (\psi^{-1}(xy^{-1}, y), xy^{-1})$ *in* $E_{(\psi,\varepsilon)}$,

6. $E_{(\psi,\varepsilon)}$ is abelian $\Leftrightarrow$ N and G are abelian, $\varepsilon \equiv 1$ and ψ is symmetric.

DEFINITION 7.4 *A factor pair (ψ_2, ε_2) of N by G is said to be* equivalent to *(ψ_1, ε_1) via ϕ, written $(\psi_2, \varepsilon_2) \sim_\phi (\psi_1, \varepsilon_1)$, if there exists a function $\phi : G \to N$ with $\phi(1) = 1$ such that, for all $x, y \in G$,*

$$\varepsilon_2(x) = \overline{\phi(x)}\varepsilon_1(x) \text{ and} \tag{7.8}$$

$$\psi_2(x, y) = \phi(x)\phi(y)^{\varepsilon_1(x)}\psi_1(x, y)\phi^{-1}(xy). \tag{7.9}$$

This equivalence relation partitions $F^2(G, N)$ into equivalence classes, and we denote the equivalence class containing (ψ, ε) by $[\psi, \varepsilon]$.

By slight abuse of Definition 6.1, we will extend its notation in the case $n = 1$ to nonabelian N, and denote by $C^1(G, N)$ the group of normalised functions

$$C^1(G, N) = \{\phi : G \to N, \phi(1) = 1\} \tag{7.10}$$

under pointwise multiplication. Thus $\phi^{-1}(x) = \phi(x)^{-1}$ for $x \in G$.

Equivalence of factor pairs may equally be interpreted as a right action by the group $C^1(G, N)$ on the set $F^2(G, N)$, if the action of ϕ on (ψ, ε) is defined by

$$(\psi, \varepsilon) \cdot \phi = (\psi', \varepsilon') \Leftrightarrow (\psi', \varepsilon') \sim_{\phi^{-1}} (\psi, \varepsilon). \tag{7.11}$$

Then the orbits of the action are precisely the equivalence classes of factor pairs.

Equivalence classes of the form $[1, \varrho]$ are especially interesting because ϱ must be a group homomorphism, by (7.1). Each element in $[1, \varrho]$ is defined by a function $\phi \in C^1(G, N)$, so in a real sense these equivalence classes represent all functions defined on groups. The importance of this relationship will become apparent in Section 7.4, where it is used to established the fifth equivalence of the Five-fold Constellation.

Factor pairs in $[1, \varrho]$ will be termed *splitting*. Splitting factor pairs generalise coboundaries, so we adopt the notational convention of (6.5) for this general case:

$$\partial\phi(x, y) = \phi^{-1}(x)(\phi^{-1}(y))^{\varrho(x)}\phi(xy), \quad x, y \in G. \tag{7.12}$$

The splitting case of (7.11) becomes

$$(\partial(\phi^{-1}), \overline{\phi}\varrho) \cdot \phi = (1, \varrho) \quad \left(\text{or, } (1, \varrho) \cdot \phi = (\partial\phi, \overline{\phi^{-1}}\varrho)\right). \tag{7.13}$$

In order to avoid confusing $\partial(\phi^{-1})$ with the inverse $(\partial\phi)^{-1}$ of $\partial\phi$, and to distinguish the mapping $\phi \mapsto \partial(\phi^{-1})$ from the mapping $\phi \mapsto \partial\phi$, *from now on, we use the notation*

$$\partial^{-1}\phi \equiv \partial(\phi^{-1}). \tag{7.14}$$

DEFINITION 7.5 *Factor pair $(\psi, \varepsilon) \in F^2(G, N)$ is a splitting factor pair if there exist $\phi \in C^1(G, N)$ and a homomorphism $\varrho : G \to \mathrm{Aut}(N)^{\mathrm{op}}$ such that $(\psi, \varepsilon) \sim_\phi$ $(1, \varrho)$. It has the form $(\psi, \varepsilon) = (\partial^{-1}\phi, \overline{\phi}\varrho)$, where*

$$(\overline{\phi}\varrho)(x) = \overline{\phi(x)}\varrho(x), \tag{7.15}$$

$$\partial^{-1}\phi(x, y) = \phi(x)\phi(y)^{\varrho(x)}\phi(xy)^{-1}, \quad x, y \in G. \tag{7.16}$$

In general, $\partial^{-1}\phi \neq (\partial\phi)^{-1}$, though if N is abelian, then $\varepsilon = \varrho$ by (7.15) and $\partial^{-1}\phi = (\partial\phi)^{-1} \in B_\varepsilon^2(G, N)$ by (7.16).

The canonical extensions (7.5) corresponding to splitting factor pairs are split extensions, and their extension groups (7.6) are semidirect products.

COROLLARY 7.6 Let $(\psi, \varepsilon) \in F^2(G, N)$. Then $E_{(\psi,\varepsilon)} \cong E_{(1,\varrho)} = N \rtimes_\varrho G$ if and only if $(\psi, \varepsilon) \sim_\phi (1, \varrho) \in F^2(G, N)$. In particular $E_{(\psi,\varepsilon)} \cong N \times G$ if and only if $(\psi, \varepsilon) \sim_\phi (1, 1)$.

We will further study splitting factor pairs in Chapters 8.2 and 9.2, where they are used to develop a new theory of equivalence and nonlinearity of functions.

The next Lemma establishes a mapping from extensions of N by G to equivalence classes of factor pairs in $F^2(G, N)$.

LEMMA 7.7 Suppose that $e : N \overset{\imath}{\rightarrowtail} E \overset{\pi}{\twoheadrightarrow} G$ is an extension of N by G and let $T = \{t_x : x \in G, \pi(t_x) = x\}$ be a transversal of $\imath(N)$ in E. Then (ψ_T, ε_T) defined by

$$\varepsilon_T(x) = \imath^{-1} \circ \overline{t_x} \circ \imath, \tag{7.17}$$

$$\psi_T(x, y) = \imath^{-1}(t_x t_y t_{xy}^{-1}), \tag{7.18}$$

for all $x, y \in G$, is a factor pair of N by G. If $T^* = \{t_x^* : x \in G, \pi(t_x^*) = x\}$ is any other transversal of $\imath(N)$ in E, then $(\psi_{T^*}, \varepsilon_{T^*}) \sim_\phi (\psi_T, \varepsilon_T)$, where $\phi(x) = \imath^{-1}(t_x^* t_x^{-1})$. Further, every factor pair in $[\psi_T, \varepsilon_T]$ derives from such a transversal.

In fact, the mapping of Lemma 7.7 is surjective.

LEMMA 7.8 Let $(\psi, \varepsilon) \in F^2(G, N)$. Then $T = \{(1, x) : x \in G\} \subseteq E_{(\psi,\varepsilon)}$ is a transversal of $\overline{N}$ in $E_{(\psi,\varepsilon)}$ with $(\psi_T, \varepsilon_T) = (\psi, \varepsilon)$.

The set of all extensions of N by G also partitions according to a basic equivalence relation.

DEFINITION 7.9 Two extensions $N \overset{\imath_1}{\rightarrowtail} E_1 \overset{\pi_1}{\twoheadrightarrow} G$ and $N \overset{\imath_2}{\rightarrowtail} E_2 \overset{\pi_2}{\twoheadrightarrow} G$ of N by G are equivalent if there exists a group homomorphism $\Phi : E_1 \rightarrow E_2$ (necessarily an isomorphism) such that the following diagram commutes:

$$
\begin{array}{ccccccccc}
1 & \longrightarrow & N & \overset{\imath_1}{\longrightarrow} & E_1 & \overset{\pi_1}{\longrightarrow} & G & & \\
 & & \mathrm{id}_N \downarrow & & \Phi \downarrow & & \mathrm{id}_G \downarrow & & . \\
 & & N & \overset{\imath_2}{\longrightarrow} & E_2 & \overset{\pi_2}{\longrightarrow} & G & \longrightarrow & 1
\end{array}
\tag{7.19}
$$

Every extension of N by G is equivalent to some canonical extension (in fact, many, in general).

COROLLARY 7.10 Let $e : N \overset{\imath}{\rightarrowtail} E \overset{\pi}{\twoheadrightarrow} G$ be a group extension and let $T = \{t_x \in E : \pi(t_x) = x\}$ be a transversal of $\imath(N)$ in E. Then e is equivalent to $N \overset{\iota}{\rightarrowtail} E_{(\psi_T,\varepsilon_T)} \overset{\kappa}{\twoheadrightarrow} G$ via the isomorphism

$$\imath(a)t_x \mapsto (a, x), \tag{7.20}$$

for all $a \in N$ and $x \in G$. Similarly, if $(\psi, \varepsilon) \sim_\phi (\psi_T, \varepsilon_T)$, then $E \cong E_{(\psi, \varepsilon)}$ via

$$\iota(a)t_x \mapsto (a\phi^{-1}(x), x), \tag{7.21}$$

with $E_{(\psi, \varepsilon)} \cong E$ via $(a, x) \mapsto \iota(a\phi(x))t_x$.

The mapping of Lemma 7.7 induces a well-defined mapping from the set of equivalence classes of extensions of N by G to the set of equivalence classes of factor pairs of N by G. Surjectivity of the induced mapping follows from Lemma 7.8 and injectivity from the following lemma.

LEMMA 7.11 *Let* $e_1 : N \overset{\iota_1}{\rightarrowtail} E_1 \overset{\pi_1}{\twoheadrightarrow} G$ *and* $e_2 : N \overset{\iota_2}{\rightarrowtail} E_2 \overset{\pi_2}{\twoheadrightarrow} G$ *be extensions of* N *by* G, *and let* $T = \{t_x : x \in G, \pi_1(t_x) = x\}$ *and* $S = \{s_x : x \in G, \pi_2(s_x) = x\}$ *be transversals of* $\iota_1(N)$ *in* E_1 *and* $\iota_2(N)$ *in* E_2, *respectively, with* $(\psi_S, \varepsilon_S) \sim_\phi$ (ψ_T, ε_T). *Then* e_1 *is equivalent to* e_2 *via the isomorphism*

$$\iota_1(a)t_x \mapsto \iota_2(a\phi^{-1}(x))s_x \tag{7.22}$$

for all $a \in N$ *and* $x \in G$.

The fact that the induced mapping on equivalence classes is bijective is a central result in the theory of group extensions [3, pp. 85–86] (a detailed proof appears in [118]).

THEOREM 7.12 *There is a bijection between the set of equivalence classes in* $F^2(G, N)$ *and the set of equivalence classes of extensions of* N *by* G, *under which* $[\psi, \varepsilon]$ *corresponds to the equivalence class containing the canonical extension* $N \rightarrowtail E_{(\psi, \varepsilon)} \twoheadrightarrow G$.

7.2 ORTHOGONALITY FOR FACTOR PAIRS

In this Section we describe Galati's characterisation of those factor pairs $(\psi, \varepsilon) \in F^2(G, N)$ for which there is an (v, w, k, λ)-RDS in the corresponding extension group $E_{(\psi, \varepsilon)}$. Assume throughout that G is finite of order v and N is finite of order w. Recall that a subset $X \subseteq G$ of a group G is identified with its sum $X = \sum_{x \in X} x$ in the group ring $\mathbb{Z}G$.

DEFINITION 7.13 [120, Definition 4.1] *Let* $(\psi, \varepsilon) \in F^2(G, N)$ *and let* D *be a* k-*subset of* G.

1. (ψ, ε) *is* (v, w, k, λ)-*orthogonal with respect to* D *if for each* $x \in G \backslash \{1\}$, *the sequence* $\{ \psi(x, y) \}_{y \in D \cap x^{-1}D}$ *lists each element of* N *exactly* λ *times, or equivalently, in the group ring* $\mathbb{Z}N$

$$\sum_{y \in D \cap x^{-1}D} \psi(x, y) = \lambda N. \tag{7.23}$$

2. *When* N *is abelian and* $\varepsilon \equiv 1$, ψ *is termed* (v, w, k, λ)-*orthogonal with respect to* D.

3. When $k = v$, so $D = G$ and $\lambda = v/w$, (ψ, ε) is termed orthogonal, and (7.23) becomes

$$\sum_{y \in G} \psi(x, y) = (v/w)N, \ \forall\, x \neq 1 \in G, \qquad (7.24)$$

so (ψ, ε) is orthogonal if and only if M_ψ is row balanced (Definition 4.6).

When N is abelian, $\varepsilon \equiv 1$ and $k = v$, Definition 7.13 specialises to the original definition of an orthogonal cocycle (Definition 6.7) due to the author and Perera [260].

Orthogonality is the property of factor pairs which characterises existence of relative difference sets. Even more useful is the fact that the characterisation is constructive, and we can identify one such relative difference set in a canonical form. Before we embark on the proof, note that, if a factor pair (ψ, ε) of N by G is (v, w, k, λ)-orthogonal with respect to $D \subseteq G$, then D is an ordinary $(v, k, w\lambda)$-difference set in G.

THEOREM 7.14 [120, Theorem 5.1] *Suppose* $e : N \overset{\imath}{\rightarrowtail} E \overset{\pi}{\twoheadrightarrow} G$ *is an extension of* N *by* G *and* $[\varphi, \tau]$ *is its associated equivalence class of factor pairs. Let* D *be a* k-subset of G. Then the following statements are equivalent:

1. there is a (v, w, k, λ)-RDS R in E relative to $\imath(N)$ lifting D;

2. there is $(\psi, \varepsilon) \in [\varphi, \tau]$ which is (v, w, k, λ)-orthogonal with respect to D;

3. there is $(\psi, \varepsilon) \in [\varphi, \tau]$ such that $R_{(\psi, \varepsilon)} = \{(1, x) : x \in D\}$ is a (v, w, k, λ) -RDS in $E_{(\psi, \varepsilon)}$ relative to $\overline{N}$ lifting D.

In this case we may take $(\psi, \varepsilon) = (\psi_T, \varepsilon_T)$, *where* T *is any transversal of* $\imath(N)$ *in* E *containing* R', *and* $R' = Rb$ *is any translate of* R *satisfying* $R' \cap \imath(N) \in \{\emptyset, \{1\}\}$ *and* $\pi(R') = D$. *Then* R *and* $R_{(\psi, \varepsilon)}$ *are equivalent (see (4.15)), with* $\beta(Rb) = R_{(\psi, \varepsilon)}$, *where* $\beta : E \to E_{(\psi, \varepsilon)}$ *is given by* $\imath(a)t_x \mapsto (a, x)$ *and* $t : G \to E$ *is the section of* π *corresponding to* T.

Proof. $1 \Leftrightarrow 3$. Let R be an (v, w, k, λ)-RDS in E relative to $\imath(N)$ lifting D. Since $R \cap \imath(N) \in \{\emptyset, \{b^{-1}\}\}$ for some $b \in \imath(N)$, one of R and Rb is a translate R' of R satisfying $R' \cap \imath(N) \in \{\emptyset, \{1\}\}$ and $\pi(R') = D$. Let $T = \{t_x : x \in G, \pi(t_x) = x\}$ be a normalised transversal of $\imath(N)$ in E containing R', so that $R' = \{t_x : x \in D\}$. Set $(\psi, \varepsilon) = (\psi_T, \varepsilon_T) \in [\varphi, \tau]$. The isomorphism $E \cong E_{(\psi, \varepsilon)}$ given by $\imath(a)t_x \mapsto (a, x)$ (see Corollary 7.10) maps R' onto $R_{(\psi, \varepsilon)}$ and $\imath(N)$ onto $\overline{N}$, and the implication follows. Conversely, by the final statement of Lemma 7.7, there is a transversal T of $\imath(N)$ in E with $(\psi_T, \varepsilon_T) = (\psi, \varepsilon)$. Take R to be the image of $R_{(\psi, \varepsilon)}$ under the inverse of the isomorphism given by (7.20).

$2 \Leftrightarrow 3$. Assume that (ψ, ε) is (v, w, k, λ)-orthogonal with respect to D. First, observe that $\{\Delta^{-1}(g) : 1 \neq g \in G\}$ gives a partition of $\{(x, y) \in D \times D : x \neq y\}$, where $\Delta : D \times D \to G$ is defined by $\Delta(x, y) = xy^{-1}$, and $\Delta^{-1}(g)$ is the preimage of $\{g\}$. Second, by Lemma 7.3.5, $(1, x)(1, y)^{-1} = (\psi^{-1}(xy^{-1}, y), xy^{-1})$ for all $x, y \in G$. Third, note that ψ^{-1} satisfies (7.23) iff ψ does, since $a \mapsto a^{-1}$ gives a permutation of N, and that $(a, x) = (a, 1)(1, x)$ in $E_{(\psi,\varepsilon)}$ for all $a \in N$ and $x \in G$. Therefore, in $\mathbb{Z}E_{(\psi,\varepsilon)}$,

$$
\begin{aligned}
R_{(\psi,\varepsilon)}R_{(\psi,\varepsilon)}^{(-1)} &= k + \sum_{x,y \in D, x \neq y}(1,x)(1,y)^{-1} \\
&= k + \sum_{1 \neq g \in G} \sum_{(x,y) \in \Delta^{-1}(g)} (\psi^{-1}(xy^{-1}, y), xy^{-1}) \\
&= k + \sum_{1 \neq g \in G} \left(\sum_{y \in D \cap g^{-1}D} (\psi^{-1}(g,y), 1) \right)(1,g) \\
&= k + \sum_{1 \neq g \in G} \lambda \overline{N}(1,g) \\
&= k + \lambda E_{(\psi,\varepsilon)} - \lambda \overline{N},
\end{aligned}
$$

which, by (4.12), gives 3. Conversely, if $R_{(\psi,\varepsilon)}R_{(\psi,\varepsilon)}^{(-1)} = k + \lambda E_{(\psi,\varepsilon)} - \lambda \overline{N}$, then in particular,

$$
\sum_{1 \neq g \in G} \sum_{y \in D \cap g^{-1}D} (\psi^{-1}(g,y), g) = \sum_{1 \neq g \in G} \sum_{a \in N} \lambda(a, g) \qquad (7.25)
$$

in $\mathbb{Z}E_{(\psi,\varepsilon)}$, or equivalently, in $(\mathbb{Z}N)E_{(\psi,\varepsilon)}$. Since $(\mathbb{Z}N)E_{(\psi,\varepsilon)}$ is a free $\mathbb{Z}N$-module, the map $E_{(\psi,\varepsilon)} \rightarrow (\mathbb{Z}N)G$ with $(a,g) \mapsto ag$ extends to a $\mathbb{Z}N$-linear map which, when applied to (7.25), yields

$$
\sum_{1 \neq g \in G} \sum_{y \in D \cap g^{-1}D} \psi^{-1}(g,y)g = \sum_{1 \neq g \in G} \sum_{a \in N} \lambda ag. \qquad (7.26)
$$

Equating coefficients in (7.26) gives $\sum_{y \in D \cap g^{-1}D} \psi^{-1}(g,y) = \lambda N$ for each $g \in G \setminus \{1\}$, so (7.23) holds as required. $\qquad \square$

The equivalence of RDSs and orthogonal factor pairs, given by Theorem 7.14, lays bare the following natural hierarchy for the class of (normal) RDSs

$$
\text{abelian} \subsetneq \text{central} \subsetneq \text{abelian kernel} \subsetneq \text{normal}, \qquad (7.27)
$$

depending on whether E is abelian, N is central in E, N is abelian and N is normal in E. Examples illustrating the proper containment of each class are given in [120]. The central and abelian kernel cases coincide if and only if $\mathrm{Aut}(N) = \{1\}$, that is, if and only if $N \cong \mathbb{Z}_2$.

When the kernel N is abelian, the coupling $\varepsilon : G \rightarrow \mathrm{Aut}(N)^{\mathrm{op}}$ of a factor pair (ψ, ε) is necessarily a homomorphism, in which case (N, ε) is a G-module. We work in the group $Z_\varepsilon^2(G, N)$ of cocycles with action ε, and are free to use any results from the cohomology theory of finite groups.

When $N(= C)$ is central in E, we are in the simpler setting of untwisted coefficients given by $\varepsilon \equiv 1$. Therefore we deal with the cocycles $Z^2(G, C)$ of Chapter 6, and we write E_ψ for $E_{(\psi,\varepsilon)}$.

Finally, E is abelian if and only if (Lemma 7.3.6) G is abelian, $N(= C)$ is central and any cocycle ψ with $E \cong E_\psi$ is symmetric, so we deal with the subgroup $S_+^2(G, C)$ of $Z^2(G, C)$ (Definition 6.5.1).

This hierarchy determines four 'orders of magnitude' of the RDS 'stars' in our Five-fold Constellation of equivalent objects, to be described in Section 7.4.

By Theorem 7.14, any (v, w, k, λ)-RDS R is equivalent to one in the canonical form $R_{(\psi,\varepsilon)} = \{(1, x) : x \in D\}$ in the group $E_{(\psi,\varepsilon)}$, where the factor pair (ψ, ε) is (v, w, k, λ)-orthogonal with respect to the ordinary $(v, k, w\lambda)$-difference set D. The next corollary shows that R is also equivalent to an RDS in $E_{(\psi',\varepsilon')}$ for any choice of $(\psi', \varepsilon') \in [\psi, \varepsilon]$.

COROLLARY 7.15 [120, Corollary 5.1] *Let* $(\varphi, \tau), (\varphi', \tau') \in F^2(G, N)$ *with* $(\varphi', \tau') \sim_\phi (\varphi, \tau)$. *Then*

1. $R_{(\varphi, \tau)} = \{(1, x) : x \in D\} \subseteq E_{(\varphi, \tau)}$ *is a* (v, w, k, λ)-*RDS in* $E_{(\varphi, \tau)}$ *relative to* $\overline{N}$ *lifting the* $(v, k, w\lambda)$-*difference set* $D \subseteq G$ *if and only if*

2. $R_{\phi^{-1}} = \{(\phi^{-1}(x), x) : x \in D\} \subseteq E_{(\varphi', \tau')}$ *is a* (v, w, k, λ)-*RDS in* $E_{(\varphi', \tau')}$ *relative to* $\overline{N}$ *lifting* D.

When this occurs, $R_{(\varphi, \tau)}$ *and* $R_{\phi^{-1}}$ *are isomorphic RDSs.*

Proof. The existence of isomorphism $\alpha : E_{(\varphi, \tau)} \to E_{(\varphi', \tau')}$ given by $\alpha(a, x) = (a\phi^{-1}(x), x)$ follows from Corollary 7.10, with (φ', τ') for (ψ, ε), $E_{(\varphi, \tau)}$ for E and $T = \{(1, x) : x \in G\} \subseteq E_{(\varphi, \tau)}$. In this case $(\psi_T, \varepsilon_T) = (\psi, \varepsilon)$ by Lemma 7.8. □

A straightforward corollary of Theorem 7.14 is its specialisation to the case of splitting RDSs (Definition 4.22).

COROLLARY 7.16 [120, p. 287] *A RDS* R *in* E *relative to* N *is a splitting RDS if and only if any corresponding* (v, w, k, λ)-*orthogonal factor pair* (ψ, ε) *is a splitting factor pair* $(\partial^{-1}\phi, \overline{\phi}\varrho)$ *for some map* $\phi \in C^1(G, N)$ *and some homomorphism* $\varrho : G \to \mathrm{Aut}(N)^{\mathrm{op}}$. *If* N *is abelian, then* $\varepsilon = \varrho$ *and* $\psi = \partial(\phi^{-1}) \in B^2_\varepsilon(G, N)$ *is a coboundary. If* N *is central in* E *then* $\varepsilon = \varrho \equiv 1$.

Combined with Theorem 4.20, Theorem 7.14 also gives us an explicit description, in terms of an (v, w, k, λ)-orthogonal factor pair (ψ, ε), for an (v, w, k, λ)-divisible design $\mathcal{D}$ with regular group E, class regular with respect to a normal subgroup N.

THEOREM 7.17 [118, Theorem 4.10] *Let* (ψ, ε) *be a factor pair of* N *by* G, *let* $D \subseteq G$ *and let* $R_{(\psi, \varepsilon)} = \{(1, d) : d \in D\} \subseteq E_{(\psi, \varepsilon)}$. *Then the following statements are equivalent:*

1. (ψ, ε) *is* (v, w, k, λ)-*orthogonal with respect to* D;

2. $\mathcal{D}_{(\psi, \varepsilon)} = \left(E_{(\psi, \varepsilon)}, \{B_{(b,y)} : b \in N, y \in G\}\right)$ *is a* (v, w, k, λ)-*divisible design with class partition* $\{\overline{N}e : e \in E_{(\psi, \varepsilon)}\}$, *where for all* $b \in N$ *and* $y \in G$, $$B_{(b,y)} = R_{(\psi, \varepsilon)}(b, y) = \{(b^{\varepsilon(d)}\psi(d, y), dy) : d \in D\}.$$

When this occurs, $E_{(\psi, \varepsilon)}$ *is a regular group for* $\mathcal{D}_{(\psi, \varepsilon)}$, *acting via right translation, and class regular with respect to* $\overline{N}$. *If* $\mathcal{D}$ *is any* (v, w, k, λ)-*divisible design with regular group* E, *class regular with respect to* N, *then* $\mathcal{D} \cong \mathcal{D}_{(\psi, \varepsilon)}$ *for a suitable* (v, w, k, λ)-*orthogonal factor pair* (ψ, ε) *of* N *by* $G = E/N$.

7.3 ALL THE COCYCLIC GENERALISED HADAMARD MATRICES

We are now in a position to describe the overarching construction of generalised Hadamard matrices available to us by developing a matrix from a factor pair. By

Theorem 4.16 this construction cannot be enlarged upon, because the $GH(w, v/w)$ over N which result are exactly those corresponding to the divisible designs, class regular with respect to N, for which N is normal in a regular group of the design. We will see in Example 7.4.2, however, that not every generalised Hadamard matrix arises this way: that is, there are divisible designs, class regular with respect to N, for which N is not normal in any regular group of the design.

DEFINITION 7.18 *Let G be a finite group of order v, let N be a group and let M be a $v \times v$ matrix with entries in N. Then M is a* coupled G-cocyclic ma-*trix over N if there exist a factor pair $(\psi, \varepsilon) \in F^2(G, N)$ and an ordering $G = \{1 = x_1, x_2, \ldots, x_v\}$ such that M is Hadamard equivalent (Definition 4.12) to the (normalised) matrix*

$$M_{(\psi,\varepsilon)} = [\psi^{-1}(x_i, x_j)^{\varepsilon(x_i)^{-1}}]_{1 \leq i,j \leq v}. \tag{7.28}$$

For brevity, or if N is abelian, we say M is G-cocyclic over N.

COROLLARY 7.19 *For $M_{(\psi,\varepsilon)}$ as in (7.28),*

$$M_{(\psi,\varepsilon)} \sim \overline{M}_{(\psi,\varepsilon)} = [\psi(x_i, x_i^{-1}x_j)]_{1 \leq i,j \leq v}. \tag{7.29}$$

Proof. [120, Theorem 10.1 proof] Lemma 7.3.4, Lemma 7.3.3 with x_i^{-1} for x and Lemma 7.3.2 with x_i^{-1}, x_j for x, y give

$$\begin{aligned}
\psi^{-1}(x_i, x_j)^{\varepsilon(x_i)^{-1}} &= \psi^{-1}(x_i, x_i^{-1})^{\varepsilon(x_i^{-1})} \, \psi^{-1}(x_i, x_j)^{\varepsilon(x_i^{-1})} \, \psi(x_i, x_i^{-1})^{\varepsilon(x_i^{-1})} \\
&= \psi^{-1}(x_i^{-1}, x_i) \, \psi^{-1}(x_i, x_j)^{\varepsilon(x_i^{-1})} \, \psi(x_i^{-1}, x_i) \\
&= \psi^{-1}(x_i^{-1}, x_i) \, \psi(x_i^{-1}, x_i x_j). \tag{7.30}
\end{aligned}$$

So $\overline{M}_{(\psi,\varepsilon)}$ may be obtained from $M_{(\psi,\varepsilon)}$ by multiplying the ith row on the left by $\psi(x_i^{-1}, x_i)$ for $i = 2, \ldots, v$, then permuting the rows according to $x_i \mapsto x_i^{-1}$. $\quad\Box$

When N is abelian and $\varepsilon \equiv 1$, the matrix $M_{(\psi,1)}$ is indeed G-cocyclic according to Definition 6.3, although at first glance this may not be apparent. For $\varepsilon(x_i)^{-1} = \mathrm{id}_N$, so $M_{(\psi,1)} = [\psi^{-1}(x_i, x_j)]$, and since inversion is an automorphism of N, by Definition 4.12 $M_{(\psi,1)}$ and the G-cocyclic matrix $M_\psi = [\psi(x_i, x_j)]$ are Hadamard equivalent.

The splitting case $(\psi, \varepsilon) \sim_\phi (1, \varrho)$ of Definition 7.18 leads to the optimal gen-eralisation of a group developed (or, after reindexing columns labelled g by g^{-1}, group-invariant) matrix having entries in a group.

DEFINITION 7.20 *Let G be a finite group of order v, let N be a group and let M be a $v \times v$ matrix with entries in N. Then M is a* coupled G-developed matrix over *N if there are an ordering $G = \{x_1, \ldots, x_v\}$, a mapping $\phi \in C^1(G, N)$ and a homomorphism $\varrho : G \to \mathrm{Aut}(N)^{\mathrm{op}}$ such that*

$$M = [\phi(x_i x_j)^{\varrho(x_i^{-1})}]_{1 \leq i,j \leq v}.$$

For brevity, or if the coupling $\varrho \equiv 1$, we say M is G-developed, as in Definition 2.17. The row of M indexed by 1 (assumed, without loss of generality, to be the top row) is always $(\phi(x_1), \phi(x_2), \ldots, \phi(x_v))$.

COROLLARY 7.21 *Let M be coupled G-developed as in Definition 7.20. Then $M_{(\partial^{-1}\phi,\overline{\phi}\varrho)} \sim M$.*

Proof. By Definition 7.5, $(\partial^{-1}\phi, \overline{\phi}\varrho) \sim_\phi (1, \varrho)$. Since $\partial^{-1}\phi(x_i^{-1}, x_i x_j) = \phi(x_i^{-1}) \phi(x_i x_j)^{\varrho(x_i^{-1})} \phi(x_j)^{-1}$, permuting the rows of $\overline{M}_{(\partial^{-1}\phi,\overline{\phi}\varrho)}$ according to $x_i \mapsto x_i^{-1}$, and multiplying the ith row of the resulting matrix on the left by $\phi(x_i^{-1})^{-1}$ and the jth column on the right by $\phi(x_j)$ produces the Hadamard equivalent matrix M, and the result follows from Corollary 7.19. $\square$

A simpler, direct proof of Galati's optimal generalisation of the cocyclic construction of generalised Hadamard matrices is presented next. This also proves it is both necessary and sufficient that the decoupled matrix M_ψ be row balanced for the coupled G-cocyclic matrix $M_{(\psi,\varepsilon)}$ to be a generalised Hadamard matrix. For abelian N and trivial action $\varepsilon \equiv 1$, the result is Lemma 6.6: the decoupled matrix is identically the G-cocyclic M_ψ of (6.8) and $M_\psi \sim M_{(\psi,1)}$.

THEOREM 7.22 [120, Theorem 10.1] *Let G be a finite group of order v, let N be a finite group of order w dividing v and let $(\psi, \varepsilon) \in F^2(G, N)$. Then the following statements are equivalent:*

1. *(ψ, ε) is orthogonal, that is, the decoupled matrix $M_\psi = [\psi(x, y)]_{x,y\in G}$ is row balanced;*

2. *the coupled G-cocyclic matrix $M_{(\psi,\varepsilon)} = [\psi^{-1}(x,y)^{\varepsilon(x)^{-1}}]_{x,y\in G}$ is a $GH(w, v/w)$ over N.*

Proof. (Compare with Lemma 6.6.) In $\mathbb{Z}N$, for each pair of elements $y, z \in G$, using (7.30) and (7.2),

$$\sum_{x\in G} \psi^{-1}(y,x)^{\varepsilon(y)^{-1}} \psi(z,x)^{\varepsilon(z)^{-1}}$$
$$= \psi^{-1}(y^{-1},y)\psi^{-1}(z^{-1}, zy^{-1})\Big(\sum_{x\in G} \psi(zy^{-1}, yx)^{\varepsilon(z^{-1})}\Big)\psi(z^{-1}, z).$$

Hence, if $d = zy^{-1} \neq 1$, then
$$\sum_{x\in G} \psi^{-1}(y,x)^{\varepsilon(y)^{-1}} \psi(z,x)^{\varepsilon(z)^{-1}} = v/w \sum_{u\in N} u \text{ if and only if}$$
$$\sum_{x\in G} \psi(zy^{-1}, yx)^{\varepsilon(z^{-1})} = v/w \sum_{u\in N} u, \text{ if and only if}$$
$$\sum_{x\in G} \psi(d,x)^{\varepsilon(z^{-1})} = v/w \sum_{u\in N} u, \text{ if and only if}$$
$$\sum_{x\in G} \psi(d,x) = v/w \sum_{u\in N} u, \text{ since } \varepsilon(z^{-1}) \text{ is an automorphism of } N. \quad \square$$

It is important to recognise that, unless both N is abelian and $\varepsilon \equiv 1$, the decoupled matrix M_ψ and the coupled cocyclic matrix $M_{(\psi,\varepsilon)}$ need not be equivalent, so that if $M_{(\psi,\varepsilon)}$ is a generalised Hadamard matrix it does not follow that M_ψ is one, and vice versa.

The splitting case of Theorem 7.22 is worth recording separately. For M as in Corollary 7.21, write the row balance condition (Theorem 7.22.1) for the corresponding decoupled matrix $M_{\partial^{-1}\phi}$ as an equation in $\mathbb{Z}N$, using (7.16). Left-multiply that equation by $\phi(x)^{-1}$.

COROLLARY 7.23 *Let G be a finite group of order v and N be a group of order w dividing v. For a mapping $\phi \in C^1(G, N)$ and a homomorphism ϱ :*

$G \to \mathrm{Aut}(N)^{\mathrm{op}}$, *the coupled G-developed matrix* $M = [\phi(xy)^{\varrho(x^{-1})}]_{x,y \in G}$ *is a $GH(w, v/w)$ over N if and only if*

$$\forall\, x \neq 1 \in G, \quad \sum_{y \in G} \phi(y)^{\varrho(x)} \phi(xy)^{-1} = (v/w)\, N. \qquad (7.31)$$

7.3.1 Cocyclic generalised Hadamard matrix constructions

By imitating construction techniques for generalised Hadamard matrices, new coupled cocyclic $GH(w, v/w)$ may be derived from known ones. Unsurprisingly, sometimes a factor pair construction may exist only under additional constraints.

First, to state the obvious, any coupled G-developed $GH(w, v/w)$ over N is coupled cocyclic.

LEMMA 7.24 *A coupled G-developed $GH(w, v/w)$ over N is coupled G-cocyclic.*

Proof. Let M be as in Definition 7.20 with $(\partial^{-1}\phi, \overline{\phi}\varrho)$ the resulting splitting factor pair of Corollary 7.21. Then M is a $GH(w, v/w)$ if and only if $M_{(\partial^{-1}\phi, \overline{\phi}\varrho)}$ is a $GH(w, v/w)$. □

Second, every factor pair has a dual (cf. Example 6.2.16); the corresponding coupled cocyclic matrices are both $GH(w, v/w)$ over N or neither is.

THEOREM 7.25 (Galati [118, Theorem 6.6]) *The dual $(\psi^*, \varepsilon^*) \in F^2(G, N)$ of a factor pair $(\psi, \varepsilon) \in F^2(G, N)$ is defined to be*

$$\varepsilon^*(x) = \varepsilon(x^{-1})^{-1}, \quad \psi^*(x, y) = \psi^{-1}(y^{-1}, x^{-1})^{\varepsilon^*(xy)}. \qquad (7.32)$$

Note $(\psi^{**}, \varepsilon^{**}) = (\psi, \varepsilon)$. Then $M_{(\psi, \varepsilon)}$ is a $GH(w, v/w)$ over N if and only if $M_{(\psi^*, \varepsilon^*)}$ is a $GH(w, v/w)$ over N.

Proof. Verification that $(\psi^*, \varepsilon^*) \in F^2(G, N)$ is by direct checking that the pair corresponds to the transversal $T = \{t_x = (1, x^{-1})^{-1} : x \in G\}$ in $E_{(\psi, \varepsilon)}$. Then $(\psi^*, \varepsilon^*) \sim_\phi (\psi, \varepsilon)$, where $\phi(x) = \psi^{-1}(x, x^{-1})$ for all $x \in G$. By (7.7), Theorem 7.14 and Corollary 7.15, (ψ, ε) is orthogonal if and only if $R_{\phi^{-1}} = \{(1, x)^{-1} : x \in G\}$ is a $(v, w, v, v/w)$-RDS in $E_{(\psi^*, \varepsilon^*)}$ if and only if $R_{\phi^{-1}}^{(-1)} = \{(1, x) : x \in G\} = R_{(\psi^*, \varepsilon^*)}$ is a $(v, w, v, v/w)$-RDS in $E_{(\psi^*, \varepsilon^*)}$, and the result follows . □

As shown next, the coupled cocyclic matrix $M_{(\psi^*, \varepsilon^*)}$ for the dual is equivalent to the transinverse $M^*_{(\psi, \varepsilon)}$, from which it follows that the transpose $M^\top_{(\psi, \varepsilon)}$ of $M_{(\psi, \varepsilon)}$ is a $GH(w, v/w)$ over N whenever $M_{(\psi, \varepsilon)}$ is a $GH(w, v/w)$ over N. For entries from an abelian group N, this has been known since 1988 [37], but it has not been shown for entries from nonabelian N (Lemma 4.10). This extension to coupled cocyclic $GH(w, v/w)$ over arbitrary N is new.

LEMMA 7.26 *If $(\psi, \varepsilon) \in F^2(G, N)$ let $M^*_{(\psi, \varepsilon)}$ be the transinverse of $M_{(\psi, \varepsilon)}$.*

1. *$M^*_{(\psi, \varepsilon)} \sim M_{(\psi^*, \varepsilon^*)} = [\psi(y^{-1}, x^{-1})^{\varepsilon(y^{-1})^{-1}}]_{x, y \in G}$;*

2. *$M_{(\psi, \varepsilon)}$ is a $GH(w, v/w)$ over $N \Leftrightarrow M^*_{(\psi, \varepsilon)}$ is a $GH(w, v/w)$ over $N \Leftrightarrow M^\top_{(\psi, \varepsilon)}$ is a $GH(w, v/w)$ over N;*

3. *If H is a coupled G-cocyclic $GH(w, v/w)$ over N then in $\mathbb{Z}N$*

$$HH^* = H^*H = vI_v + v/w \left(\sum_{u \in N} u \right) (J_v - I_v).$$

Proof. For part 1, by (7.28), $M_{(\psi,\varepsilon)} = [\psi^{-1}(x,y)^{\varepsilon(x)^{-1}}]_{x,y \in G}$ and

$M_{(\psi^*,\varepsilon^*)} = [(\psi^*(x,y)^{-1})^{\varepsilon^*(x)^{-1}}]_{x,y \in G} = [(\psi(y^{-1},x^{-1})^{\varepsilon^*(xy)})^{\varepsilon^*(x)^{-1}}]_{x,y \in G}$.

By (7.32) and (7.1),

$$\varepsilon^*(x)^{-1} \circ \varepsilon^*(xy) = \varepsilon(x^{-1}) \circ \varepsilon(y^{-1}x^{-1})^{-1} = \varepsilon(y^{-1})^{-1} \circ \overline{\psi(y^{-1}, x^{-1})},$$

so $M_{(\psi^*,\varepsilon^*)}$ has the stated form. Under the row permutation $y^{-1} \mapsto y$ and column permutation $x^{-1} \mapsto x$, $M_{(\psi^*,\varepsilon^*)} \sim [\psi(y,x)^{\varepsilon(y)^{-1}}]_{x,y \in G} = M^*_{(\psi,\varepsilon)}$. Then parts 2 and 3 follow from (4.8), Lemma 4.10.1 and Theorem 7.25. □

Although by Lemma 7.26 the transinverse $M^*_{(\psi,\varepsilon)}$ is a coupled G-cocyclic matrix over N, the transpose $M^\top_{(\psi,\varepsilon)}$ is not necessarily coupled cocyclic itself (cf. (6.10)). This underscores our contention that the *dual*, rather than the transpose, is the proper object of study for generalised Hadamard matrices.

Significantly, $M_{(\psi,\varepsilon)}$ and the matrix $M_{(\psi^*,\varepsilon^*)}$ for its dual can be inequivalent over N, even when they are both $GH(w, v/w)$.

Example 7.3.1   *There exist G of order 16 and $\psi \in Z^2(G, \{\pm 1\})$ such that both M_ψ and M_{ψ^*} are G-cocyclic Hadamard matrices but $M_\psi \not\sim M_{\psi^*}$.*

Proof. There exist inequivalent Hadamard matrices H and $H^\top$ of order 16 (though they are Q-equivalent). By Example 7.4.1 in Section 7.4 following, $H \sim M_\psi$ for some G and $\psi \in Z^2(G, \{\pm 1\})$, so by Lemma 7.26, $H^\top = H^* \sim (M_\psi)^* = (M_\psi)^\top \sim M_{\psi^*}$. □

Third, from (Drake's) Lemma 4.13, the image of a $GH(w, v/w)$ over N under an epimorphism $\sigma : N \twoheadrightarrow N'$ with $|N'| = w'$ is a $GH(w', v/w')$ over N'. Such a projection also preserves coupled G-cocyclic $GH(w, v/w)$, provided that the kernel of σ is $\text{Im}(\varepsilon)$-invariant in N, that is, every automorphism in $\text{Im}(\varepsilon)$ fixes $\text{Ker}(\sigma)$. This extra condition ensures that projection of a factor pair is a factor pair [118, Theorem 6.2], which has already been noted for cocycles with trivial action in Example 6.2.12. Combined with Theorem 7.22, this result generalises [260, Theorem 5.2], where the case of abelian N and trivial action is proved.

THEOREM 7.27 *Let $\sigma : N \twoheadrightarrow N'$ with $|N| = w'$ be an epimorphism of groups for which $\text{Ker}(\sigma)$ is $\text{Im}(\varepsilon)$-invariant in N. The projection $(\sigma \circ \psi, \varepsilon') \in F^2(G, N')$ of a factor pair $(\psi, \varepsilon) \in F^2(G, N)$ is defined by $\varepsilon'(x) = \sigma \circ \varepsilon(x) \circ s$ for all $x \in G$ and any section $s : N' \to N$ of σ.*
If $M_{(\psi,\varepsilon)}$ is a $GH(w, v/w)$ over N then $M_{(\sigma \circ \psi, \varepsilon')}$ is a $GH(w', v/w')$ over N'.

Finally, by Lemma 4.10 the tensor product of a $GH(w, v/w)$ and a $GH(w, v'/w)$ over N is a $GH(w, vv'/w)$ over N. By Theorem 6.9, if $N = C$ is central, so $\varepsilon \equiv 1$, then $M_{\psi_1 \otimes \psi_2} = M_{\psi_1} \otimes M_{\psi_2}$ is generalised Hadamard if and only if M_{ψ_i} is generalised Hadamard, $i = 1, 2$. No tensor product construction corresponding to Example 6.2.14 has been found for more general factor pairs.

Research Problem 44 *Generalise Theorem 6.9 to cocycles with nontrivial action, or to more general factor pairs.*

Galati has found a particular solution in a skewed tensor square construction, which for $\varepsilon \equiv 1$ equals $\psi \otimes \psi$ if and only if ψ is symmetric, and is otherwise inequivalent.

LEMMA 7.28 (Galati [120, Lemma 7.1]) *Let G and N be finite abelian groups of orders v and w, respectively, and let $(\psi, \varepsilon) \in F^2(G, N)$. The skew tensor square $(\psi \circledast \psi, \varepsilon \circledast \varepsilon)$ of (ψ, ε), defined for all $(x_1, x_2), (y_1, y_2) \in G \times G$ by $(\varepsilon \circledast \varepsilon)(x_1, x_2) = \varepsilon(x_1 x_2)$ and*

$$(\psi \circledast \psi)\big((x_1, x_2), (y_1, y_2)\big)$$
$$= \Big[\psi(x_1, y_2)\, \psi^{-1}(y_2, x_1) \Big]^{\varepsilon(x_2)} \psi(x_1, y_1)^{\varepsilon(x_2 y_2)} \, \psi(x_2, y_2)$$

is in $F^2(G \times G, N)$. Then (ψ, ε) is $(v, w, v, v/w)$-orthogonal if and only if $(\psi \circledast \psi, \varepsilon \circledast \varepsilon)$ is $(v^2, w, v^2, v^2/w)$-orthogonal.

7.4 THE FIVE-FOLD CONSTELLATION

This Section delivers the full flowering of our theory. In it, mutual equivalences of coupled cocyclic generalised Hadamard matrices and four other objects of interest — 'stars' — are proved, emphasising the power and pervasiveness of the group extensions approach to Hadamard matrices.

First, pull Theorems 7.14, 4.20 (or 7.17) and 7.22 together, for the case $k = v$.

THEOREM 7.29 *(Four-fold equivalence) Let G and N be finite groups of order v and w, respectively, where w divides v. Let $N \overset{\iota}{\rightarrowtail} E \overset{\pi}{\twoheadrightarrow} G$ be an extension of N by G and let $[\varphi, \tau]$ be its associated equivalence class of factor pairs. Then the following statements are equivalent:*

1. *there exists a coupled G-cocyclic $GH(w, v/w)$ over N;*

2. *there exists an orthogonal factor pair in $[\varphi, \tau]$;*

3. *there exists a (normal) $(v, w, v, v/w)$-RDS in E relative to $\iota(N)$;*

4. *there exists a $(v, w, v, v/w)$-divisible design with regular group E, class regular with respect to $\iota(N)$.*

The splitting case is itself important, and will prove useful for extraction of the fifth equivalence.

THEOREM 7.30 *(Splitting equivalence) Let G and N be finite groups of order v and w, respectively, where w divides v. Let $N \overset{\iota}{\rightarrowtail} E \overset{\pi}{\twoheadrightarrow} G$ be a split extension of N by G, so $E \cong N \rtimes_\varrho G$, a semidirect product of N by G, and let $[1, \varrho]$ be its associated equivalence class of splitting factor pairs. Then the following statements are equivalent:*

1. *there exists a coupled G-developed GH$(w, v/w)$ over N;*

2. *there exists an orthogonal factor pair in $[1, \varrho]$;*

3. *there exists a splitting $(v, w, v, v/w)$-RDS in E relative to $\imath(N)$;*

4. *there exists a $(v, w, v, v/w)$-divisible design with regular group $N \rtimes_\varrho G$, class regular with respect to $\imath(N)$.*

Partial results towards this complete theory have been obtained by many other authors.

The splitting equivalences of Theorem 7.30, in the special case $\varrho \equiv 1$, are the most familiar. For subsequent ease of reference, they are restated in Theorem 9.13. They correspond to the original G-developed case (Corollary 4.23) of Jungnickel [189], using the traditional definition of a splitting RDS. As we now see, Definition 4.22 is the more appropriate definition of a splitting RDS.

The general equivalences of Theorem 7.29, in the special case when N is abelian and the action ε is trivial, so $N = C$ is central in E, correspond to the original G-cocyclic case, due to Perera and the author [260].

COROLLARY 7.31 *(Central equivalence)* [260, Theorem 4.1] *Let G be a finite group of order v and C be a finite abelian group of order w such that $w|v$. Let $C \xrightarrow{\imath} E \xrightarrow{\pi} G$ be a central extension of C by G and let $[\varphi] \in H^2(G, C)$ be its associated cohomology class of cocycles. Then the following statements are equivalent:*

1. *there exists a G-cocyclic GH$(w, v/w)$ over C;*

2. *there exists an orthogonal cocycle in $[\varphi]$;*

3. *there exists a central relative $(v, w, v, v/w)$-difference set in E, relative to $\imath(C)$;*

4. *there exists a divisible $(v, w, v, v/w)$-design with regular group E, class regular with respect to $\imath(C)$.*

Restriction of Corollary 7.31 to abelian extensions E of C by G is the most familiar case in the literature on RDSs, corresponding to abelian semiregular RDSs [266, 267]. This forces G to be abelian and $[\varphi]$ to contain only symmetric cocycles, that is, by (6.15), $[\varphi] \in \mathrm{Ext}_\mathbb{Z}(G, C)$.

Restriction of Corollary 7.31 to $C = \{\pm 1\} \cong \mathbb{Z}_2$ gives the corresponding set of equivalences, first demonstrated by de Launey in 1993, for G-cocyclic Hadamard matrices. Necessarily $\varepsilon \equiv 1$, since $\mathrm{Aut}(C) = \{1\}$. Apart from the case $G \cong \mathbb{Z}_2$ — for which the appropriate equivalences also all hold — we may assume v is divisible by 4. Subsequently, Flannery [112] proved that the existence of a G-cocyclic Hadamard matrix is equivalent to the existence of a *Hadamard group*, a term coined earlier by Ito [180] to describe the group containing a *Hadamard set* relative to a central involution, in other words a $(4t, 2, 4t, 2t)$-RDS.

COROLLARY 7.32 *(Binary equivalence)* [88] *Let G be a finite group of order $4t$. Let $\mathbb{Z}_2 \overset{\imath}{\rightarrowtail} E \overset{\pi}{\twoheadrightarrow} G$ be a central extension of $\mathbb{Z}_2$ by G and let $[\varphi]$ be its associated cohomology class of cocycles. Then the following statements are equivalent:*

1. *there exists a G-cocyclic Hadamard matrix;*

2. *there exists an orthogonal cocycle in $[\varphi]$;*

3. *there exists a central relative $(4t, 2, 4t, 2t)$-difference set in E, relative to $\imath(\mathbb{Z}_2)$;*

4. *there exists a divisible $(4t, 2, 4t, 2t)$-design with regular group E, class regular with respect to $\imath(\mathbb{Z}_2)$;*

5. [180, 112] *there exists a Hadamard group E with respect to the central involution $\imath(1)$.*

For our final specialisation, we come full circle, to the splitting binary case which returns us the family of Menon Hadamard matrices (Definition 2.20). That is, $C \cong \{\pm 1\} \cong \mathbb{Z}_2$; necessarily $\varepsilon = \varrho \equiv 1$; and $\varphi = \partial\phi$ is a coboundary. We also recover the additional equivalence with Menon-Hadamard difference sets, the proof of which was promised in Lemma 2.19.

COROLLARY 7.33 *(Splitting binary equivalence) Let G be a finite group of order $4u^2$, let $E \cong \mathbb{Z}_2 \times G$ and let $B^2(G, \mathbb{Z}_2)$ be the associated cohomology class of coboundaries. Then the following statements are equivalent:*

1. *there exists a G-developed Menon Hadamard matrix;*

2. *there exists an orthogonal coboundary in $B^2(G, \mathbb{Z}_2)$;*

3. *there exists a relative $(4u^2, 2, 4u^2, 2u^2)$-difference set in $\mathbb{Z}_2 \times G$, relative to $\mathbb{Z}_2 \times \{1\}$;*

4. *there exists a divisible $(4u^2, 2, 4u^2, 2u^2)$-design with regular group $\mathbb{Z}_2 \times G$, class regular with respect to $\mathbb{Z}_2 \times \{1\}$;*

5. *there exists a Menon-Hadamard difference set in G.*

Proof. All that is missing is the equivalence 5 ⇔ 3. This is provided by Jungnickel's proof [189, Theorem 3.7, Proposition 3.9] that a subset D in G is a Menon-Hadamard difference set if and only if $R = \{-1\} \times D \cup \{1\} \times \overline{D}$ is a $(4u^2, 2, 4u^2, 2u^2)$-RDS in $\{\pm 1\} \times G$, relative to $\{\pm 1\} \times \{1\}$. □

As a small illustration, Corollary 7.32 has been used by de Launey [85] to show that all Hadamard matrices of orders 16 and 20 are cocyclic. Using MAGMA, he checked that, for a representative in each of the 5 equivalence classes of order 16 and 3 of order 20, there is a regular automorphism group of the associated divisible design with a normal class regular subgroup of order 2.

Example 7.4.1 *All Hadamard matrices of orders ≤ 20 are cocyclic.*

We have located four stars in our Five-fold Constellation of equivalences: coupled cocyclic generalised Hadamard matrices; orthogonal factor pairs (or their group extensions); semiregular relative difference sets; and semiregular class regular divisible designs with regular group of automorphisms.

Where's the fifth?

The fifth equivalence arises from applications in signal correlation and in cryptography, through confluence of the concepts of perfect array and perfect nonlinear function. From this point of view we are able to encompass and reconcile many diverse and occasionally inconsistent attempts at definition in the literature.

Recall the perfect binary arrays PBA (Definition 3.22) and their quaternary counterparts PQA (Definition 4.25), defined for *abelian* groups G and for $C = \{\pm 1\}$ and $C = \{\pm 1, \pm i\}$, respectively. The existence of a PBA is equivalent to the existence of a Menon-Hadamard difference set in G (Theorem 3.23) and to the existence of a Menon Hadamard matrix with the PBA as top row (Lemma 3.25), giving an additional equivalence with those of Corollary 7.33. Similarly, the existence of a PQA is equivalent to the existence of a G-developed quaternary complex Hadamard matrix with the PQA as top row (Lemma 4.27). The top rows of G-developed $GH(4, v/4)$ are the *flat* PQAs (in the coinage of Hughes [175]), giving an additional equivalence with the splitting case of Corollary 7.31 for abelian G and $C = \{\pm 1, \pm i\}$.

When G and N are abelian, the top rows of G-developed $GH(w, v/w)$ over N are also familiar within the cryptographic community, where, following Nyberg, the defining functions $G \to N$ are called *perfect nonlinear (PN)*. Nyberg's original definition [251, Definition 3.1] of PN functions has $G = \mathbb{Z}_r^n$ and $N = \mathbb{Z}_r^m$, $n \geq m$, and for $r = 2$ they are precisely the vectorial bent functions (see Chapter 3.5.2).

That is, when $N = \{\pm 1\} \cong \mathbb{Z}_2$, a PN function is the same as a PBA, and when $N = \{\pm 1, \pm i\} \cong \mathbb{Z}_4$, a PN function is the same as a flat PQA, or equivalently by Lemma 4.29, its square is a PBA.

So it is obvious that the function defining the top row of a coupled G-developed $GH(w, v/w)$ is a most interesting object for study. We adopt Nyberg's nomenclature for this most general case. Perfect nonlinear functions will appear again in Chapter 9.2.1.

DEFINITION 7.34 *Let G and N be finite groups of order v and w, respectively, where $w|v$ and let $\phi \in C^1(G, N)$. For a homomorphism $\varrho : G \to \mathrm{Aut}(N)^{\mathrm{op}}$, let M be the coupled G-developed matrix $[\phi(xy)^{\varrho(x^{-1})}]_{x,y \in G}$.*

The function ϕ is perfect nonlinear (PN) *relative to ϱ if M is a $GH(w, v/w)$ over N. If $\varrho \equiv 1$ we say ϕ is* perfect nonlinear (PN).

Equivalently, on inverting each term in (7.31), ϕ is PN relative to ϱ if and only if, in the group ring $\mathbb{Z}N$,

$$\forall\, x \neq 1 \in G, \quad \sum_{y \in G} \phi(xy)(\phi(y)^{-1})^{\varrho(x)} = (v/w)\, N. \tag{7.33}$$

The function $\Delta_x(\phi) \in C^1(G, N)$ defined by $\Delta_x(\phi)(y) = \phi(xy)(\phi(y)^{-1})^{\varrho(x)}$ will be termed the directional derivative *of ϕ in direction x with twist ϱ.*

We have found our fifth star, in the splitting case.

THEOREM 7.35 *The splitting equivalences of Theorem 7.30 are further equivalent to the following statement:*

5. *there exists a PN function* ϕ *relative to* ϱ.

How do perfect arrays fit here? Remember, they are called 'perfect' because they have ideal autocorrelation: all off-peak correlations are zero and the on-peak correlation equals the *energy* $|G|$ of the signal. It is plain now how to link group developed $GH(w, v/w)$ to a right notion of *flat* perfect array. Of course, we must enlarge the idea of correlation appropriately, to include sequences defined over arbitrary groups.

A flat perfect array must be the top row of a coupled G-developed $GH(w, v/w)$ over N, which, in order for its top row to have the ideal autocorrelation property, must be invertible (Definition 4.14) over a suitable ring R with $N \leq R^*$. By Lemma 7.26.3 invertibility over R for coupled G-developed $GH(w, v/w)$ is equivalent to the condition that $\sum_{u \in N} u = 0$ in R, just as it is for any $GH(w, v/w)$ when N is abelian (cf. Example 4.3.3).

DEFINITION 7.36 *With the terminology of Definition 7.34, let R be a ring with unity for which* char R *does not divide* v, $N \leq R^*$ *and* $\sum_{u \in N} u = 0$ *in* R.
The sequence $(\phi(x), x \in G)$ *is a* flat perfect array (FPA) *over* R *relative to* ϱ *if* ϕ *is PN relative to* ϱ. *If* $\varrho \equiv 1$ *we say it is a flat perfect array (FPA) over* R.

To lift Theorem 7.35 to full generality — a fifth equivalence with Theorem 7.29 — we look to a generalisation developed for perfect arrays rather than for PN functions.

Jedwab's GPBAs successfully extend the PBAs, dramatically overcoming their scarcity. Since a GPBA is equivalent to an abelian $(4t, 2, 4t, 2t)$-RDS [186, Theorem 3.2], there is an appropriate fifth equivalence with Corollary 7.32 in the event that E is abelian.

Hughes exploits this equivalence and Corollary 7.31 to link PQAs and GPBAs. Essentially he shows that the link arises from imposing the condition of group development of a quaternary complex Hadamard matrix (required by Lemma 4.27) onto the corresponding Hadamard matrix (given by Theorem 4.8.1).

LEMMA 7.37 [175, Theorem 3.1] *Let* $\gamma \in Z^2(\mathbb{Z}_2, \mathbb{Z}_2)$ *be the cocycle of Example 6.2.2 for* $v = 2$ *and* $\omega = -1$. *Let G be an abelian group and let φ and ϕ be related by (4.18). Then φ is a PQA if and only if $M_{(\gamma \otimes 1)\partial \phi}$ is a $(\mathbb{Z}_2 \times G)$-cocyclic Hadamard matrix.*

It follows from Lemma 7.37 and Corollary 7.32 that any PQA is equivalent to an abelian $(4t, 2, 4t, 2t)$-RDS of a certain type. This is also proved directly in the 1-D (cyclic) case $G = \mathbb{Z}_{2t}$ in [9] and in the 2-D (bicyclic) case $G = \mathbb{Z}_{s_1} \times \mathbb{Z}_{s_2}$ in [8].

In [173], Hughes generalises Jedwab's GPBAs to 'base sequences' in the central case. For $\varphi \in Z^2(G, C)$ and $\phi : G \to C$, Hughes defines ϕ to be φ-*correlated* if, in the group ring $\mathbb{Z}C$,

$$\sum_{y \in G} \varphi(x, y)\phi(y)\phi(xy)^{-1} = (v/w)\, C, \ \forall\, x \neq 1 \in G.$$

He thinks of this definition of correlation as being the usual autocorrelation function of ϕ 'twisted' by φ. It follows that $\varphi\partial(\phi^{-1})$ is orthogonal if and only if ϕ is φ-correlated. If $\psi \in Z^2(G, C)$ and $\nu_1 = 1, \ldots, \nu_s$ is a list of representatives for the cohomology classes in $H^2(G, C)$, then $\psi = \nu_i \partial(\phi^{-1})$ for a unique i and some 1-cochain ϕ. If ψ is orthogonal, Hughes calls any such ν_i-correlated ϕ a *base sequence with respect to* ν_i. The map ϕ is not unique, but if we can identify such a ϕ we have, so to speak, a 1-D representation of the cocycle ψ.

This is another equivalence with Corollary 7.31.2. When $C = \{\pm 1\}$, G is abelian and ψ is symmetric, a base sequence is the same as a GPBA.

LEMMA 7.38 *The equivalent statements of Corollary 7.31 are further equivalent to the following statement:*

5. [173] *there is a base sequence with respect to some $\psi \in [\varphi]$.*

When E is abelian, the equivalent statements of Corollary 7.32 are further equivalent to the following statement:

6. [174, Theorem 5.3] *there is a generalised perfect binary array (GPBA) coordinatised by G.*

It is now apparent how to generalise Hughes' base sequences to orthogonal (ψ, ε) in $F^2(G, N)$, using (7.11).

DEFINITION 7.39 *Let G be a group of order v and let N be a group of order w, where $w|v$. Let $(\nu_1, \eta_1) = (1, 1), \ldots, (\nu_s, \eta_s)$ be a list of representatives for the equivalence classes in $F^2(G, N)$. If $(\psi, \varepsilon) \in F^2(G, N)$, write $(\psi, \varepsilon) = (\nu_i, \eta_i) \cdot \phi^{-1}$, where i is unique and $\phi \in C^1(G, N)$.*

If (ψ, ε) is orthogonal, the sequence $(\phi(x), x \in G)$ (or, equally, the function ϕ) is called a (generalised) base sequence with respect to (ν_i, η_i).

Equivalently, if $(\psi, \varepsilon) = (\nu_i, \eta_i) \cdot \phi^{-1}$ and (ψ, ε) is orthogonal, multiplying equation (7.24) by $\phi(x)^{-1}$ gives

$$\sum_{y \in G} \phi(y)^{\eta_i(x)} \nu_i(x, y) \phi(xy)^{-1} = (v/w) \, N, \; \forall \, x \neq 1 \in G, \tag{7.34}$$

and we will say the base sequence ϕ has (ν_i, η_i)-*twisted autocorrelation*.

Definition 7.39 is consistent with Definition 7.34. For when $\phi : G \to N$ is PN relative to ϱ, then by Definition 7.34, Corollary 7.21 and Lemma 7.24, $M_{(\partial^{-1}\phi, \overline{\phi}\varrho)}$ is a coupled G-cocyclic $GH(w, v/w)$, and $(\partial^{-1}\phi, \overline{\phi}\varrho) \sim_\phi (1, \varrho)$. Thus (7.31) is (7.34) for the choice of representative $(\nu_i, \eta_i) = (1, \varrho)$. Hence ϕ is a base sequence with respect to $(1, \varrho)$, and vice versa.

We have found our fifth star, in the general case.

THEOREM 7.40 *The equivalences of Theorem 7.29 are further equivalent to the following statement:*

5. *there exists a base sequence ϕ with respect to $(\psi, \varepsilon) \in F^2(G, N)$ for some $(\psi, \varepsilon) \in [\varphi, \tau]$. In the splitting case (Theorem 7.30), ϕ is a base sequence with respect to $(1, \varrho)$ if and only if ϕ is PN relative to ϱ.*

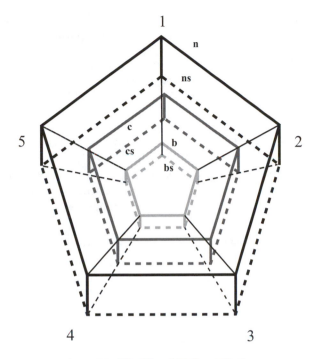

Figure 7.1 **The Five-fold Constellation:**
1. Coupled cocyclic generalised Hadamard matrix
2. Orthogonal factor pair
3. Semiregular relative difference set
4. Semiregular divisible design with regular group
and class regular normal subgroup
5. Base sequence
n = normal, **c** = central, **b** = binary, **s** = splitting

The interequivalence of these five areas can be visually represented as the *Five-fold Constellation* which is pictured in Figure 7.1.

In it, the 'orders of magnitude' are determined by successively broader classes of cocyclic Hadamard matrices. The outermost and principal pentagon, labelled **n**, represents Theorems 7.29 and 7.40, the most general, or 'normal N' case. Inside pentagon **n** are three concentric copies, labelled **a**, **c** and **b**, representing the 'abelian kernel N', 'central N' and 'binary $N = \{\pm 1\}$' cases, respectively (compare with (7.27)). For simplicity's sake, pentagon **a** has been suppressed. Pentagon **c** represents Corollary 7.31 and Lemma 7.38 and pentagon **b** represents Corollary 7.32 and the binary case of Lemma 7.38. Shadowing these should be envisaged a second level with identical concentric pentagonal structure, labelled **ns**, **as** (suppressed), **cs** and **bs**, denoting the respective splitting cases.

The fifth equivalence extracts a 1-D representation $\phi : G \to N$ from what is essentially a 2-D construction ($\psi : G \times G \to N, \varepsilon$). In the process, the perfect autocorrelation of PBAs is modified and twisted so much that the resulting definition of autocorrelation (7.34) may have limited value for applications.

Research Problem 45 *Are there useful implementations of FPAs over R relative to ϱ (Definition 7.36) in signal processing?*

Are there useful implementations of twisted autocorrelation (as defined in (7.34)) in signal processing?

Another approach, different from that of the Five-fold Constellation, to the problem of extending a 1-D perfect nonlinear function, is taken in Chapter 9.5.3. It also begins with 1-D PN functions, but asks what perfect nonlinearity of 2-D arrays might mean. The *total differential uniformity* introduced there (for the central case $N = C$) does not coincide with orthogonality for cocycles.

7.4.1 Restrictions on existence of cocyclic generalised Hadamard matrices

We can exploit the equivalence of a coupled G-cocyclic $GH(w, v/w)$ over N and a semiregular RDS in an extension E of N by G (Theorem 7.29) to determine restrictions on E necessary for existence of $GH(w, v/w)$. The first of the results below is well known for abelian semiregular RDSs (see [265, Theorem 3.1]), and the proof there works with the weaker hypothesis of centrality on N.

LEMMA 7.41 [119, Proposition 2.1] *Suppose a group $E \not\cong \mathbb{Z}_4$ of order vw contains a semiregular RDS relative to a central subgroup C of order w. Then $\exp(E)$ divides v.*

If, in addition, the orthogonal cocycle corresponding to such a central semiregular RDS is multiplicative — so, by Theorem 6.10 there is a prime p such that both G and C are elementary abelian p-groups — the restrictions on E are far more severe.

LEMMA 7.42 *For p prime, $G = \mathbb{Z}_p^n$ and $C = \mathbb{Z}_p^m$, $n \geq m \geq 1$, suppose $\psi \in Z^2(G, C)$ is multiplicative and orthogonal. Let E_ψ be the corresponding extension group (7.6). Then*

    1. if $p = 2$, $\exp(E_\psi) = 4$;

    2. if $p > 2$, $\exp(E_\psi) = p$;

    3. if $p > 2$ and ψ is symmetric then $\psi = \partial\phi \in B^2(G, C)$ and ϕ is PN.

Proof. Direct computation gives parts 1 and 2 (cf. proof of [168, Theorem 3.3.ii]). If $p > 2$ and ψ is symmetric, then E_ψ is abelian and by part 2, is an elementary abelian p-group. Therefore $E_\psi \cong C \times G$ splits and $\psi = \partial\phi$ for some $\phi : G \to C$. By Definition 7.34, ϕ is PN. $\square$

The second restriction on E generalises observations made by Flannery, Ito and others in the case $C = \{\pm 1\}$ that, if $E/\{\pm 1\}$ has a cyclic Sylow 2-subgroup then any cocyclic Hadamard matrix M_ψ must have ψ a coboundary. That result covers the cases where G is a cyclic group $\mathbb{Z}_{4t}$ or quaternion Q_{4t} for t odd, mentioned in Chapter 6.4.6. In neither this result nor Lemma 7.41 can the centrality condition be weakened to allow an abelian normal forbidden subgroup N [119, §4].

THEOREM 7.43 [119, Theorem 1.1] *Suppose a group $E \not\cong \mathbb{Z}_4$ of order vw contains a semiregular RDS relative to a central subgroup C of order w. If E/C has cyclic Sylow p-subgroups for each prime p dividing w, then E splits over C.*

Proof. Gaschütz' Theorem (cf. [13, 10.4]) states that for a prime p, if N is an abelian normal p-subgroup of a finite group E, and P is a Sylow p-subgroup of E, then E splits over N if and only if P splits over N. A corollary is that E splits over C if and only if for each prime $p \mid w$, each Sylow p-subgroup P of E splits over the Sylow p-subgroup C_p of C (cf. [134, Theorem 15.8.6, p. 246]).   □

Under these conditions, the central splitting case of Theorem 7.30 applies and the corresponding $GH(w, v/w)$ over C must be (the normalised version of) a G-developed $GH(w, v/w)$.

COROLLARY 7.44 *Suppose a G-cocyclic $GH(w, v/w)$ M_ψ over C exists, under the conditions of Corollary 7.31. If G has cyclic Sylow p-subgroups for each prime p dividing w, then ψ is a coboundary and M_ψ is Hadamard equivalent to a G-developed $GH(w, v/w)$.*

It is a fact [106, Theorem 6.2] that cyclic groups other than $\mathbb{Z}_4$ do not contain semiregular RDSs. (This also follows from Lemma 7.41.) The third new nonexistence result to follow from Theorem 7.29 generalises this. The finite cyclic groups are characterised amongst the finite abelian groups as those with all Sylow subgroups cyclic. More generally, suppose E has all Sylow subgroups cyclic. Then (cf. [276, 10.1.10, p. 290]) E has a presentation

$$E(r, s, t) = \langle\, a, b \mid a^s = 1 = b^t,\ b^{-1}ab = a^r \,\rangle, \qquad (7.35)$$

where $r^t \equiv 1 \bmod s$, s is odd, $0 \le r < s$, and s and $t(r-1)$ are coprime. Conversely, every such $E(r, s, t)$ has all its Sylow subgroups cyclic. Instances are $E(0, 1, m) \cong \mathbb{Z}_m$, $E(n-1, n, 2) \cong D_{2n}$ for n odd and $E(8, 9, 2j)$ for $(3, j) = 1$ [118, Example 8.13].

Suppose $E(r, s, t)$ contains a semiregular RDS R. Galati uses order arguments and projection of R to derive a contradiction. Since no such R can exist, by Theorem 7.29 no coupled cocyclic $GH(w, v/w)$ with entries in any normal subgroup of $E(r, s, t)$ can exist.

THEOREM 7.45 [121, Theorem 1] *Let E be a finite group of order $vw > 4$ with all Sylow subgroups cyclic, let $N \lhd E$ with $|N| = w$ and let $G \cong E/N$. Then there is no coupled G-cocyclic $GH(w, v/w)$ over N.*

The fourth nonexistence result for coupled cocyclic $GH(w, v/w)$ is due to Elvira and Hiramine [108], who prove that no dihedral group contains a semiregular RDS. (The odd n case also follows from Theorem 7.45 and another proof for arbitrary n appears in [121, Theorem 2].)

LEMMA 7.46 [108] *Let $E = D_{2n}$ be the dihedral group of order $2n > 4$, let $N \triangleleft E$ and let $G \cong E/N$. Then there is no coupled G-cocyclic $GH(w, v/w)$ over N.*

Whilst the factor pair construction for generalised Hadamard matrices given in Theorem 7.22 is very general, our final example shows that not every generalised Hadamard matrix arises in this way.

Example 7.4.2 [120, Example 10.1] *There are no cocyclic $GH(3, 2)$. In particular, the normalised $GH(3, 2)$ of Example 4.3.1 is not cocyclic.*

Proof. The only groups G of order 6 are $\mathbb{Z}_6$ and D_6. Since $\mathrm{Aut}(\mathbb{Z}_3) \cong \mathbb{Z}_2$, any homomorphism $\varepsilon : G \to \mathrm{Aut}(\mathbb{Z}_3)$ has kernel of order either 3 or 6. There are two possible actions of $\mathbb{Z}_6$ on $\mathbb{Z}_3$, the trivial action and a nontrivial action ε_1. Similarly, there are two possible actions of D_6 on $\mathbb{Z}_3$, the trivial action and a nontrivial action ε_2. An exhaustive computer search of the groups $Z^2(\mathbb{Z}_6, \mathbb{Z}_3)$, $Z_{\varepsilon_1}^2(\mathbb{Z}_6, \mathbb{Z}_3)$, $Z^2(D_6, \mathbb{Z}_3)$, and $Z_{\varepsilon_2}^2(D_6, \mathbb{Z}_3)$ revealed there are no orthogonal cocycles from G to $\mathbb{Z}_3$ and thus by Theorem 7.22, no cocyclic $GH(3, 2)$. $\square$

7.4.2 Two approaches

The Five-fold Constellation creates two possible approaches to the search for generalised Hadamard matrices.

The first, which might be considered a 'bottom-up' approach, starts with a group G of order v, and focusses on discovery of orthogonal factor pairs for some coefficient group N. This approach, the essence of this monograph, concentrates more on the pair of groups (G, N) and uses results from finite group cohomology, the theory of group extensions and ring theory to investigate the sets $F^2(G, N)$. Once detailed knowledge of $F^2(G, N)$ is obtained, the hunt is on to isolate its orthogonal elements (ψ, ε) and their corresponding generalised Hadamard matrices $M_{(\psi, \varepsilon)}$.

The second, which might be considered a 'top-down' approach, starts with an extension group E, and focusses on discovery of designs having regular groups E which are class regular relative to some normal subgroup N. This approach, taken for instance by Ito in his search for Hadamard matrices, concentrates more on the pair of groups (E, N), and uses tools from finite group theory and character theory. Once such a divisible design $\mathcal{D}$ is found, by Theorem 7.17 a block may be selected from which to derive (ψ, ε). It partners the converse design theoretic approach which starts with a known divisible design $\mathcal{D}$, then attempts to determine its automorphism group $\mathrm{Aut}(\mathcal{D})$ and the isotypes E of those subgroups of $\mathrm{Aut}(\mathcal{D})$ which act regularly. These are more probable candidates E from which to begin the top-down search for other designs on which they act regularly (since they are already known to do so for at least one class of designs).

Both approaches naturally also give a new weapon for attacking the problem regarded as the heart of research in relative difference sets (see, for example, [267, p. 203]): that of finding all groups E containing semiregular RDSs and of classifying the RDSs.

Warwick de Launey has used this converse process to great effect in his investigations of the regular action by different groups on the 'expanded design' of a huge range of combinatorial designs. For example, with Stafford [92, 87] he determines the isotypes of all the regular automorphism groups of the divisible designs associated with the Paley Type I Hadamard matrices P_q and Paley Type II Hadamard matrices P'_q, respectively. In the former case, they show that for $q + 1 > 8$, every regular action is normal [87, Corollary 5.2.8], and that, except for the 5 values $q = 3, 7, 11, 23, 59$, the only group isotype acting regularly on the divisible design is the generalised quaternion group [87, Corollary 5.2.4] of order $2(q + 1)$. In the latter case [87, Section 5.3.8], they show that for $q > 5$ all the regular automorphism groups correspond to Dickson near-fields, and regular actions are all normal. With Smith [91, Theorem 1.2.2] he shows that the regular automorphism groups of the divisible design associated with the Sylvester Hadamard matrix S_t include every extension of $\mathbb{Z}_2$ by $\mathbb{Z}_2^t$ except, if $t = 2s - 1$, the split extension $\mathbb{Z}_2^{2s}$.

Moreover, de Launey has developed a general theory for the study of designs which admit a cocyclic development, for the central case — cf. (5.15) — under the rubric of *pairwise combinatorial designs*. These include not merely generalised Hadamard matrices, but generalised weighing matrices and orthogonal designs. The reader is referred to his monograph with Flannery [87], which extends the power of the cocyclic approach from generalised Hadamard matrices to more general designs, and consequently has wider application, for instance to perfect *ternary* arrays (with entries in $\{0, \pm1\} \subset \mathbb{Z}$) [10].

However, we are hunting orthogonal factor pairs. Our next task is to examine the relationship between the various concepts of equivalence inherent to the Five-fold Constellation, and the extent to which they preserve orthogonality.

Chapter Eight

Bundles and Shift Action

If orthogonality were an easily identified property of factor pairs, the quest for orthogonal factor pairs with which to construct coupled cocyclic generalised Hadamard matrices might be simple.

The correspondence between semiregular relative difference sets and orthogonal factor pairs (Theorem 7.29) is a special case of the correspondence between transversals and factor pairs (Lemmas 7.7, 7.8). The natural equivalence relation on transversals preserves semiregular relative difference sets, and it could be hoped that the natural equivalence relation on factor pairs (Definition 7.4) would preserve orthogonality.

However, even for cocycles, the natural equivalence relation (cohomology) does not preserve orthogonality. Over the past decade, exploitation of the correspondence in the central case (Corollary 7.31) has therefore largely consisted of searches for orthogonal cocycles, both theoretical and experimental (using the algorithms of Chapter 6.3) [18, 88, 112, 162, 163, 173, 260]. This gave rise to the perception that orthogonality is an essentially combinatorial property, with no natural cohomological interpretation.

This perception is largely false — there is a natural atomic structure within equivalence classes of factor pairs, which discriminates between orthogonal and nonorthogonal factor pairs. This atomic structure is determined by a differential action we term the *shift* action, by G on $F^2(G, N)$, which defines a stronger equivalence relation on $F^2(G, N)$ than the natural equivalence $\sim_\phi$. Our fundamental question now becomes: what makes the factor pairs in one shift orbit orthogonal and not those in another?

Orbits under the shift action lie wholly within $\sim_\phi$ classes, so their partition of each $\sim_\phi$ class is invisible from the usual cohomological point of view. This is undoubtedly why the action has not been detected earlier.

This Chapter collects what little is known about the atomic structure of equivalence classes of factor pairs. This glimpse, however brief, reveals part of a gorgeous tapestry, rich in interconnecting details and insights, ready for further unfolding.

In the first Section, the natural equivalence between transversals is transcribed to an equivalence between factor pairs (Theorem 8.5). Each resulting equivalence class is a $G \rtimes (\mathrm{Aut}(N) \times \mathrm{Aut}(G))$-orbit of factor pairs (Corollary 8.9). It results from following a shift action within $\sim_\phi$ classes (the G-action) by an $(\mathrm{Aut}(N) \times \mathrm{Aut}(G))$-action across $\sim_\phi$ classes which preserves shift orbits (or indeed, vice versa). The term *bundle* was adopted in 1999 by the author [154, 156] to capture this orbit-of-orbits structure within and across equivalence classes of factor pairs.

Then equivalence of RDSs is partially mapped around the Five-fold Constellation. Each equivalence class of normalised (v, w, k, λ)-RDSs is determined by at least one bundle of (v, w, k, λ)-orthogonal factor pairs. For semiregular RDSs, there is exactly one such bundle: Theorem 8.11 states that the set of equivalence classes of semiregular RDSs is in one-to-one correspondence with the set of bundles of orthogonal factor pairs. The equivalence imposed by orthogonal bundles on semiregular divisible designs is touched on briefly. Then the orthogonal bundles are mapped into Hadamard equivalence classes (Definition 4.12) of generalised Hadamard matrices.

More generally, the 'row' frequency distribution taken by values of a factor pair is the same within each bundle. Orthogonal bundles are an extreme case, with uniform distribution of row frequencies. This statistical invariance of bundles (Theorem 8.15) may prove their most important characteristic. It is revisited in Chapter 9.5.

Translation of RDS equivalence to the fifth star of the Five-fold Constellation, base sequences, is detailed in Section 8.2, but only for the splitting case, and again in Section 8.5, but only for shift action in the central splitting case. Section 8.2 lays the groundwork for a new theory of equivalence and nonlinearity of functions between groups (Chapter 9.2.1), which, it is argued, is the natural context for studying such functions.

The rest of the Chapter is restricted to the central case $N = C$. In Section 8.3, salient properties of bundles are extracted and a 7-parameter taxonomy for classifying equivalence classes of central semiregular RDSs is established. A classification program is announced, based on the taxonomy (Research Problem 50). Such a classification is the principal goal of the theory of central semiregular RDSs. Classification of equivalence classes of central $(p^n, p^n, p^n, 1)$-RDSs, using the taxonomy, is commenced.

The remaining two Sections focus on shift action in the central case. Section 8.4 reports what is known about shift action as an abstract group action. It is a remarkably general action and so should be recognisable in more areas of mathematics than in fact appears to be the case.

Section 8.5 deals with the main link detected so far: shift action on $B^2(G, C)$ derives from the usual left action of G on the standard RG-module, where R is a commutative ring with unity and $C = (R, +)$. It relates to a sequence of quotients of RG which generalises the Loewy series for p-groups. Shift action on $B^2(G, C)$ also crops up in cryptography: see Corollary 9.58. The principal result of this Section (Theorem 8.48) is that for certain p-groups G of large enough order, almost all the shift orbits in $B^2(G, \mathbb{Z}_p^n)$ are maximal (that is, of size $|G|$).

There is yet a vast array of unanswered questions.

8.1 BUNDLES AND THE FIVE-FOLD CONSTELLATION

8.1.1 Equivalence of transversals

Since a relative $(v, w, v, v/w)$-difference set in E relative to N is a transversal of N in E, definition (4.15) of equivalence for RDSs is adopted for transversals.

Note that, if T is a transversal of the normal subgroup N in E, $d, e \in E$ and $\alpha \in \mathrm{Aut}(E)$, then eTd is a transversal of N in E, $\alpha(T)$ is a transversal of the normal subgroup $\alpha(N)$ in E and further, the groups E/N and $E/\alpha(N)$ are isomorphic. Two transversals T, T' of the isomorphic normal subgroups N, N', respectively, in E are *equivalent* if there exist $\alpha \in \mathrm{Aut}(E)$ and $d, e \in E$ such that $\alpha(N) = N'$ and $T' = e\, \alpha(T)\, d$.

We can refine this definition somewhat. First, note that $T' = e\,\alpha(T)\,d = ed\,(\overline{d^{-1}} \circ \alpha)(T)$, where $\overline{d^{-1}}$ is the inner automorphism determined by d^{-1}. Thus equivalence need involve only one-sided translation (on either side).

Second, a transversal is *normalised* if it intersects N in 1. Since uT is normalised for some $u \neq 1 \in N$ if and only if T is not normalised, any equivalence class contains at least one normalised representative and the set of equivalence classes of transversals of N in E coincides with the set of equivalence classes of normalised transversals of N in E. Thus, with no loss of generality, we may restrict the study of equivalence to equivalence of normalised transversals.

DEFINITION 8.1 *Let T, T' be normalised transversals of the isomorphic normal subgroups N, N', respectively, in E. Define T and T' to be* equivalent *if there exist $\alpha \in \mathrm{Aut}(E)$ and $e \in E$ such that $\alpha(N) = N'$ and $e T' = \alpha(T)$. Denote the equivalence class of T in E by $[T]$.*

Each equivalence class either consists entirely of RDSs or contains no RDSs. These equivalence classes are further specified by the isomorphism type of E/N.

DEFINITION 8.2 *Let N be a normal subgroup of order w in a group E of order vw with $G \cong E/N$ a group of order v. Denote by $\mathcal{T}(N, E, G)$ the set of equivalence classes of normalised transversals of N in E, and by $\mathcal{R}(N, E, G) \subseteq \mathcal{T}(N, E, G)$ the set of equivalence classes of normalised relative $(v, w, v, v/w)$-difference sets in E relative to N.*

In [156], the author translates equivalence of normalised transversals to a relationship between the corresponding cocycles (Lemma 7.7), when $N = C$ is central in E.

THEOREM 8.3 *[156, Theorem 3.2] Let T and T' be normalised transversals in E of the central subgroups N, N' isomorphic to C, respectively, for which $E/N \cong E/N' \cong G$. Then $[T] = [T']$ if and only if there exist automorphisms $\gamma \in \mathrm{Aut}(C)$ and $\theta \in \mathrm{Aut}(G)$ and an element $a \in G$ such that*

$$\psi_T = \gamma \circ (\psi_{T'}\, \partial\phi) \circ (\theta \times \theta), \qquad (8.1)$$

where $\phi(g) = \psi_{T'}(a^{-1}, g)$, $g \in G$.

Proof. Suppose $\alpha(T) = eT'$ for $e \in E$ and $\alpha \in \mathrm{Aut}(E)$, where $N' = \alpha(N)$. Set $T^* = \alpha(T)$. Then α induces an automorphism δ of C and an automorphism θ of G. Let $T = \{t_g : g \in G\}$, $T' = \{t'_g : g \in G\}$ and $T^* = \{t^*_g : g \in G\}$. Then $\alpha(t_g) = t^*_{\theta(g)}$. Hence $\psi_T = \gamma \circ \psi_{T^*} \circ (\theta \times \theta)$, where $\gamma = \delta^{-1}$. Express $e = x t'_a$ uniquely for $x \in N'$ and $a \in G$, and let σ be the permutation on G

such that $xt'_a t'_g = t^*_{\sigma(g)}$, $g \in G$. By Lemma 7.7, there is a mapping $\phi : G \to C$ such that $t^*_g = \iota'(\phi(g)^{-1})t'_g$, $g \in G$. Under the isomorphism $E \to E_{\psi_{T'}}$, we obtain $(\iota'^{-1}(x), a)(1, g) = (\phi(\sigma(g))^{-1}, \sigma(g))$. Set $c = \iota'^{-1}(x)$, so $\sigma(g) = ag$ and $\phi(ag)^{-1} = c\psi_{T'}(a, g)$, $\forall g \in G$. After application of (6.3) this becomes $\phi(g)^{-1} = c\psi_{T'}(a, a^{-1})\psi_{T'}(a^{-1}, g)^{-1}$. But $\phi(1) = 1$, so $\phi(g) = \psi_{T'}(a^{-1}, g)$, and the result follows. The converse follows without difficulty, on reversing the argument and setting $c = \psi_{T'}(a, a^{-1})^{-1}$, $e = \iota'(c)t'_a$ and $\alpha(\iota(d)t_g) = \iota'(\gamma^{-1}(d))t'_{\theta(g)}$, $d \in C$, $g \in G$. $\square$

The set of cocycles corresponding to an equivalence class of transversals under (8.1) is called a bundle. That is, the *bundle* $\mathcal{B}(\psi)$ of $\psi \in Z^2(G, C)$ is

$$\mathcal{B}(\psi) = \{\gamma \circ (\psi \, \partial\psi_{a^{-1}}) \circ (\theta \times \theta) : \gamma \in \mathrm{Aut}(C), \theta \in \mathrm{Aut}(G), a \in G\}.$$

where for each $d \in G$, the function $\psi_d : G \to C$ is given by the dth row of M_ψ,

$$\psi_d(g) = \psi(d, g), \ g \in G. \tag{8.2}$$

Note that

$$\psi \, \partial\psi_{a^{-1}}(g, h) = \psi(a^{-1}, h)^{-1}\psi(a^{-1}g, h). \tag{8.3}$$

LEMMA 8.4 *If $\psi \in Z^2(G, C)$, $a \in G$ and $\varphi = \psi \, \partial\psi_a$ as in (8.3) then φ is orthogonal if and only if ψ is orthogonal.*

Proof. Suppose $a, g \in G$, $g \neq 1$. Then, by (6.3),

$$\sum_{h \in G} \varphi(g, h) = \sum_{h \in G} \psi(ag, h)\psi(a, h)^{-1} = \psi(aga^{-1}, a)^{-1} \sum_{h \in G} \psi(aga^{-1}, h),$$

and, since $\psi(aga^{-1}, a)^{-1} \in C$ is fixed, the result follows from (7.24). $\square$

In [156, Theorems 3.5, 4.10], the author proves that, if $\mathcal{B}(C, E, G)$ is the set of bundles of cocycles in $Z^2(G, C)$ having $E_\psi \cong E$ by an isomorphism preserving the images of C, then the mapping $\mathcal{B}$ from $\mathcal{T}(C, E, G)$ to $\mathcal{B}(C, E, G)$ given by $\mathcal{B}([T]) = \mathcal{B}(\psi_T)$ is a well-defined bijection which maps $\mathcal{R}(C, E, G)$ onto the set of bundles of orthogonal cocycles.

8.1.2 Bundles of factor pairs

Galati's extension to factor pairs [118] of the author's results (given in Section 8.1.1 for cocycles with $\varepsilon \equiv 1$) is now presented. Proofs involve mostly straightforward checking.

THEOREM 8.5 *Let T and T' be normalised transversals in E of the normal subgroups K and K' isomorphic to N, respectively, for which $E/K \cong E/K' \cong G$. Let $(\psi, \varepsilon), (\psi', \varepsilon')$ be the corresponding factor pairs of Lemma 7.7, respectively. Then there exist $\alpha \in \mathrm{Aut}(E)$ and $e \in E$ such that $\alpha(K) = K'$ and $\alpha(T) = eT'$ if and only if there exist $\gamma \in \mathrm{Aut}(N)$, $\theta \in \mathrm{Aut}(G)$ and $a \in G$ such that*

$$\varepsilon = \gamma \circ ((\varepsilon' \cdot a^{-1}) \circ \theta) \circ \gamma^{-1}, \tag{8.4}$$

$$\psi = \gamma \circ (\psi' \cdot a^{-1}) \circ (\theta \times \theta), \tag{8.5}$$

where $e = \iota'(\psi'(a, a^{-1})^{-1})t'_a$, $(\psi' \cdot a^{-1}, \varepsilon' \cdot a^{-1}) = (\psi', \varepsilon') \cdot \phi_{a^{-1}}^{-1}$ and $\phi_{a^{-1}}(x) = \psi'(a, a^{-1})^{-1}\psi'(a, a^{-1}x)$, $x \in G$.

Proof. [161, Theorem 1] Straightforward adaptation of the proof of Theorem 8.3 using left translation $T^* = e\,T'$, where $e = k't'_a$, to the normal case, suffices. □

The action of $\phi^{-1}_{a^{-1}}$ on (ψ', ε') in Theorem 8.5, derived from translation by e and renormalisation, is our *shift action* by the element a^{-1} of G. Now we separate it from the action resulting from automorphism α.

DEFINITION 8.6 *Let* $(\psi, \varepsilon) \in F^2(G, N)$. *For each* $s \in G$, $\psi_s \in C^1(G, N)$ *is defined by*

$$\psi_s(x) = \psi(s^{-1}, sx)^{-1}\psi(s^{-1}, s) = \psi(s, x)^{\varepsilon(s)^{-1}}, \quad x \in G. \tag{8.6}$$

The shift action *of* s *on* (ψ, ε) *is* $(\psi, \varepsilon) \cdot s = (\psi, \varepsilon) \cdot \psi_s \in F^2(G, N)$ *(see (7.11))*. *That is,* $(\psi, \varepsilon) \cdot s = (\psi \cdot s, \varepsilon \cdot s)$ *is*

$$(\varepsilon \cdot s)(x) = \varepsilon(s)^{-1}\varepsilon(sx), \tag{8.7}$$

$$(\psi \cdot s)(x, y) = \left(\psi^{-1}(s, y)^{\varepsilon(sx)\varepsilon(s)^{-1}} \psi(sx, y)\right)^{\varepsilon(s)^{-1}}. \tag{8.8}$$

The map $s \mapsto \left((\psi, \varepsilon) \mapsto (\psi, \varepsilon) \cdot s\right)$ gives a right action of G on $F^2(G, N)$, with orbits partitioning the equivalence classes $[\psi, \varepsilon] \subseteq F^2(G, N)$. When $\varepsilon \equiv 1$ and N is abelian, we recover (8.3) on setting $s = a^{-1}$.

It is easy to show [118, Theorem 5.5, Corollary 5.7] that the mapping defined on factor pairs according to the next definition gives a right action of $\mathrm{Aut}(N) \times \mathrm{Aut}(G)$ on $F^2(G, N)$ which preserves (v, w, k, λ)-orthogonality. It generalises the $\mathrm{Aut}(C) \times \mathrm{Aut}(G)$ action on cocycles, implicit in Theorem 8.3, to factor pairs.

DEFINITION 8.7 *Let* $(\psi, \varepsilon) \in F^2(G, N)$ *and let* $\gamma \in \mathrm{Aut}(N)$, $\theta \in \mathrm{Aut}(G)$. *The* automorphism action *of* (γ, θ) *on* (ψ, ε) *is* $(\psi, \varepsilon)^{(\gamma, \theta)} = (\psi^{(\gamma, \theta)}, \varepsilon^{(\gamma, \theta)}) \in F^2(G, N)$, *where*

$$\varepsilon^{(\gamma, \theta)}(x) = \gamma \circ \varepsilon(x^{\theta^{-1}}) \circ \gamma^{-1}, \quad x \in G, \tag{8.9}$$

$$\psi^{(\gamma, \theta)} = \gamma \circ \psi \circ (\theta^{-1} \times \theta^{-1}). \tag{8.10}$$

Galati [118, Theorem 5.2] derives a variant shift action by adapting the proof of Theorem 8.3 directly to the normal case, using *right* translation. A simpler function $\phi_{a^{-1}}(x) = \psi'(x, a^{-1})$, $x \in G$ results, because of the asymmetry inherent in the representation $e = k't'_a$. That is, he obtains the variant shift action $(\psi, \varepsilon) \odot s = (\psi^s, \varepsilon^s)$, where $\varepsilon^s : G \to \mathrm{Aut}(N)^{\mathrm{op}}$ and $\psi^s : G \times G \to N$ are

$$\varepsilon^s(x) = \varepsilon(xs^{-1})\varepsilon(s^{-1})^{-1}, \tag{8.11}$$

$$\psi^s(x, y) = \psi^{-1}(x, s^{-1})\psi(x, ys^{-1}), \tag{8.12}$$

for all $x, y \in G$. It is plain to see that

$$(\psi, \varepsilon) \cdot s^{-1} = \left((\psi, \varepsilon) \odot s\right)^{(\gamma_s, \theta_s)}, \tag{8.13}$$

where $\gamma_s = \varepsilon(s^{-1})^{-1}$ and $\theta_s = \bar{s}$ are the automorphisms of N and G, respectively, induced by the inner automorphism $\bar{e} = \imath'(\psi'(s, s^{-1})^{-1})t'_s$ of E.

Galati extends Lemma 8.4 to show that his variant shift action preserves (v, w, k, λ)-orthogonality, and also that the RDSs corresponding to (v, w, k, λ)-orthogonal factor pairs in the same shift orbit are equivalent.

LEMMA 8.8 [118, Lemma 5.4] *Let* $(\psi, \varepsilon) \in F^2(G, N)$ *and* $s \in G$, *and let* D *be a* k-*subset of* G. *Then* (ψ, ε) *is* (v, w, k, λ)-*orthogonal with respect to* D *if and only if* (ψ^s, ε^s) *is* (v, w, k, λ)-*orthogonal with respect to* Ds. *When this occurs, the RDS* $R_{(\psi,\varepsilon)} = \{(1, x) : x \in D\} \subseteq E_{(\psi,\varepsilon)}$ *is equivalent to* $R_{(\psi^s,\varepsilon^s)} = \{(1, y) : y \in Ds\} \subseteq E_{(\psi^s,\varepsilon^s)}$, *with* $\alpha_s(R_{(\psi,\varepsilon)})(1, s) = R_{(\psi^s,\varepsilon^s)}$, *where* $\alpha_s : E_{(\psi,\varepsilon)} \to E_{(\psi^s,\varepsilon^s)}$ *is defined by* $\alpha_s(a, x) = (a\psi(x, s^{-1}), x)$.

The variant shift and $\mathrm{Aut}(N) \times \mathrm{Aut}(G)$ actions are combined into a single action on $F^2(G, N)$. Let $\tau : \mathrm{Aut}(N) \times \mathrm{Aut}(G) \to \mathrm{Aut}(G)^{\mathrm{op}}$ be defined by $x^{\tau(\gamma,\theta)} = x^{\theta^{-1}}$. It is readily checked that τ gives a left action of $\mathrm{Aut}(N) \times \mathrm{Aut}(G)$ on G.

COROLLARY 8.9 [118, Lemma 5.8, Corollary 5.9] *There is a right action of the semidirect product* $H = G \rtimes_\tau (\mathrm{Aut}(N) \times \mathrm{Aut}(G))$ *on* $F^2(G, N)$ *given by*

$$(\psi, \varepsilon)^{(s,(\gamma,\theta))} = \left((\psi^s)^{(\gamma,\theta)}, (\varepsilon^s)^{(\gamma,\theta)} \right) = \left((\psi^{(\gamma,\theta)})^{\theta(s)}, (\varepsilon^{(\gamma,\theta)})^{\theta(s)} \right) \quad (8.14)$$

for all $(\psi, \varepsilon) \in F^2(G, N)$ *and* $(s, (\gamma, \theta)) \in H$. *If* (ψ, ε) *is* (v, w, k, λ)-*orthogonal, then* $(\psi, \varepsilon)^{(s,(\gamma,\theta))}$ *is* (v, w, k, λ)-*orthogonal for each* $(s, (\gamma, \theta)) \in H$.

We call the orbits of this action *bundles* of factor pairs. By (8.13) they may equally be expressed in terms of shift action as in terms of variant shift action. If so, we call the resulting action *bundle action* on $F^2(G, N)$.

DEFINITION 8.10 *Let* $(\psi, \varepsilon) \in F^2(G, N)$ *and* H *be as in Corollary 8.9. The bundle* $\mathcal{B}((\psi, \varepsilon))$ *of* (ψ, ε) *is the* H-*orbit* $\mathcal{B}((\psi, \varepsilon)) = \{(\psi, \varepsilon)^h : h \in H\}$; *or, equally,*

$$\mathcal{B}((\psi, \varepsilon)) = \left\{ \left((\psi, \varepsilon) \cdot s \right)^{(\gamma,\theta)} : s \in G, \gamma \in \mathrm{Aut}(N), \theta \in \mathrm{Aut}(G) \right\}. \quad (8.15)$$

Each bundle in $F^2(G, N)$ therefore consists entirely of (v, w, k, λ)-orthogonal factor pairs, for some fixed k and λ, or it contains none, for any parameters k and λ satisfying (4.13).

Research Problem 46 *For what* G *and* N *are there NO bundles of* (v, w, k, λ)-*orthogonal factor pairs, for any* k *and* $\lambda = k(k - 1)/(w(v - 1))$?

It is not known, in general, whether the bundle action on $F^2(G, N)$ gives a finer partition than does equivalence of partial transversals. For full transversals, however, the author's results extend with no surprises, since bundle action is defined exactly by equivalence of transversals (Definition 8.1). There is a bijective correspondence between the *orthogonal* bundles in $F^2(G, N)$ (bundles containing orthogonal factor pairs) and the equivalence classes of semiregular RDSs in a given extension E of N by G.

Since all factor pairs in an equivalence class of $F^2(G, N)$ determine isomorphic extension groups, we define

$$F^2(G, N, E) = \{(\psi, \varepsilon) \in F^2(G, N) : E_{(\psi,\varepsilon)} \cong E\}. \quad (8.16)$$

THEOREM 8.11 [118, Theorem 5.10] *Let N, G and E be finite groups and let $N_1, \ldots, N_r$ be all the distinct normal subgroups of E with $N_i \cong N$ and $E/N_i \cong G$. Let $\mathcal{O}_1, \ldots, \mathcal{O}_t \subseteq F^2(G, N, E)$ be the distinct bundles which consist of (v, w, k, λ)-orthogonal factor pairs. For each $j \in \{1, \ldots, t\}$, fix a representative $(\psi_j, \varepsilon_j) \in \mathcal{O}_j$ and an isomorphism $f_j : E_{(\psi_j, \varepsilon_j)} \to E$. Then $f_j(\overline{N}) = N_{i_j}$ for some $1 \leq i_j \leq r$. Define $\Phi(\mathcal{O}_j) = [f_j(R_{(\psi_j, \varepsilon_j)})]$. Then*

$$\Phi : \{\mathcal{O}_1, \ldots, \mathcal{O}_t\} \to \bigcup_{i=1}^{r} \{[R] \ : \ R \subseteq E \text{ is a } (v, w, k, \lambda) - \text{RDS relative to } N_i\}$$

is surjective.

If $k = v$ and $\mathcal{O}(N, E, G)$ denotes the set of orthogonal bundles in $F^2(G, N, E)$, then $\Phi : \mathcal{O}(N, E, G) \twoheadrightarrow \mathcal{R}(N, E, G)$ is bijective.

Research Problem 47 *For which groups G, N, E and which $1 < k < v$ is Φ in Theorem 8.11 injective?*

As an immediate consequence of Theorem 7.17, each bundle $\mathcal{O}$ of (v, w, k, λ)-orthogonal factor pairs in $F^2(G, N, E)$ defines a set of (v, w, k, λ)-divisible designs $\{\mathcal{D}_{(\psi, \varepsilon)} : (\psi, \varepsilon) \in \mathcal{O}\}$ with regular group E and class regular with respect to N. Conversely, for every (v, w, k, λ)-divisible design $\mathcal{D}$ with regular group E and class regular with respect to N, there exist an RDS R in E relative to N and a $(\psi, \varepsilon) \in F^2(G, N, E)$ such that $[R] = [R_{(\psi, \varepsilon)}]$ and $\mathcal{D} \cong \text{dev}(R) \cong \mathcal{D}_{(\psi, \varepsilon)}$ as designs. For consistency we will call $\{\mathcal{D}_{(\psi, \varepsilon)} : (\psi, \varepsilon) \in \mathcal{O}\}$ a *bundle* of (v, w, k, λ)-divisible designs.

Theorem 8.11 then gives us the following one-to-one mapping of bundles around three stars of the Five-fold Constellation.

COROLLARY 8.12 *Let N, G and E be finite groups of orders w, v (where $w|v$) and vw, respectively. Then the mappings*

$$\mathcal{B}((\psi, \varepsilon)) \leftrightarrow [R_{(\psi, \varepsilon)}] \leftrightarrow \{\mathcal{D}_{(\psi', \varepsilon')} : (\psi', \varepsilon') \in \mathcal{B}((\psi, \varepsilon))\}$$

define one-to-one correspondences between the corresponding sets of bundles of orthogonal factor pairs in $F^2(G, N, E)$, equivalence classes of semiregular RDSs in E relative to N, and bundles of semiregular divisible designs with regular group E and class regular with respect to N.

By using the equivalent form of a coupled G-cocyclic matrix over N given in Corollary 7.19, it is easy to verify from Definition 4.12 and (8.14) that factor pairs in the same bundle determine Hadamard equivalent coupled cocyclic matrices. That is,

$$\mathcal{B}((\psi, \varepsilon)) = \mathcal{B}((\psi', \varepsilon')) \Rightarrow [M_{(\psi, \varepsilon)}] = [M_{(\psi', \varepsilon')}]. \tag{8.17}$$

The equivalence operations on $M_{(\psi, \varepsilon)}$ determined by bundle action on (ψ, ε) are restricted (for instance, not all possible row or column permutations are applied) so that a single Hadamard equivalence class of coupled cocyclic matrices could contain the images of two, or more, distinct bundles of factor pairs.

COROLLARY 8.13 *Set*

$$\mathcal{H}(N,E,G) = \bigcup_{(\psi,\varepsilon)\in F^2(G,N,E)} \left\{ [\, M_{(\psi,\varepsilon)} \,] : M_{(\psi,\varepsilon)} \text{ is a } GH(w,v/w) \right\}.$$

Then $\mathcal{B}((\psi,\varepsilon)) \mapsto [\, M_{(\psi,\varepsilon)} \,]$ *defines a set surjection* $\mathcal{O}(N,E,G) \twoheadrightarrow \mathcal{H}(N,E,G)$.

Research Problem 48 *Under what conditions is the set surjection of Corollary 8.13 an injection?*

Although every equivalence class of coupled G-cocyclic generalised Hadamard matrices over N is the image of at least one bundle of orthogonal factor pairs, Example 7.4.2 proves that not every equivalence class of generalised Hadamard matrices over N contains a coupled G-cocyclic matrix for some G.

In order to study generalised Hadamard matrices over N from the cocyclic point of view, it is clear we must deal with bundles of orthogonal factor pairs. However, orthogonality is an extreme condition to impose on factor pairs. Relaxing it to (v,w,k,λ)-orthogonality still gives a condition preserved by bundles.

In fact, more general statistical properties, of considerable significance for our applications, are invariants of bundles. First, we define the distribution of a normalised function $\Phi : G \times G \to N$, in terms of its first coordinate. As with $C^1(G,N)$, by a slight abuse of terminology we adopt the cochain notation of Definition 6.1 for such functions.

DEFINITION 8.14 *Set* $C^2(G,N) = \{\Phi : G \times G \to N, \Phi(x,1) = \Phi(1,x) = 1, x \in G\}$ *and, for each* $x \in G$, $a \in N$, *set* $N_\Phi(x,a) = |\{y \in G : \Phi(x,y) = a\}|$. *The (row) distribution* $\mathcal{D}(\Phi)$ *of* $\Phi \in C^2(G,N)$ *is the* multiset *(that is, including repetitions) of all frequencies*

$$\mathcal{D}(\Phi) = \{N_\Phi(x,a) : x \in G, a \in N\}. \tag{8.18}$$

Example 8.1.1 *The matrix of Example 4.1.4 represents a mapping from* $\mathbb{Z}_7$ *to the group of* 6th *roots of unity with distribution (by row, where* n^k *means that k values appear n times each in a row)* $\{7; 1^3, 2^2; 1^3, 2^2; 1^3, 2^2; 1^3, 2^2; 1^3, 2^2; 1^3, 2^2\}$.

Similarly, the matrix of (4.4) represents a mapping from any group of order 8 to the group of 4th *roots of unity with distribution* $\{8; 2^4; 2^4; 2^4; 4^2; 2^4; 2^4; 2^4\}$.

The distribution of a factor pair is an invariant of its bundle. This important result explains the significance of bundles in cryptographic applications, discussed next in Section 8.2, and again in Chapter 9.2.1 and Chapter 9.5.

THEOREM 8.15 *Let* $(\psi,\varepsilon) \in F^2(G,N)$. *Then* $\mathcal{D}(\varphi) = \mathcal{D}(\psi)$ *for all* $(\varphi,\epsilon) \in \mathcal{B}((\psi,\varepsilon))$.

Proof. Since $N_{\psi(\gamma,\theta)}(x,a) = N_\psi(\theta^{-1}(x),\gamma^{-1}(a))$ for $\theta \in \mathrm{Aut}(G)$ and $\gamma \in \mathrm{Aut}(N)$, and $N_{\psi^s}(x,a) = N_\psi(x,\psi(x,s^{-1})\,a)$ for $s \in G$, the result follows from Corollary 8.9. $\square$

In the next section (from Horadam [161]) we study bundles of splitting factor pairs. These in turn define bundles of functions, which we argue are the natural equivalence classes for functions between groups. Subsequently, we map bundles of splitting factor pairs to the fifth star of the Five-fold Constellation.

8.2 BUNDLES OF FUNCTIONS — THE SPLITTING CASE

From Definition 7.5 for splitting factor pairs, we see that for each homomorphism $\varrho : G \to \mathrm{Aut}(N)^{\mathrm{op}}$ there is a surjection $\partial_\varrho^{-1} : C^1(G, N) \twoheadrightarrow [1, \varrho] \subset F^2(G, N)$ (if $\varrho \equiv 1$, denoted simply ∂^{-1}) defined by

$$\partial_\varrho^{-1}(\phi) = (\partial^{-1}\phi, \overline{\phi}\varrho). \tag{8.19}$$

The preimage of $(1, \varrho)$ under ∂_ϱ^{-1} is a group which is important for our analysis. For $\varrho \equiv 1$, the preimage of $(1, 1)$ is the group of homomorphisms $\mathrm{Hom}(G, N)$. (In the special case that G is abelian with exponent m and N is the cyclic group of complex m^{th} roots of unity, $\mathrm{Hom}(G, N)$ is the character group $\widehat{G}$ of G.) For N abelian, the preimage of $(1, \varrho)$ is the group of 1-cocycles $Z_\varrho^1(G, N)$ (cf. (6.2)), often called the *crossed homomorphisms*, which name we adopt for the general case.

DEFINITION 8.16 *Let $\varrho : G \to \mathrm{Aut}(N)^{\mathrm{op}}$ be a homomorphism. Then $\chi \in C^1(G, N)$ is a ϱ-crossed homomorphism if $\overline{\chi}\varrho = \varrho$ and*

$$\chi(xy) = \chi(x)\, \chi(y)^{\varrho(x)} \ \left(= \chi(y)^{\varrho(x)}\, \chi(x) \right), \quad x, y \in G.$$

Denote the subgroup of ϱ-crossed homomorphisms in $C^1(G, N)$ by $\mathrm{Hom}_\varrho(G, N)$.

Though $\mathrm{Hom}_\varrho(G, N)$ may not be a normal subgroup of $C^1(G, N)$, its (left) cosets $\phi\,\mathrm{Hom}_\varrho(G, N)$ are the preimages of the distinct elements in $[1, \varrho]$. The coset mapping $\widehat{\partial_\varrho^{-1}}(\phi\,\mathrm{Hom}_\varrho(G, N)) = \partial_\varrho^{-1}(\phi)$ induced by (8.19) is a set isomorphism,

$$\widehat{\partial_\varrho^{-1}} : \{\phi\,\mathrm{Hom}_\varrho(G, N) : \phi \in C^1(G, N)\} \overset{\cong}{\longrightarrow} [1, \varrho]. \tag{8.20}$$

When N is abelian, $C^1(G, N)$ is abelian and

$$\widehat{\partial_\varrho^{-1}} : C^1(G, N)/\mathrm{Hom}_\varrho(G, N) \rightarrowtail [1, \varrho] = B_\varrho^2(G, N)$$

is a group isomorphism.

The splitting case of Theorem 8.5 may now be extracted without much difficulty, using (8.15).

THEOREM 8.17 *Let $\phi \in C^1(G, N)$ and let $\varrho : G \to \mathrm{Aut}(N)^{\mathrm{op}}$ be a homomorphism. Let $s \in G$, $\theta \in \mathrm{Aut}(G)$ and $\gamma \in \mathrm{Aut}(N)$. Define the shift $\phi \cdot s \in C^1(G, N)$ of ϕ by s to be*

$$(\phi \cdot s)(x) = \left(\phi(s)^{-1}\phi(sx)\right)^{\varrho(s^{-1})}, \quad x \in G, \tag{8.21}$$

and define $\phi^{(\gamma, \theta)} \in C^1(G, N)$ to be

$$\phi^{(\gamma, \theta)}(x) = (\gamma \circ \phi \circ \theta^{-1})(x), \quad x \in G. \tag{8.22}$$

Suppose $(\psi, \varepsilon) = (\partial^{-1}\phi, \overline{\phi}\varrho) \sim_\phi (1, \varrho)$. Then

1. *$(\psi, \varepsilon) \cdot s = (\partial^{-1}(\phi \cdot s), \overline{(\phi \cdot s)}\,\varrho) \sim_{\phi \cdot s} (1, \varrho)$;*

2. *$(\psi, \varepsilon)^{(\gamma, \theta)} = (\partial^{-1}\phi', \overline{\phi'}\,\varrho') \sim_{\phi'} (1, \varrho')$, where $\phi' = \phi^{(\gamma, \theta)}$, $\varrho' = \varrho^{(\gamma, \theta)}$.*

Proof. Since $(\psi, \varepsilon) \sim_\phi (1, \varrho)$ and $(\psi, \varepsilon) \cdot s \sim_{\psi_s^{-1}} (\psi, \varepsilon)$ by Theorem 8.5, so $(\psi, \varepsilon) \cdot s \sim_{\psi_s^{-1}\phi} (1, \varrho)$. Therefore, by (7.16),

$$(\psi_s^{-1}\phi)(x) = (\phi(s)^{-1}\phi(sx))^{\varrho(s^{-1})} = (\phi \cdot s)(x),$$

giving part 1, and part 2 follows from (8.9) since $\overline{\phi'(x)} = (\overline{\phi})^{(\gamma,\theta)}$. $\square$

Hence the bundle of a splitting factor pair consists entirely of splitting factor pairs:

$$\mathcal{B}((\partial^{-1}\phi, \overline{\phi}\varrho)) = \left\{ (\partial^{-1}\phi', \overline{\phi'}\,\varrho') \; : \; \varrho' = \varrho^{(\gamma,\theta)}, \, \phi' = (\phi \cdot s)^{(\gamma,\theta)}, \right.$$
$$\left. \gamma \in \mathrm{Aut}(N), \, \theta \in \mathrm{Aut}(G), \, s \in G \right\}. \quad (8.23)$$

Consequently, the set of splitting factor pairs partitions into disjoint bundles. However, it is important to recognise that a bundle of splitting factor pairs cuts across equivalence classes of splitting factor pairs, and vice versa. In fact, for a particular homomorphism $\varrho : G \to \mathrm{Aut}(N)^{\mathrm{op}}$ we have

$$\bigvee_{\theta,\gamma} [1, \varrho^{(\gamma,\theta)}] = \bigvee_\phi \mathcal{B}(\partial_\varrho^{-1}(\phi)).$$

Putting (8.20) and (8.23) together, we say two functions $\varphi, \phi \in C^1(G, N)$ are *equivalent relative to ϱ* if there exist $\theta \in \mathrm{Aut}(G)$ and $\gamma \in \mathrm{Aut}(N)$ such that $\mathcal{B}(\partial_{\varrho'}^{-1}(\varphi)) = \mathcal{B}(\partial_\varrho^{-1}(\phi))$, where $\varrho' = \varrho^{(\gamma,\theta)}$. Theorem 8.17 gives the following more workable definition.

DEFINITION 8.18 *Let $\varrho : G \to \mathrm{Aut}(N)^{\mathrm{op}}$ be a homomorphism. Two functions $\varphi, \phi \in C^1(G, N)$ are equivalent relative to ϱ if there exist $s \in G$, $\theta \in \mathrm{Aut}(G)$, $\gamma \in \mathrm{Aut}(N)$ and $f \in \mathrm{Hom}_{\varrho^{(\gamma,\theta-1)}}(G, N)$ such that*

$$\varphi = (\gamma \circ (\phi \cdot s) \circ \theta) f. \quad (8.24)$$

The shift action of s on $C^1(G, N)$ is defined by $\phi \mapsto \phi \cdot s$. The equivalence class $\mathbf{b}(\phi, \varrho)$ of ϕ relative to ϱ is called its bundle relative to ϱ. *That is, $\mathbf{b}(\phi, \varrho) =$*

$$\left\{ (\gamma \circ (\phi \cdot s) \circ \theta) f \; : \; f \in \mathrm{Hom}_{\varrho^{(\gamma,\theta-1)}}(G, N), \theta \in \mathrm{Aut}(G), \gamma \in \mathrm{Aut}(N), s \in G \right\}.$$
$$(8.25)$$

In particular, if $\varrho \equiv 1$, so $\varrho^{(\gamma,\theta^{-1})} \equiv 1$, the bundle $\mathbf{b}(\phi) = \mathbf{b}(\phi, 1)$ of ϕ is

$$\mathbf{b}(\phi) = \left\{ (\gamma \circ (\phi \cdot s) \circ \theta) f \; : \; f \in \mathrm{Hom}(G, N), \, \theta \in \mathrm{Aut}(G), \, \gamma \in \mathrm{Aut}(N), \, s \in G \right\}.$$
$$(8.26)$$

As in the general case (8.14), bundle action on functions is a shift action followed by an automorphism action, or vice versa.

COROLLARY 8.19 *Let $\varrho : G \to \mathrm{Aut}(N)^{\mathrm{op}}$ be a homomorphism and let $\phi \in C^1(G, N)$. For every $s \in G$, $\theta \in \mathrm{Aut}(G)$ and $\gamma \in \mathrm{Aut}(N)$,*

$$\mathbf{b}(\phi, \varrho) = \mathbf{b}(\phi \cdot s, \varrho),$$
$$\mathbf{b}(\phi, \varrho) = \mathbf{b}(\gamma \circ \phi \circ \theta, \gamma \circ \varrho(\theta) \circ \gamma^{-1}).$$

For each homomorphism $\varrho : G \to \mathrm{Aut}(N)^{\mathrm{op}}$, the group of all normalised functions $C^1(G, N)$ therefore partitions into disjoint bundles relative to ϱ,

$$C^1(G, N) = \bigvee_\phi \mathbf{b}(\phi, \varrho).$$

COROLLARY 8.20 *Let* $\varrho : G \to \mathrm{Aut}(N)^{\mathrm{op}}$ *be a homomorphism. The mapping* $\widehat{\partial}_\varrho^{-1} : \mathbf{b}(\phi, \varrho) \mapsto \mathcal{B}(\partial_\varrho^{-1}(\phi))$ *induced by (8.20) on bundles is a set isomorphism*

$$\widehat{\partial}_\varrho^{-1} : \{\mathbf{b}(\phi, \varrho) : \phi \in C^1(G, N)\} \xrightarrow{\cong} \{\mathcal{B}(\partial_\varrho^{-1}(\phi)) : \phi \in C^1(G, N)\}.$$

When N is abelian, a more appealing 'positive' version of (8.20) and Corollary 8.20 is available to us. Define the *coboundary operator* on $C^1(G, N)$ to be ∂_ϱ : $\phi \mapsto (\partial\phi, \varrho)$, so there is an induced surjection $\partial_\varrho : C^1(G, N) \twoheadrightarrow [1, \varrho]$ and the coset mapping

$$\widehat{\partial}_\varrho : C^1(G, N)/\mathrm{Hom}_\varrho(G, N) \rightarrowtail [1, \varrho] = B_\varrho^2(G, N) \qquad (8.27)$$

is a group isomorphism. Furthermore, $\partial_\varrho(\phi^{-1}) = \partial_\varrho^{-1}(\phi)$. (When $\varrho \equiv 1$, simply write ∂ and $\widehat{\partial}$.)

COROLLARY 8.21 *Suppose N is abelian and $\varrho : G \to \mathrm{Aut}(N)^{\mathrm{op}}$ is a homomorphism. Then* $\mathbf{b}(\phi, \varrho) = \mathbf{b}(\phi^{-1}, \varrho)$, *and the coboundary operator* $\widehat{\partial}_\varrho : \mathbf{b}(\phi, \varrho) \mapsto \mathcal{B}(\partial_\varrho(\phi))$ *induced by ∂_ϱ on bundles, is the isomorphism* $\widehat{\partial}_\varrho^{-1}$ *of Corollary 8.20.*

Proof. Since the inversion permutation $a \mapsto a^{-1}$, $a \in N$, is an automorphism $\gamma \in \mathrm{Aut}(N)$, $\phi^{-1} = \gamma \circ \phi$ and by (8.25) $\phi^{-1} \in \mathbf{b}(\phi, \varrho)$. Moreover, $\partial_\varrho^{-1}(\phi) = (\partial^{-1}\phi, \varrho) = (\partial(\phi^{-1}), \varrho) = ((\partial\phi)^{-1}, \varrho) = (\gamma \circ \partial\phi, \varrho) \in \mathcal{B}(\partial_\varrho(\phi))$. $\qquad\square$

In other words, when N is abelian it is unnecessary to distinguish between ϕ and ϕ^{-1}, up to bundle equivalence.

In the Five-fold Constellation (see pentagon **ns** in Figure 7.1, Theorems 7.30 and 7.35), Corollary 8.20 determines a one-to-one correspondence between equivalence classes of PN functions relative to ϱ (the fifth star) and bundles of orthogonal splitting factor pairs (the second star). The latter have already been equated to equivalence classes of splitting semiregular RDSs (the third star) and their corresponding bundles of divisible designs (the fourth star) in Corollary 8.12. By (8.17) they all map onto equivalence classes of coupled G-developed generalised Hadamard matrices (the first star).

This is the first of many reasons to conclude that bundles are the natural equivalence classes for normalised functions.

The careful reader may be wondering what happens for un-normalised functions $f : G \to N$, where $f(1) \neq 1$. Necessarily any un-normalised f determines the normalised function $\phi \in C^1(G, N)$ where $\phi(x) = f(1)^{-1}f(x)$. In fact, when the definition of shift action in (8.21) is extended to un-normalised functions; that is to say, if for $s \in G$, the *shift action* $f \cdot s$ of s on f is given by

$$(f \cdot s)(x) = \left(f(s)^{-1}f(sx)\right)^{\varrho(s^{-1})}, \ x \in G, \qquad (8.28)$$

then the normalisation of f is clearly $f \cdot 1$. If f is normalised, then $f \cdot 1 = f$.

Thus, we use (8.28) to extend equivalence of normalised functions (Definition 8.18) to *affine* equivalence of un-normalised functions.

DEFINITION 8.22 *Define* $UC^1(G, N) = \{f : G \to N\}$ *and let* $\varrho : G \to \mathrm{Aut}(N)^{\mathrm{op}}$ *be a homomorphism. The normalisation of $f \in UC^1(G, N)$ is $f \cdot 1 \in$*

$C^1(G, N)$. *Two functions* f, $f' \in UC^1(G, N)$ *are* affinely equivalent relative to ϱ *if their normalisations* $f \cdot 1$, $f' \cdot 1 \in C^1(G, N)$ *are equivalent relative to ϱ. The affine bundle* $\mathbf{b}(f, \varrho)$ *of* $f \in UC^1(G, N)$ *relative to ϱ is*

$$\mathbf{b}(f, \varrho) = \{f' \in UC^1(G, N), \ f' \cdot 1 \in \mathbf{b}(f \cdot 1, \ \varrho)\}.$$

The second reason to conclude that (affine) bundles are the appropriate equivalence classes for functions is that the Five-fold Constellation is an optimal expression of a fundamental statistical invariance within bundles, inherited from Theorem 8.15. Consider the difference distribution of f relative to ϱ, which is significant in several nonlinearity measures for functions.

DEFINITION 8.23 *Let* $f : G \to N$, $\varrho : G \to \operatorname{Aut}(N)^{\mathrm{op}}$ *be a homomorphism, and for each* $x \in G$ *and* $a \in N$, *set*

$$n_{(f, \varrho)}(x, a) = |\{y \in G : f(xy)(f(y)^{-1})^{\varrho(x)} = a\}|.$$

The difference distribution *of* f *relative to ϱ is the multiset of all frequencies*

$$\mathcal{D}(f, \varrho) = \{n_{(f, \varrho)}(x, a) : x \in G, a \in N\}. \tag{8.29}$$

The difference distribution of a function relative to ϱ equals that of its normalisation, which equals that of its image under ∂_ϱ^{-1} and is an invariant of its affine bundle relative to ϱ.

THEOREM 8.24 *Let* $\varrho : G \to \operatorname{Aut}(N)^{\mathrm{op}}$ *be a homomorphism and* $\phi \in C^1(G, N)$, *so that* $\partial_\varrho^{-1}(\phi) = (\partial^{-1}\phi, \overline{\phi}\varrho)$. *Then*

1. *if* $f : G \to N$ *and* $\phi = f \cdot 1$, *then* $\mathcal{D}(f, \varrho) = \mathcal{D}(\phi, \varrho)$;

2. $\mathcal{D}(\phi, \varrho) = \mathcal{D}(\partial^{-1}\phi)$;

3. *if* $\mathbf{b}(\phi, \varrho) = \mathbf{b}(\varphi, \varrho')$, *then* $\mathcal{D}(\phi, \varrho) = \mathcal{D}(\varphi, \varrho')$;

4. *if* N *is abelian, then* $\mathcal{D}(\phi, \varrho) = \mathcal{D}(\partial\phi)$.

Proof. For part 1, $n_{(f, \varrho)}(x, a) = n_{(\phi, \varrho)}(x, f(1)^{-1}af(1)^{\varrho(x)})$. Set

$$\hat{n}_{(\phi, \varrho)}(x, a) = |\{y \in G : \phi(y)^{\varrho(x)}\phi(xy)^{-1} = a\}| = n_{(\phi, \varrho)}(x, a^{-1}),$$

so that $\mathcal{D}(\phi, \varrho) = \{\hat{n}_{(\phi, \varrho)}(x, a) : x \in G, a \in N\}$.
For part 2, if $f \in \operatorname{Hom}_\varrho(G, N)$, then $\hat{n}_{(\phi f, \varrho)}(x, a) = \hat{n}_{(\phi, \varrho)}(x, f(x)\,a)$, so $\mathcal{D}(\phi, \varrho)$ $= \mathcal{D}(\phi f, \varrho)$. Then, $\hat{n}_{(\phi, \varrho)}(x, a) = |\{y \in G : \phi(x)\phi(y)^{\varrho(x)}\phi(xy)^{-1} = \phi(x)a\}| = N_{\partial^{-1}\phi}(x, \phi(x)a)$, by (7.16) and Definition 8.14. Thus $\mathcal{D}(\phi, \varrho) = \mathcal{D}(\partial^{-1}\phi)$.
For part 3, ϕ is equivalent to φ relative to ϱ if and only if there exist $\theta \in \operatorname{Aut}(G)$ and $\gamma \in \operatorname{Aut}(N)$ such that $\mathcal{B}(\partial_{\varrho'}^{-1}(\varphi)) = \mathcal{B}(\partial_\varrho^{-1}(\phi))$, where $\varrho' = \gamma \circ \varrho(\theta) \circ \gamma^{-1}$, if and only if $\mathcal{D}(\partial^{-1}\phi) = \mathcal{D}(\partial^{-1}\varphi)$, by Theorem 8.15.
Corollary 8.21 gives part 4. $\qquad \square$

A third reason to presume that affine bundles are the natural equivalence classes for functions is their familiarity: affine bundle equivalence is recognisable when $\varrho \equiv 1$ and $G = N$. It includes equivalence of planar functions, affine equivalence of Boolean functions and linear equivalence of cryptographic functions. This familiar case is detailed in Chapter 9.2, where it is used to argue that the results of this Section form the correct framework for studying nonlinearity of functions between groups.

8.3 BUNDLES OF COCYCLES — THE CENTRAL CASE

Clearly, in order to identify bundles in $F^2(G, N, E)$, it is necessary to fix the isotypes N, E and G, such that $N \overset{\iota}{\rightarrowtail} E \overset{\pi}{\twoheadrightarrow} G$ is an extension of N by G.

A further subtlety in classification of these bundles is glossed over in Theorem 8.11. We know equivalent factor pairs in $F^2(G, N)$ determine equivalent extensions (Lemma 7.11) and thus extension groups with the same isotype E, but it is also possible for inequivalent factor pairs to belong to $F^2(G, N, E)$.

The purpose of this Section is to characterise several significant properties of bundles, for the central case with abelian $N = C$ and $\varepsilon \equiv 1$. First, a deeper analysis of the interconnection between the two components of bundle action, by automorphisms and by shifts, is undertaken.

8.3.1 Automorphism action versus shift action

Now suppose $N \cong C$ is central in E, with $E/N \cong G$. Theorem 8.17.2 may be applied to show that automorphism action preserves cohomology classes; more particularly, if $(\gamma, \theta) \in \mathrm{Aut}(C) \times \mathrm{Aut}(G)$ and $\varphi = \psi \, \partial \phi$, then $\varphi^{(\gamma, \theta)} = \psi^{(\gamma, \theta)} \, \partial(\phi^{(\gamma, \theta)})$.

DEFINITION 8.25 *For $[\psi] \in H^2(G, C)$, denote its $(\mathrm{Aut}(C) \times \mathrm{Aut}(G))$-orbit by $\mathcal{A}([\psi])$. Denote by $\mathcal{A}(C, E, G)$ the set of $\mathcal{A}([\psi])$ which determine extension groups isomorphic to E.*

The number $|\mathcal{A}(C, E, G)|$ depends on the way in which C is embedded in E. By [112, Theorem 2.2], extension groups determined by the equivalence classes of central extensions corresponding to $[\psi]$, $[\varphi] \in H^2(G, C)$ are isomorphic by an isomorphism which preserves the images of C if and only if $\mathcal{A}([\psi]) = \mathcal{A}([\varphi])$. Furthermore, $\mathcal{A}(B^2(G, C)) = \{B^2(G, C)\}$, so there is an isomorphism $E_\psi \cong C \times G$ which preserves the images of C if and only if ψ is a coboundary.

COROLLARY 8.26 *Let E be a group containing a central subgroup N isomorphic to C. Then $|\mathcal{A}(C, E, G)|$ is equal to the number ν of central subgroups N_i of E such that $N_i \cong C$, $E/N_i \cong G$, $1 \leq i \leq \nu$, but no automorphism of E maps N_i to N_j for any $i \neq j$. If every isomorphism from N to a central subgroup M of E with $E/N \cong E/M$ extends to an automorphism of E, then $|\mathcal{A}(C, E, G)| = 1$.*

Probably the smallest E with $|\mathcal{A}(C, E, G)| \geq 2$ are of order 64. An abelian example is given in [156, Example 2.9]. A nonabelian example with $G = C = \mathbb{Z}_2^3$ and $|\mathcal{A}(C, E, G)| = 2$ is given in Example 8.3.3.1 below. In [112, Example 2.3], an example of a nonabelian group E of order 128 with $|\mathcal{A}(C, E, G)| \geq 2$ is given.

However, if E is an abelian p-group and C is an elementary abelian p-group or if E is an abelian group and C is cyclic, then $|\mathcal{A}(C, E, G)| = 1$ [156, §5].

Next, we untangle the relationship between shift action and automorphism action in the construction of bundles. An equivalence between transversals has two components: one derived from an action of $\mathrm{Aut}(E, C)$, the automorphisms of E which restrict to automorphisms on the image of C, and one an E-action defined by translation. Similarly, a bundle equivalence between cocycles as in (8.1) has two components: one an automorphism action of $\mathrm{Aut}(C) \times \mathrm{Aut}(G)$ and one the shift action by G. However, the $\mathrm{Aut}(E, C)$ action on transversals does not correspond solely to the $\mathrm{Aut}(C) \times \mathrm{Aut}(G)$ action on cocycles, nor the shift action by G solely to the translation action by E, so the two components do not act independently.

In the central case, we use (8.8) to extend the definition of shift action to 2-cochains. Shift action on 1-cochains is defined by (8.21), and on arbitrary functions $G \to C$ by (8.28).

DEFINITION 8.27 *The shift action of G on $C^2(G, C)$ is defined for $a \in G$ and $\Phi \in C^2(G, C)$ to be $\Phi \cdot a \in C^2(G, C)$, where*

$$(\Phi \cdot a)(g, h) = \Phi(ag, h)\, \Phi(a, h)^{-1}, \; g, h \in G. \tag{8.30}$$

The shift action of G on $UC^1(G, C)$, for $a \in G$ and $f \in UC^1(G, C)$, is

$$(f \cdot a)(g) = f(ag)\, f(a)^{-1}, \; g \in G. \tag{8.31}$$

Note from (8.3) that if $\psi \in Z^2(G, C)$ then $\psi \cdot a = \psi \, \partial \psi_a$. If $\partial \phi \in B^2(G, C)$ then $(\partial \phi) \cdot a = \partial(\phi \cdot a)$. We could just as easily work with a left action of G on $Z^2(G, C)$. If $(a \cdot \psi)(g, h) = (\psi \cdot a)(a^{-1}ga, a^{-1}ha)$, so that $(a \cdot \psi)(g, h) = \psi(g, ha)\psi(g, a)^{-1}$, we return the right action $\psi^{a^{-1}}$ of (8.12). Similarly, for $f \in UC^1(G, C)$, we could work with $(a \cdot f)(g) = (f \cdot a)(a^{-1}ga) = f(ga)f(a)^{-1}$, $\forall a, g \in G$ and then for $\phi \in C^1(G, C)$, $a \cdot \partial \phi = \partial(a \cdot \phi)$.

When G is abelian, there is no need to choose between right and left versions of shift action, since they coincide for cocycles and also for 1-cocycles. We might even ask if higher dimensional shift actions exist.

Research Problem 49 *Suppose G is abelian. Is there an analogue of shift action for n-cocycles (Definition 6.1) if $n \geq 3$?*

The exact conversion from shift-equivalent cocycles to equivalent transversals can now be specified.

LEMMA 8.28 [159, Lemma 2.2] *In the central extension $C \overset{\iota}{\rightarrowtail} E \overset{\pi}{\twoheadrightarrow} G$ let $T = \{t_g, \; g \in G\}$ and $T' = \{t'_g, \; g \in G\}$ be normalised transversals of $\iota(C)$ in E with $\pi(t_g) = \pi(t'_g) = g$, $g \in G$. Let ψ and ψ' be the cocycles defined in (7.18) by T and T', respectively. Let $\zeta : G \to C$ be a homomorphism and let e be an element in E. Write e uniquely as $e = \iota(c)t'_a$, $c \in C$, $a \in G$, and define $\phi : G \to C$ by $\phi(g) = \psi'(a^{-1}, g)^{-1}$, $g \in G$.*
Then $\iota(\zeta)T = eT' \Leftrightarrow \psi' = \psi \cdot a$, $c = \phi(a)$ and $\zeta(g) = \phi(g)\iota^{-1}(t_g^{-1}t'_g)$, $g \in G$.

Proof. Write $T^* = \iota(\zeta)T$, that is, let $t_g^* = \iota(\zeta(g))\, t_g$, $g \in G$, so T^* and T determine the same cocycle $\psi^* = \psi$, and since the mapping $(c, g) \mapsto (c\, \zeta(g), g)$ is an automorphism from E_ψ to E_{ψ^*}, with inverse $(d, h) \mapsto (d\, \zeta(h)^{-1}, h)$, it defines the automorphism α of E given by $\alpha(\iota(c)t_g) = \iota(c\, \zeta(g))t_g = \iota(c)t_g^*$, with $\alpha(T) = T^*$, which induces the identity of $\mathrm{Aut}(C) \times \mathrm{Aut}(G)$. The result follows on setting $\gamma = \theta = 1$ in Theorem 8.3. $\qquad\square$

The elements of $\mathrm{Aut}(E)$ which leave $\iota(C)$ invariant and induce the identity of $\mathrm{Aut}(C) \times \mathrm{Aut}(G)$ are precisely those of the form α in the above proof, and together form a subgroup IdAut(E, C) of Aut(E, C). This means that the shift action by G on cocycles corresponds exactly to the combination of the translation action by E and the IdAut(E, C) action on transversals.

8.3.2 A taxonomy for central semiregular RDSs

Very little information is available on whether known RDSs with the same parameters are equivalent. The literature contains a huge number of inventive construction methods for RDSs but often no analysis of whether or not genuinely new examples have been found. Theorem 8.11 shows that the problem of listing equivalence classes of RDSs is the same as the problem of listing bundles of orthogonal factor pairs.

A complete classification scheme or taxonomy for equivalence classes of central semiregular RDSs — and of course for the corresponding sets in the Five-fold Constellation (Corollary 7.31 and Lemma 7.38) — derives from the preceding sections. It confirms that the study of semiregular RDSs is not sufficiently defined by the pair E and C, but depends intrinsically on knowledge of the triple of groups (C, E, G) in a central extension $1 \to C \to E \to G \to 1$.

The mere existence of the classification scheme does not mean it is populated! For only one infinite family $\{C = \mathbb{Z}_p,\ G = \mathbb{Z}_p,\ p \text{ prime}\}$ for which central semiregular RDSs are known always to exist, is the classification complete. The challenge remains to complete the classification for any other (family of) triples (C, E, G).

The only known $(v, w, v, v/w)$-RDSs in groups E which are not p-groups have $|C| = w = 2^n$ or $w = 3$ [75, p. 70]. When $w = 2$ we are in the familiar situation of $(4t, 2, 4t, 2t)$-RDSs (Corollary 7.32 and the binary case of Lemma 7.38 — pentagon **b** in Figure 7.1).

Now to describe the classification: a transversal T of a central subgroup N of E may be partly specified by five parameters (v, w, C, E, G), where $v = |T| = |G|$; w, where $w|v$, is the order of C (the isotype of N); E is the (isotype of the) group of order vw containing T; and G is the (isotype of the) group E/N of order v.

Two further variables are needed to identify uniquely the equivalence class containing T, which will consist entirely of central semiregular RDSs, or contain none. The first of these, by Section 8.3.1, is an index specifying which $(\text{Aut}(C) \times \text{Aut}(G))$-orbit in $H^2(G, C)$ contains $[\psi_T]$. Suppose $\mathcal{A}(C, E, G) = \{\mathcal{A}_1, \ldots, \mathcal{A}_\nu\}$, where $\nu = |\mathcal{A}(C, E, G)|$. Thus $\mathcal{A}([\psi_T]) = \mathcal{A}_i$ for a uniquely specified index i. Finally, because shift orbits lie entirely within cohomology classes, the set of all cocycles within the cohomology classes in each orbit $\mathcal{A}_i$ itself partitions into bundles of cocycles, say $\mathcal{B}_{ij}$, $1 \le j \le b_i$. With a slight abuse of notation, write $\mathcal{A}_i = \bigcup_{j=1}^{b_i} \mathcal{B}_{ij}$, $1 \le i \le \nu$.

THEOREM 8.29 [156, §6] *Let T be a transversal of a central subgroup N of E. The equivalence class $[T]$ is uniquely specified by the 7-variable parameter set*

$$\langle v, w, C, E, G, i, j \rangle,\tag{8.32}$$

where $v = |T| = |G|$; w, where $w|v$, is the order of C (the isotype of N); E is the (isotype of the) group of order vw containing T; G is the (isotype of the) group

E/N of order v; i indexes the $(\mathrm{Aut}(C) \times \mathrm{Aut}(G))$-orbit $\mathcal{A}_i = \mathcal{A}([\psi_T])$ containing $[\psi_T]$ within $\mathcal{A}(C, E, G)$; and j indexes the bundle $\mathcal{B}_{ij} = \mathcal{B}(\psi_T)$ containing ψ_T within $\mathcal{A}_i$. If $|\mathcal{A}(C, E, G)| = 1$, the sixth parameter i will be deleted from the classification.

This taxonomy suggests the following research program for classification of equivalence classes of central semiregular RDSs.

Research Problem 50 *Given parameters (v, w), where $w|v$, fix C in (8.32). Suppose a central $(v, w, v, v/w)$-RDS relative to C, or an equivalent object from the Five-fold Constellation, is known to exist. Complete the classification of (8.32), determining*

1. *the isotypes possible for each of E and G;*

2. *for each triple (C, E, G), the number of automorphism orbits $|\mathcal{A}(C, E, G)|$;*

3. *for $1 \le i \le |\mathcal{A}(C, E, G)|$, the number b_i of bundles in $\mathcal{A}_i$;*

4. *for $1 \le j \le b_i$, an RDS R such that $\psi_R \in \mathcal{B}_{ij}$ or, equivalently, a cocycle $\psi_{ij} \in \mathcal{B}_{ij}$, from which R can be calculated by Lemma 7.8.*

We know a few basic results. Since any group E of order p^2, where p is prime, must be abelian, either $E = \mathbb{Z}_p^2$ or $E = \mathbb{Z}_{p^2}$ and any semiregular RDS in E must be abelian (7.27).

Example 8.3.1 [232, Results 2.1, 2.2] *For each prime p, there is a single equivalence class of $(p, p, p, 1)$-RDSs,*

$$\langle p, p, \mathbb{Z}_p, E, \mathbb{Z}_p, \mu \rangle,$$

where $E = \mathbb{Z}_4$ if $p = 2$; $E = \mathbb{Z}_p^2$ if p is odd; and μ is the multiplication cocycle for $GF(p)$ of Example 6.2.7.

Example 8.3.2 [156, Corollary 4.12] *There are exactly 2 equivalence classes of central $(4, 4, 4, 1)$-RDSs,*

$$\langle 4, 4, \mathbb{Z}_2^2, \mathbb{Z}_4^2, \mathbb{Z}_2^2, \mu \rangle \quad and \quad \langle 4, 4, \mathbb{Z}_2^2, \mathbb{Z}_4 \ltimes \mathbb{Z}_4, \mathbb{Z}_2^2, \mu_1 \rangle,$$

where μ is the multiplication cocycle for $GF(4)$ of Example 6.2.7 and $\mu_1(g, h) = g^2 h$, $g, h \in GF(4)$. The first is abelian and the second is nonabelian.

The classification program for equivalence classes of central RDSs with parameters $(p^n, p^n, p^n, 1)$, for $C = G = \mathbb{Z}_p^n$, continues in Section 8.3.3.1 and Chapter 9.3.2.

Before proceeding to it, consider the classification program for the other parameter family of critical interest to us: the equivalence classes of the central $(4t, 2, 4t, 2t)$-RDSs. Here we are in the familiar territory of equivalence classes of Hadamard matrices.

8.3.2.1 Example: Equivalence classes of $(4t, 2, 4t, 2t)$-RDSs and Hadamard matrices

Since $C = \mathbb{Z}_2$ is fixed and $\text{Aut}(C) = 1$, the bundle of any $\psi \in Z^2(G, \mathbb{Z}_2)$ is $\mathcal{B}(\psi) = \{(\psi \cdot a) \circ (\theta \times \theta) : \theta \in \text{Aut}(G), a \in G\}$. If $E \cong E_\psi$ is abelian (so G is abelian and ψ is symmetric), then $|\mathcal{A}(\mathbb{Z}_2, E, G)| = 1$ by [156, §5] and the sixth parameter may be deleted.

The classification program therefore asks for a complete list $\langle 4t, 2, \mathbb{Z}_2, E, G, j \rangle$, $1 \leq j \leq b$ when E is abelian, and $\langle 4t, 2, \mathbb{Z}_2, E, G, i, j \rangle$, $1 \leq j \leq b_i, 1 \leq i \leq |\mathcal{A}(\mathbb{Z}_2, E, G)|$ when E is nonabelian. For simplicity, the first three parameters may be suppressed, and the equivalence class $\langle 4t, 2, \mathbb{Z}_2, E, G, i, j \rangle$ of RDSs may be represented by $\langle E, G, \psi_{ij} \rangle$, $\mathcal{B}(\psi_{ij}) = \mathcal{B}_{ij} \in \mathcal{A}_i, \mathcal{A}_i \in \mathcal{A}(\mathbb{Z}_2, E, G)$.

By Corollary 8.13, for each allowable pair of isotypes $\langle E, G \rangle$ there is a set surjection

$$S\langle E, G \rangle : \{\langle E, G, \psi_{ij} \rangle, \mathcal{B}_{ij} \in \mathcal{A}(\mathbb{Z}_2, E, G)\} \twoheadrightarrow \{[M_{\psi_{ij}}] : \mathcal{B}_{ij} \in \mathcal{A}(\mathbb{Z}_2, E, G)\},$$

where $S\langle E, G \rangle(\mathcal{B}_{ij}) = [M_{\psi_{ij}}]$, from the set of equivalence classes of RDSs to the corresponding set of equivalence classes of cocyclic Hadamard matrices.

Under these surjections we know, from Chapter 6.5.1, that there exist bundles for which $S\langle E_1, D_{4t} \rangle(\mathcal{B}_1) = S\langle E_2, \mathbb{Z}_2^2 \times \mathbb{Z}_t \rangle(\mathcal{B}_2)$, but also for which $S\langle E, D_{4t} \rangle(\mathcal{B}) \neq S\langle E', D_{4t} \rangle(\mathcal{B}')$ for any $\mathcal{B}'$. So the problem of identifying distinct equivalence classes of cocyclic Hadamard matrices (cf. Research Problem 43) is very complex, but the classification program gives us tools to make sense of it. The D_{4t}-cocyclic matrices, being prolific sources of (Ito) Hadamard matrices, seem the best place to start.

Research Problem 51 *Determine the isotypes of all extension groups E of $\mathbb{Z}_2$ by D_{4t}. For each such E, what are the distinct Hadamard equivalence classes in the image of $S\langle E, D_{4t} \rangle$?*

8.3.3 Bundles with trivial shift action — the multiplicative cocycles

The simplicity of the shift action means that it is easy to characterise the cocycles in the fixed point (or, G-stable) subgroup

$$\mathcal{C}_Z(G) = \{\psi \in Z^2(G, C) : \psi \cdot g = \psi, \forall g \in G\} \subseteq Z^2(G, C)$$

as the multiplicative cocycles and similarly for $\mathcal{C}_B(G) \subseteq B^2(G, C)$. Therefore, the bundle of a multiplicative cocycle is just its $(\text{Aut}(C) \times \text{Aut}(G))$-orbit.

LEMMA 8.30 *Given groups G and abelian C, let $M^2(G, C) \subset Z^2(G, C)$ denote the subgroup of multiplicative cocycles (Definition 6.5.3). Then*

1. *$\mathcal{C}_Z(G) = M^2(G, C)$, $\mathcal{C}_B(G) = B^2(G, C) \cap M^2(G, C)$ and*

2. *if $\psi \in M^2(G, C)$ then $\mathcal{B}(\psi) = \{\gamma \circ \psi \circ (\theta \times \theta), \gamma \in \text{Aut}(C), \theta \in \text{Aut}(G)\}$.*

Proof. For $g \in G$, $\psi \cdot g = \psi \Leftrightarrow \partial \psi_g = 1 \Leftrightarrow \psi(g, kh) = \psi(g, k)\psi(g, h)$, $\forall k, h \in$
G. This holds $\forall g \in G \Leftrightarrow \psi$ is multiplicative, giving part 1. $\qquad\square$

The power of the ideas above is illustrated by application to a single multiplicative orthogonal cocycle: the multiplication μ of Example 6.2.7 in a finite field $GF(p^n)$. Thus $G = C = (GF(p^n), +) \cong \mathbb{Z}_p^n$ and the bundle of $\mu \in M^2(G, G)$ is its $(\mathrm{Aut}(G) \times \mathrm{Aut}(G))$-orbit.

8.3.3.1 Example: Equivalence classes of central $(p^n, p^n, p^n, 1)$-RDSs from field multiplication

For our purposes, if $G = (GF(p^n), +) \cong \mathbb{Z}_p^n$ there is a preferred description of $\mathrm{Aut}(G)$. A *linearized polynomial* of $GF(p^n)$ is a polynomial of the form

$$\lambda(x) = \sum_{i=0}^{n-1} l_i x^{p^i} \in GF(p^n)[x] \qquad (8.33)$$

and is a *linearized permutation polynomial* (LPP) if the function $\lambda : GF(p^n) \to GF(p^n)$ is one-to-one [223, 3.49, 7.1]. Then $\lambda \in \mathrm{Aut}(G)$. The set of LPPs of $GF(p^n)$ constitutes a group, the *Betti-Mathieu* group, under the operation of composition modulo $x^{p^n} - x$. This group is isomorphic to the general linear group $GL(n, p)$ [223, 7.24] and consequently may be identified with $\mathrm{Aut}(G)$.

The field multiplication cocycle $\mu \in Z^2(G, G)$ is orthogonal, and if $\lambda \in \mathrm{Aut}(G)$ then it is easy to check that $\mu_\lambda \in Z^2(G, G)$ defined by $\mu_\lambda(g, h) = \lambda(g) h$ is both multiplicative and orthogonal.

DEFINITION 8.31 *Let* $\lambda(x) = \sum_{i=0}^{n-1} l_i x^{p^i}$ *be an LPP of* $GF(p^n)$ *and let* $G = (GF(p^n), +)$. *The* linearized permutation (LP) *cocycle* $\mu_\lambda : G \times G \to G$ *is defined to be* $\mu_\lambda(g, h) = \lambda(g) h$, *for all* $g, h \in G$. *The cases of monomial* λ *will be* termed power *cocycles and abbreviated* $\mu_i(g, h) = g^{p^i} h$, $i = 0, \ldots, n-1$. *We set* $\mu = \mu_0 = \mu_{\mathrm{Id}}$.

The power cocycle μ_1 for $p = 2$ has already appeared in Example 8.3.2.

We can identify which LP cocycles lie in the same bundle, and therefore list inequivalent classes of central $(p^n, p^n, p^n, 1)$-RDSs, providing a partial solution to Research Problem 50, when $v = w = p^n$ and $G = C = \mathbb{Z}_p^n$, that is, to list the $\langle p^n, p^n, \mathbb{Z}_p^n, E, \mathbb{Z}_p^n, i, j \rangle$.

THEOREM 8.32 [168, Theorem 3.5, Corollary 3.6] *Let* μ_λ *be an LP cocycle and let* $\mathcal{D}(\mu_\lambda)$ *be the subset of* $\mathcal{B}(\mu_\lambda)$ *containing only LP cocycles. Then* $\mathcal{D}(\mu_\lambda) =$

$$\left\{ \mu_\tau \mid \tau(x) = \alpha\, \theta\big(\lambda(\beta\, \theta^{-1}(x))\big),\ \alpha, \beta \in GF(p^n)^\star,\ \theta \in \{\sigma_i, 0 \leq i < n\} \right\},$$
$$(8.34)$$

where $\sigma_i(x) = x^{p^i}$ *is a Frobenius automorphism. For a power cocycle* μ_i,

$$\mathcal{D}_{\mu_i} = \{\mu_\tau \mid \tau(x) = \alpha x^{p^i},\ \alpha \in GF(p^n)^\star\},\ 0 \leq i \leq n-1.$$

Distinct power cocycles represent distinct equivalence classes of RDSs, so that, for any n *and* p, *there are at least* n *equivalence classes of central* $(p^n, p^n, p^n, 1)$-*RDSs, of which one is abelian and* $n-1$ *are nonabelian.*

RDS Class	LPP representative $\lambda(x)$	$|\mathcal{D}_{\mu_\lambda}|$
$p = 2; n = 1$		
1	x	1
$p = 2; n = 2$		
1	x	3
2	x^2	3
$p = 2; n = 3$		
1	x	7
2	x^2	7
3	x^4	7
4	$x^4 + \alpha^2 x^2 + x$	147
$p = 3; n = 1$		
1	x	2
$p = 3; n = 2$		
1	x	8
2	x^3	8
3	$x^3 + \alpha x$	32

Table 8.1 [168, Table 1] Complete list of equivalence classes of central $(p^n, p^n, p^n, 1)$-
RDSs for $p = 2, 3, \ p^n < 16$

Thus at most $n(p^n - 1)^2$ LP cocycles are in the bundle of each LP cocycle.

Using the abelian group Algorithm 1 of Chapter 6.3, combined with programs
for computing $\text{Aut}(\mathbb{Z}_p^n)$, the author and Udaya [168] checked every cocycle in
$Z^2(\mathbb{Z}_p^n, \mathbb{Z}_p^n)$ for $p = 2$ and $n \leq 3$, and for $p = 3$ and $n \leq 2$, for orthogonality, and
so obtained a complete list of all equivalence classes of central $(p^n, p^n, p^n, 1)$-
RDSs. This appears in Table 8.1, in which α refers to a primitive element in
$GF(p^n)$. For $n = 1$ and $p = 2, \ n = 2$ the computational results confirm Examples 8.3.1 and 8.3.2, respectively.

Immediately it is obvious that the lower bound n for the number of equivalence
classes, given by Theorem 8.32, is not tight.

Example 8.3.3 *Let $G = C = \mathbb{Z}_p^n$.*

1. *There are exactly 4 equivalence classes of central $(8, 8, 8, 1)$-RDSs,*

   $\langle 8, 8, \mathbb{Z}_2^3, \mathbb{Z}_4^3, \mathbb{Z}_2^3, \mu \rangle$;

   $\langle 8, 8, \mathbb{Z}_2^3, E_1, \mathbb{Z}_2^3, \mu_1 \rangle$; $\langle 8, 8, \mathbb{Z}_2^3, E_2, \mathbb{Z}_2^3, \mu_2 \rangle$; *and* $\langle 8, 8, \mathbb{Z}_2^3, E_3, \mathbb{Z}_2^3, \mu_\lambda \rangle$,

   where λ is the LPP for RDS Class 4. The first is abelian and the rest are
   nonabelian. By Lemma 7.42, $\exp(E_i) = 4$ for $i = 1, 2, 3$.

2. *There are exactly 3 equivalence classes of central $(9, 9, 9, 1)$-RDSs,*

   $\langle 9, 9, \mathbb{Z}_3^2, \mathbb{Z}_3^4, \mathbb{Z}_3^2, \mu \rangle$; $\langle 9, 9, \mathbb{Z}_3^2, E_1, \mathbb{Z}_3^2, \mu_1 \rangle$; *and* $\langle 9, 9, \mathbb{Z}_3^2, E_2, \mathbb{Z}_3^2, \mu_\lambda \rangle$,

   where λ is the LPP for RDS Class 3. The first is abelian and the others are
   nonabelian. By Lemma 7.42, $\exp(E_i) = 3$ for $i = 1, 2$.

Smith [22] informs us that for $p^n = 8$ all RDSs are central and the 3 nonabelian groups E_i of exponent 4 have only 2 isotypes. These are $\langle x, y, z, w : x^4 = y^4 = z^4 = w^2 = 1, x^2 = y^2, w = [x, y], [x, z^2] = [y, z^2] = [w, z] = [y, z] = [x^2, z] = 1 \rangle$, containing 1 equivalence class of $(8, 8, 8, 1)$-RDSs, and $\langle x, y, z, w : x^4 = y^4 = z^4 = w^2 = 1, w = [x, y] = z^2, [x, w] = [y, w] = 1, [x, z] = y^2 w, [y, z] = x^2 y^2 \rangle$, containing 2 equivalence classes of $(8, 8, 8, 1)$-RDSs. So, there are two automorphism orbits $\mathcal{A}_i$ with nonabelian extension groups, one of which contains 2 bundles and the other only 1.

To complete the classification of these equivalence classes according to Research Problem 50, it remains to sort the the isotypes of E_i into automorphism orbits for $p^n = 8$ and identify the isotypes of E_i for $p^n = 9$.

Research Problem 52 *Identify the automorphism orbit of each of the 3 nonabelian groups E_i of exponent 4 for $p^n = 8$ and the isotype of each of the 2 nonabelian groups E_i of exponent 3 for $p^n = 9$, defined by Example 8.3.3.*

For $p = 2$ and $n = 4$, we sorted the $20,160 = |GL(4, 2)|$ LP cocycles defined from field multiplication on $GF(16)$ into distinct bundles according to Theorem 8.32. The resulting list of orthogonal bundles is given in Table 8.2, in which α refers to a primitive element in $GF(16)$. As will be shown in Chapter 9.3.2, this is not a complete list of orthogonal bundles in $Z^2(\mathbb{Z}_2^4, \mathbb{Z}_2^4)$.

Nonetheless, observe that in each case we know of, the number of bundles listed is a power of p.

Research Problem 53 *(See also Research Problem 74) If $G = C = \mathbb{Z}_p^n$, is the number of equivalence classes of central $(p^n, p^n, p^n, 1)$-RDSs always a power of p?*

Again, to complete the classification of the equivalence classes defined in Table 8.2, each extension group E must be identified.

Research Problem 54 *Identify the isotypes of each of the 31 nonabelian 2-groups E_{μ_λ} of exponent 4 defined by Table 8.2 and hence sort the RDS Classes into automorphism orbits.*

Analysis of multiplicative orthogonal cocycles continues in Chapter 9.3.2.

8.4 SHIFT ACTION — THE CENTRAL CASE

Shift action is defined for *any* pair of groups G and N as an action by G on the set of factor pairs $F^2(G, N)$. As such, it is an astonishingly general action and should be more widely known and far better understood. Analysed as an abstract group action, it has so far received attention only in the central case. We will see in Section 8.5 that in fact shift action on coboundaries does appear, in disguise, within the G-module structure of a group ring RG.

Most of the initial results reported in this section appear in [159]. Here, if the groups G and C are obvious from context, we will write Z for $Z^2(G, C)$, B for

RDS Class	LPP representative $\lambda(x)$	$|\mathcal{D}_{\mu_\lambda}|$
	$p = 2; n = 4;\ \mathbb{F} = GF(16)$	
1	x	15
2	x^2	15
3	x^4	15
4	x^8	15
5	$x^4 + \alpha^7\, x$	150
6	$x^8 + \alpha^4\, x^2$	150
7	$x^4 + \alpha^{12}\, x^2 + \alpha^{14}\, x$	900
8	$x^4 + \alpha^3\, x^2 + \alpha^{12}\, x$	225
9	$x^8 + x^4 + \alpha^{11}\, x$	900
10	$x^8 + \alpha^{14}\, x^4 + \alpha^2\, x^2$	900
11	$x^8 + \alpha^4\, x^4 + \alpha^6\, x^2$	225
12	$x^8 + \alpha^7\, x^2 + \alpha^8\, x$	900
13	$x^8 + \alpha^6\, x^2 + \alpha^{12}\, x$	225
14	$x^8 + \alpha^8\, x^4 + \alpha^{14}\, x$	225
15	$x^8 + \alpha^7\, x^4 + \alpha^{14}\, x^2 + \alpha^6\, x$	900
16	$x^8 + \alpha^9\, x^4 + \alpha^6\, x^2 + x$	900
17	$x^8 + \alpha^{14}\, x^4 + \alpha^{11}\, x^2 + \alpha^2\, x$	450
18	$x^8 + \alpha^5\, x^4 + \alpha^{11}\, x^2 + \alpha^{12}\, x$	900
19	$x^8 + \alpha^{10}\, x^4 + \alpha^5\, x^2 + \alpha^{12}\, x$	900
20	$x^8 + \alpha^{11}\, x^4 + \alpha^9\, x^2 + \alpha^{13}\, x$	450
21	$x^8 + \alpha^9\, x^4 + \alpha^9\, x^2 + \alpha^2\, x$	900
22	$x^8 + \alpha^{10}\, x^4 + \alpha^6\, x^2 + \alpha^4\, x$	900
23	$x^8 + \alpha^2\, x^4 + \alpha^{12}\, x^2 + \alpha^{11}\, x$	900
24	$x^8 + \alpha^{14}\, x^4 + \alpha\, x^2 + \alpha^5\, x$	900
25	$x^8 + \alpha^8\, x^4 + \alpha^8\, x^2 + \alpha^{13}\, x$	900
26	$x^8 + \alpha^2\, x^4 + \alpha^{11}\, x^2 + \alpha^3\, x$	900
27	$x^8 + \alpha^5\, x^4 + \alpha^5\, x^2 + \alpha^3\, x$	900
28	$x^8 + \alpha^{13}\, x^4 + \alpha^{11}\, x^2 + \alpha^{12}\, x$	900
29	$x^8 + \alpha^{13}\, x^4 + \alpha^3\, x^2 + \alpha\, x$	900
30	$x^8 + \alpha^5\, x^4 + \alpha^2\, x^2 + \alpha^{14}\, x$	900
31	$x^8 + \alpha^6\, x^4 + \alpha^{10}\, x^2 + \alpha^8\, x$	900
32	$x^8 + \alpha^{14}\, x^4 + \alpha^9\, x^2 + \alpha^4\, x$	900

Table 8.2 [168, Table 1] Classification of equivalence classes of central $(16, 16, 16, 1)$-RDSs derived from field multiplication μ

$B^2(G,C)$ and M for the subgroup $M^2(G,C)$ of multiplicative cocycles. It is immediate from the definition that shift action is a permutation action of G on Z which preserves the group structure of Z. Since $1 \cdot a = 1$ for any $a \in G$, shift action is never transitive on Z (unless C is the trivial group). The subgroup B is closed under shift action, so the same characteristics apply when the action is restricted to B. These properties are merely recorded; their proof is left to the reader.

LEMMA 8.33 *For any group G and abelian group C, the shift action of G on Z has the following properties.*

1. $\psi \cdot 1 = \psi, \ \forall \, \psi \in Z$;

2. $\psi \cdot (ab) = (\psi \cdot a) \cdot b, \ \forall \, \psi \in Z, \ a, b \in G$;

3. $\psi \cdot a = \varphi \cdot a$ *if and only if* $\psi = \varphi, \ \forall \, \psi, \varphi \in Z, \ a \in G$;

4. $(\psi\varphi) \cdot a = (\psi \cdot a)(\varphi \cdot a), \ \forall \, \psi, \varphi \in Z, \ a \in G$;

5. $1 \cdot a = 1, \ \forall \, a \in G$;

6. $\partial\phi \cdot a = \partial(\phi \cdot a), \ \forall \, \partial\phi \in B, \ a \in G$.

By parts 2, 3 and 4 above, G acts by automorphisms of Z, and by 2 and 6, as a group of automorphisms of Z which leave B fixed setwise.

LEMMA 8.34 *For any group G and abelian group C, define* $\mathrm{Aut}(Z, B) \leq \mathrm{Aut}(Z)$ *to be the subgroup of automorphisms of Z which leave B fixed setwise. Define* $\zeta : G \to \mathrm{Aut}(Z)$ *to be* $\zeta_a(\psi) = \psi \cdot a, \ a \in G, \ \psi \in Z$. *Then* $\zeta(G) \leq \mathrm{Aut}(Z, B)$.

It is a simple matter to check that for abelian G, shift action also fixes setwise the subgroup $S^2_+(G,C)$ of symmetric cocycles.

Similarly, some elementary results relating the shift actions for different groups are easily determined.

LEMMA 8.35 *Let* $\psi \in Z^2(G,C)$ *and let* $\varphi \in Z^2(H,C)$. *Let* $\theta : H \to G$ *and* $\gamma : C \to C'$ *be homomorphisms.*

1. *For* $a \in G$, $(\gamma \circ \psi) \cdot a = \gamma \circ (\psi \cdot a)$ *in* $Z^2(G, C')$.

2. *For* $b \in H$, $\big(\psi \circ (\theta \times \theta)\big) \cdot b = \big(\psi \cdot \theta(b)\big) \circ (\theta \times \theta)$ *in* $Z^2(H, C)$. *In particular, for* $b \in H \leq G$, $\psi|_H \cdot b = (\psi \cdot b)|_H$.

3. *For* $a \in G$ *and* $b \in H$, $(\psi \otimes \varphi) \cdot (a, b) = (\psi \cdot a) \otimes (\varphi \cdot b)$ *in* $Z^2(G \times H, C)$.

If the additively written abelian group C carries an R-module structure for some ring R, then $Z^2(G,C)$ is an R-module (Example 6.2.9). It is immediate from the definition that R-action preserves shift orbits (but not necessarily orbit sizes).

LEMMA 8.36 *Let R be a ring and C be a (left) R-module. For any* $a \in G$, $\varphi \in Z^2(G,C)$ *and* $r \in R$, *if* $\psi = \varphi \cdot a$ *then* $(r\psi) = (r\varphi) \cdot a$. *If* $r \neq 0$ *is (left) cancellable in C, the converse holds.*

Very little is known about shift action. The following list of research problems opens a broad frontier to future exploration.

Research Problem 55   *1. Identify the fixed point subgroups $M^2(G, C)$ and $M^2(G, C) \cap B^2(G, C)$.*

2. Identify the stabiliser of $Z^2(G, C)$ in G.

3. What is the shift orbit structure of $B^2(G, C)$?

4. How are shift orbit structures within the cosets of $B^2(G, C)$ in $Z^2(G, C)$ related?

5. For abelian G, how are shift orbit structures within the cosets of $B^2(G, C)$ in $S^2_+(G, C)$ related?

6. What characteristics of the shift orbit structure are determined by different families of groups G, such as simple, cyclic, abelian, dihedral or p-groups?

Of course, we are still faced with the fundamental question.

Research Problem 56 *Which shift orbits in $Z^2(G, C)$ are (v, w, k, λ)-orthogonal (contain only (v, w, k, λ)-orthogonal cocycles) for some k and λ satisfying (4.13)? In particular, which orbits are orthogonal?*

This latter question might be weakened to ask:

Research Problem 57 *What proportion of the shift orbits in $Z^2(G, C)$ are orthogonal?*

8.4.1 Orbit structure for cyclic groups

The simplest class of groups G for which to explore the shift orbit structure (Research Problem 55.6) is the class of cyclic groups. This analysis is begun at an elementary level in [159], but further progress requires more sophisticated tools. The only result of note is that, in the vast majority of cases, all the multiplicative cocycles on a cyclic group are coboundaries.

LEMMA 8.37 [159, Corollary 4.13] *If $G = \mathbb{Z}_v$ and C has order w, then $M \subseteq B$ in any of the following cases:*

1. v is odd;

2. w is odd;

3. v is even, $v = 2^k u$ for u odd, w is even and C has no elements of order 2^k.

However, by Theorem 6.10, we know that there are no orthogonal cocycles in $M^2(\mathbb{Z}_v, C)$, unless v is a prime.

Research Problem 58 *What is the shift orbit structure of the cyclic group $\mathbb{Z}_v$?*

This problem is solved below for $B^2(\mathbb{Z}_{p^r}, \mathbb{Z}_p^r)$, p prime — see Example 8.5.1 and Theorem 8.45.

8.4.2 Relationship between orbit structures in distinct cohomology classes

A preliminary analysis of Research Problem 55.4 follows from the fact that any $\psi \in Z$ can be represented uniquely with respect to B and M.

Set $J = Z/MB$, $K = MB/B \cong M/(B \cap M)$, $L = B/(B \cap M)$, let $R = \{r_j, j \in J, r_1 = 1\}$ be a normalised transversal of MB in Z, let $S = \{s_k, k \in K, s_1 = 1\}$ be a normalised transversal of $B \cap M$ in M and let $T = \{t_l, l \in L, t_1 = 1\}$ be a normalised transversal of $B \cap M$ in B. Each $\psi \in Z$ has a unique representation in the form

$$\psi = r_j \, s_k \, t_l \, \psi_0, \quad j \in J, \; k \in K, \; l \in L, \; \psi_0 \in B \cap M. \tag{8.35}$$

In this form, by Lemma 8.33.4,

$$\psi \cdot a = (r_j s_k t_l \psi_0) \cdot a = (r_j \cdot a) \, s_k \, (t_l \cdot a) \, \psi_0 = r_j \, s_k \, \partial(r_j)_a \, \partial(t_l)_a \, t_l \, \psi_0$$

(since s_k and ψ_0 are fixed by a). Since $Z = \bigvee_{j \in J} \bigvee_{k \in K} r_j s_k B$ and multiplication of $r_j T$ by $s_k \psi_0$ is a bijection, the shift actions on $r_j T$ and on $r_j s_k T \psi_0$ are *similar* G-actions (see Kerber [200, p. 31]).

Thus it is necessary only to determine, for each $j \in J$, the orbit structure of the set $r_j T$. We formalise this in the following theorem.

THEOREM 8.38 [159, Theorem 3.11] *For any group G and abelian group C, the orbit structure of $Z^2(G, C)$ is wholly defined by the orbit structures of the sets $r_j T$, $j \in J$, defined in (8.35). Let $\psi \in Z^2(G, C)$ and denote its shift orbit by $\psi \cdot G = \{\psi \cdot a : a \in G\}$. If $\psi = r_j \, s_k \, t_l \, \psi_0$ for unique $j \in J$, $k \in K$, $l \in L$, $\psi_0 \in B \cap M$, then*

$$\psi \cdot G = s_k \, \psi_0((r_j \, t_l) \cdot G) = \psi \, \{\partial(r_j \, t_l)_a : a \in G\}.$$

Plainly, the simplest of the sets $r_j T$ has $j = 1$: the place to start is with shift orbits in $B^2(G, C)$.

8.5 SHIFT ORBITS — THE CENTRAL SPLITTING CASE

Assume throughout this Section that $N = C$ is central, G and C are finite and C is written *additively*. Recall from Section 8.2 (Corollary 8.21) that bundles in $B^2(G, C)$ can just as easily be thought of as bundles of normalised functions (1-cochains). The first results on shift orbits within bundles of coboundaries (Research Problem 55.3) are due to LeBel [217].

LeBel's critical observation is that, because arbitrary functions $f : G \to C$ may be represented as elements of a group ring RG, the shift action on coboundaries exists, in another guise, within the standard G-module structure on RG.

We must convert from G-actions on $B^2(G, C)$ to RG-modules, where R is a commutative ring with unity and $C = (R, +)$. This involves no loss of generality, since any finite abelian group C is isomorphic to a direct product of finite cyclic groups $C \cong \mathbb{Z}_{n_1} \times \cdots \times \mathbb{Z}_{n_k}$, say. Regarding each cyclic factor as a commutative ring with unity means C carries the direct product ring structure, so C is always isomorphic to the underlying additive group of at least one commutative ring R with unity. Furthermore, C is an R-module for each such R.

Let $UC^1(G, R)$ be the R-module of all mappings from G to R (cf. Definitions 6.1 and 8.22). It is readily verified that $UC^1(G, R)$ is an R-algebra under the convolution multiplication (cf. Lemma 3.6.5)

$$f_1 f_2(x) = \sum_{y \in G} f_1(y) f_2(y^{-1}x), \ x \in G, \tag{8.36}$$

for all $f_1, f_2 \in UC^1(G, R)$. Since G is finite, there is (see, for example, [71]) an R-algebra isomorphism $\vartheta : RG \to UC^1(G, R)$, where $\vartheta(\sum_{x \in G} a_x x)$ is defined by $x \mapsto a_x$, with inverse $f \mapsto \sum_{x \in G} f(x)x$.

The *standard* left RG-module, denoted $RG^{(0)}$, has underlying additive group RG, with left G-action given by left multiplication:

$$g \cdot \left(\sum_{x \in G} a_x x \right) = \sum_{x \in G} a_x(gx) = \sum_{x \in G} a_{(g^{-1}x)} x, \ \forall g \in G, \ \sum_{x \in G} a_x x \in RG.$$

The R-module isomorphism $\vartheta^{(0)} : RG^{(0)} \to UC^1(G, R)$ defined by $\alpha \mapsto \vartheta(\alpha)$ imposes a natural left RG-module structure on $UC^1(G, R)$, with

$$(g \cdot f)(x) = f(g^{-1}x), \ \forall \, x \in G, \tag{8.37}$$

for all $g \in G$ and $f \in UC^1(G, R)$. Consequently, $\vartheta^{(0)}$ is an RG-module isomorphism.

LeBel defines a sequence of quotient algebras inductively from $RG^{(0)}$.

DEFINITION 8.39 [217, Definition 3.2] *Let $RG^{(0)}$ be as above and $j \in \mathbb{N}$. Define $RG^{(j)}$ to be the quotient of $RG^{(j-1)}$ by its submodule of G-stable elements:*

$$RG^{(j)} = RG^{(j-1)} \Big/ \left({}^G RG^{(j-1)} \right). \tag{8.38}$$

Call $RG^{(j)}$ the jth quotient group algebra of RG.

Flannery [113] points out that this sequence of quotients generalises the theory of Loewy and socle series for finite p-groups (see, for example, [7]).

For our purposes the most important quotient is $RG^{(2)}$, which we proceed to show is isomorphic to $B^2(G, R)$.

Under $\vartheta^{(0)}$, the G-stable set in $UC^1(G, R)$ corresponding to ${}^G RG^{(0)}$ is the set of all constant maps $G \to R$. The quotient of $UC^1(G, R)$ by the set of constant maps is easily identified with $C^1(G, R)$. There is an imposed left G-action on $C^1(G, R)$: if $\phi \in C^1(G, R)$, then

$$(g \cdot \phi)(x) = \phi(g^{-1}x) - \phi(g^{-1}) = (\phi \cdot g^{-1})(x), \ \forall x \in G, \tag{8.39}$$

so its orbits are precisely those induced by shift action (8.31) on $C^1(G, R)$. Hence there is an RG-module isomorphism

$$\vartheta^{(1)} : RG^{(1)} \xrightarrow{\cong} C^1(G, R).$$

Moreover, $\vartheta^{(1)} \left({}^G RG^{(1)} \right)$ is the set of group homomorphisms $\mathrm{Hom}(G, R)$, so the induced isomorphism $\widehat{\partial}$ of (8.27) in turn induces an R-module isomorphism

$$\vartheta^{(2)} : RG^{(2)} \to B^2(G, R).$$

As with $\vartheta^{(0)}$ and $\vartheta^{(1)}$, $\vartheta^{(2)}$ imposes the RG-module structure of $RG^{(2)}$ onto $B^2(G, R)$. Again, the induced left G-action on $B^2(G, R)$ is $g \cdot \partial\phi = \partial\phi \cdot g^{-1}$, so its orbits are *exactly* those induced by shift action (8.30) on $B^2(G, R)$.

THEOREM 8.40 [217, Theorem 5.1] *The RG-module homomorphism*

$$\vartheta^{(2)} : RG^{(2)} \xrightarrow{\cong} B^2(G, R)$$

is an isomorphism which induces a size-preserving bijection between the set of G-orbits in $RG^{(2)}$ and the set of shift orbits in $B^2(G, R)$. Similarly,

$$\vartheta^{(2)}\left(^G RG^{(2)}\right) \cong B^2(G, R) \cap M^2(G, R).$$

Therefore, Research Problem 55.3 may be reformulated in terms of the second quotient group algebra $RG^{(2)}$. Any solution must be independent of the multiplicative structure of the commutative ring R with unity, since the underlying set is $B^2(G, C)$ (cf. Lemma 8.36).

Research Problem 59 *(= Research Problem 55.3) For $C = (R, +)$, what is the G-orbit structure of $RG^{(2)}$?*

LeBel studies Research Problem 59 in the case that $C = (R, +)$ is an elementary abelian p-group and R is a finite field. His results are outlined in the next two subsections.

8.5.1 When C is an elementary abelian p-group

Throughout this subsection and the next, the ring R is a finite field $\mathbb{F}_q = GF(p^n)$ of order $q = p^n$, so that $C \cong \mathbb{Z}_p^n$ is an elementary abelian p-group.

LeBel summarises the orbit structure of the modules $\mathbb{F}_q G^{(j)}$ in a matrix.

DEFINITION 8.41 *Let G be a finite group of order v. Let m be the least integer $m \in \mathbb{N}$ such that $\mathbb{F}_q G^{(m)} \neq \{0\}$ and $^G \mathbb{F}_q G^{(m)} = \{0\}$. If no such integer exists, then let $m \in \mathbb{N}$ be the least integer such that $^G \mathbb{F}_q G^{(m)} = \mathbb{F}_q G^{(m)} \neq \{0\}$. Let $\hat{v}$ be the number of positive integer divisors of v.*

The orbit size table *for $\mathbb{F}_q G$ is the $(m + 1) \times \hat{v}$ matrix with rows labelled 0 to m and columns labelled j, $1 \leq j \leq v$, $j|v$, whose (i, j) entry is the number of elements in $\mathbb{F}_q G^{(i)}$ with G-orbit size j. If G is a finite p'-group of order $(p')^r$, the columns will be labelled $0 \leq j \leq r$ for simplicity, so that the (i, j) entry means the $(i, (p')^j)$ entry.*

Denote by $\kappa^{(i)}(G)$ the (i, v) entry of the orbit size table for $\mathbb{F}_q G$, $0 \leq i \leq m+1$.

If all the subgroups of G are normal (for example, if G is abelian) the 0th row of the orbit size table for $\mathbb{F}_q G$ is enumerable.

THEOREM 8.42 [217, Theorem 4.3] *Let G be a finite group of order v such that all its subgroups are normal. Then the number of elements with orbit size j in $\mathbb{F}_q G^{(0)}$ is $\sum \kappa^{(0)}(G/H)$, where the sum is taken over all possible subgroups H of order v/j. In particular,*

$$\kappa^{(0)}(G) = q^v - \sum_{\{1\} \neq H \leq G} \kappa^{(0)}(G/H). \tag{8.40}$$

LeBel's approach to finding subsequent rows of the orbit size table is to study the quotient group algebras $\{\mathbb{F}_q G^{(j)}\}_{j \geq 0}$ in terms of powers of the augmentation ideal $\omega(\mathbb{F}_q G)$. Recall that the *augmentation ideal* of $\mathbb{F}_q G$ is

$$\omega(\mathbb{F}_q G) = \left\{ \sum_{x \in G} a_x x \in \mathbb{F}_q G : \sum_{x \in G} a_x = 0 \right\}. \qquad (8.41)$$

It is an $\mathbb{F}_q G$-submodule of $\mathbb{F}_q G^{(0)}$, generated as an $\mathbb{F}_q$-module by $\{g - 1 : g \in G\}$.

When p does not divide $|G|$, determining the quotient group algebras of $\mathbb{F}_q G$ is straightforward, since by Maschke's theorem [258, Theorem 2.4.2] this condition is equivalent to $\mathbb{F}_q G$ being semisimple. Refer to Passman [258], for instance, for any unfamiliar terms or concepts.

LEMMA 8.43 [217, Lemma 3.4] *Let G be a finite group of order v and suppose p does not divide v. Then*

$$\mathbb{F}_q G^{(0)} = {}^G \mathbb{F}_q G^{(0)} \oplus \omega(\mathbb{F}_q G), \qquad (8.42)$$

and for all $j \geq 1$, $\mathbb{F}_q G^{(j)} \cong \omega(\mathbb{F}_q G)$ as $\mathbb{F}_q G$-modules.

In this case, by Theorem 8.40 there are no nontrivial multiplicative coboundaries (we already know there are no orthogonal multiplicative coboundaries by Theorem 6.10).

COROLLARY 8.44 *Suppose $p \nmid |G|$. Then*

$$B^2(G, \mathbb{Z}_p^n) \cong_{\mathbb{F}_q G} \omega(\mathbb{F}_q G) \quad and \quad B^2(G, \mathbb{Z}_p^n) \cap M^2(G, \mathbb{Z}_p^n) = \{1\}.$$

In this semisimple case the orbit size table for $\mathbb{F}_q G$ is $2 \times \hat{v}$. By (8.42), $\mathbb{F}_q G^{(0)} \cong \mathbb{F}_q \oplus_{\mathbb{F}_q G} \mathbb{F}_q G^{(1)}$, where $\mathbb{F}_q$ has trivial G-action. Therefore, in this case, the $(0, j)$ entry of the orbit size table is always q times the $(1, j)$ entry.

The other extreme, when G is itself a p-group, is particularly interesting. First, by Theorem 6.10, multiplicative orthogonal cocycles can exist only when G is an elementary abelian p-group. Second, by [258, Lemma 3.1.6], $\omega(\mathbb{F}_q G)$ is nilpotent if and only if G is a finite p-group. Third, as pointed out by Flannery [113], in this case the socle of each quotient group algebra is a quotient in the socle series of $\mathbb{F}_q G$, and the duality of the Loewy and socle series might be used to determine more general results.

Research Problem 60 *(Flannery) When G is a p-group, what information does the duality between the Loewy and socle series give us about $\mathbb{F}_q G^{(j)}$? Does the duality have an extension to the general quotient group algebra sequence (8.38) ?*

8.5.2 When C is an elementary abelian p-group and G is a p-group

When $\omega(\mathbb{F}_q G)$ is nilpotent, the sequence $\{E_t = \omega(\mathbb{F}_q G)^t\}_{t \geq 1}$ is a filtration of $\mathbb{F}_q G$ called the *power filtration*. By writing $\mathbb{F}_q G^{(j)}$ and ${}^G \mathbb{F}_q G^{(j)}$ as quotients of the power filtration, LeBel determines the $\mathbb{F}_q$-dimension of $\mathbb{F}_q G^{(j)}$. For various p-groups, in particular for $G = \mathbb{Z}_{p^r}$ and $G = \mathbb{Z}_p^r$, he derives the G-orbit structure of $\mathbb{F}_q G^{(j)}$ iteratively, by counting elements of each G-orbit size in subgroups of G.

In the cyclic p-group case $G = \mathbb{Z}_{p^r}$ he shows that if $0 \leq j < p^r - p^{r-1}$, then elements in $\mathbb{F}_q\mathbb{Z}_{p^r}{}^{(j)}$ which are stable under the action of some nontrivial subgroup must be in a submodule isomorphic to $\mathbb{F}_q\mathbb{Z}_{p^{r-1}}{}^{(0)}$. If $j \geq p^r - p^{r-1}$, then he shows $\mathbb{F}_q\mathbb{Z}_{p^r}{}^{(j)} \cong \mathbb{F}_q\mathbb{Z}_{p^{r-1}}{}^{(j-p^r+p^{r-1})}$. The entire orbit size table for $\mathbb{F}_q\mathbb{Z}_{p^r}$ can thus be computed from that of $\mathbb{F}_q\mathbb{Z}_{p^{r-1}}$ and the fact that $|\mathbb{F}_q\mathbb{Z}_{p^r}{}^{(j)}| = q^{p^r-j}$. Initially, row 0 is calculated from Theorem 8.42.

Example 8.5.1 [217, Lemma 4.6] *The orbit size table for* $\mathbb{F}_q\mathbb{Z}_p$ *is the* $p \times 2$ *matrix*

$$
\begin{pmatrix}
q & q^p - q \\
q & q^{p-1} - q \\
q & q^{p-2} - q \\
\vdots & \vdots \\
q & q^2 - q \\
q & 0
\end{pmatrix}.
$$

THEOREM 8.45 [217, Theorem 4.7] *For all integers* $r \geq 2$ *and all primes* p, *the orbit size table for* $\mathbb{F}_q\mathbb{Z}_{p^r}$ *is the* $p^r \times (r+1)$ *matrix with* (i, j) *entry*

1. *the* $(0, j)$ *entry of the orbit size table for* $\mathbb{F}_q\mathbb{Z}_{p^{r-1}}$ *if* $i \leq p^r - p^{r-1}$ *and* $j < r + 1$;

2. *the* $(i - (p^r - p^{r-1}), j)$ *entry of the orbit size table for* $\mathbb{F}_q\mathbb{Z}_{p^{r-1}}$ *if* $i \geq p^r - p^{r-1}$ *and* $j < r + 1$;

3. $q^{-i+p^r} - q^{p^{r-1}}$, *if* $i \leq p^r - p^{r-1}$ *and* $j = r + 1$; *and*

4. 0, *if* $i \geq p^r - p^{r-1}$ *and* $j = r + 1$.

In the elementary abelian p-group case $G = \mathbb{Z}_p^r$, the power filtration of $\mathbb{F}_q\mathbb{Z}_p^r$ has length $r(p-1)$, so after applying Theorem 8.42 to enumerate row 0, there are another $r(p-1)$ rows to compute.

DEFINITION 8.46 *For* $t \geq 0$, *let* $\beta_p(t, n)$ *be the number of distinct elements* $(a_1, \ldots, a_n)$ *in* $\{0, \ldots, p-1\}^n$ *such that* $a_1 + \cdots + a_n = t$. *For* $t < 0$, *define* $\beta_p(t, n) = 0$.

LeBel shows that the number of elements of orbit size p^j which are stable under a given subgroup of order p^{r-j} in $\mathbb{F}_q\mathbb{Z}_p^r{}^{(i)}$ is

$$
q^{\beta_p(r(p-1)-i,r) - \beta_p(j(p-1)-i,j)} \cdot \kappa^{(i)}(\mathbb{Z}_p^j),
$$

so his next result follows on repeating the work for all the distinct subgroups of order p^{r-j} in $\mathbb{Z}_p^r$. There are

$$
\binom{r}{j}_p = \prod_{i=0}^{j-1} (1 - p^{r-i}) \Big/ \prod_{i=1}^{j} (1 - p^i)
$$

(the p-binomial, or Gaussian, coefficient) of these.

THEOREM 8.47 [217, Theorem 4.11] *For $0 \leq j < r$ and for $0 \leq i \leq r(p-1)$, the (i,j) entry of the orbit size table for $\mathbb{F}_q \mathbb{Z}_p^r$ is*

$$q^{\beta_p(r(p-1)-i,r) - \beta_p(j(p-1)-i,j)} \cdot \binom{r}{j}_p \cdot \kappa^{(i)}(\mathbb{Z}_p^j).$$

In $B^2(G, \mathbb{Z}_p^n)$ when $|G| = p^r$, the number of coboundaries in orbits of each possible size p^j is read off from the row indexed 2 (the third row) of the orbit size table for $\mathbb{F}_q G$.

Theorems 8.45 and 8.47 thus solve Research Problem 59 for $C = \mathbb{Z}_p^n$ and for $G = \mathbb{Z}_{p^r}$ and $\mathbb{Z}_p^r$, respectively.

Example 8.5.2 [217, §4.4.2] *Let $G = \mathbb{Z}_2^r$, $1 \leq r \leq 5$. The orbit size table for $\mathbb{F}_2 \mathbb{Z}_2$ is* $\begin{pmatrix} 2 & 2 \\ 2 & 0 \end{pmatrix}$. *Then for $\mathbb{F}_2 \mathbb{Z}_2^2$ and $\mathbb{F}_2 \mathbb{Z}_2^3$ it is*

$$\begin{pmatrix} 2 & 6 & 8 \\ 4 & 0 & 4 \\ 2 & 0 & 0 \end{pmatrix} \quad and \quad \begin{pmatrix} 2 & 14 & 56 & 184 \\ 8 & 0 & 56 & 64 \\ 8 & 0 & 0 & 8 \\ 2 & 0 & 0 & 0 \end{pmatrix},$$

respectively, for $\mathbb{F}_2 \mathbb{Z}_2^4$ it is

$$\begin{pmatrix} 2 & 30 & 280 & 2760 & 62,464 \\ 16 & 0 & 560 & 1920 & 30,272 \\ 64 & 0 & 0 & 960 & 1024 \\ 16 & 0 & 0 & 0 & 16 \\ 2 & 0 & 0 & 0 & 0 \end{pmatrix},$$

and for $\mathbb{F}_2 \mathbb{Z}_2^5$ it is

$$\begin{pmatrix} 2 & 62 & 1240 & 28,520 & 1,936,384 & \approx 4.3 \times 10^9 \\ 32 & 0 & 4960 & 39,680 & 15,014,912 & \approx 2.1 \times 10^9 \\ 1024 & 0 & 0 & 158,720 & 507,904 & 66,441,216 \\ 1024 & 0 & 0 & 0 & 31,744 & 32,768 \\ 32 & 0 & 0 & 0 & 0 & 32 \\ 2 & 0 & 0 & 0 & 0 & 0 \end{pmatrix},$$

where in each case the third row gives the shift orbit structure of $B^2(\mathbb{Z}_2^r, \mathbb{Z}_2)$.

Observation of computational results such as Example 8.5.2 for small values of p led LeBel to conclude that as r increases, almost all coboundaries in $B^2(G, \mathbb{Z}_p^n)$ lie in shift orbits of maximum size p^r. He proves this from Theorems 8.45 and 8.47 for several classes of p-groups.

THEOREM 8.48 [217, Theorem 4.12, Corollary 4.14] *Suppose a p-group G of order p^r satisfies either*

    1. G is cyclic, or

    2. G satisfies the inequality

$$\frac{|\mathbb{F}_q G^{(2)}| - \kappa^{(2)}(G)}{|\mathbb{F}_q G^{(2)}|} \leq \frac{|\mathbb{F}_q \mathbb{Z}_p^{r\,(2)}| - \kappa^{(2)}(\mathbb{Z}_p^r)}{|\mathbb{F}_q \mathbb{Z}_p^{r\,(2)}|}.$$

For every $\epsilon > 0$, there exists $m \in \mathbb{N}$ of order $O(\log(1/\epsilon))$ such that, for all $p^r > m$, the proportion of coboundaries with a less than maximal shift orbit in $B^2(G, \mathbb{Z}_p^n)$ is less than ϵ.

In other words, for a large enough finite p-group G satisfying either condition of Theorem 8.48, the probability that a randomly chosen coboundary in $B^2(G, \mathbb{Z}_p^n)$ has a shift orbit of size $|G|$ is arbitrarily close to 1.

Shift action raises far more questions than we know how to answer at present. Nonetheless, the orthogonality measure and distribution of a factor pair are invariant under shift action (Corollary 8.9 and Theorem 8.15), and so shift action is one key to all of the questions we have asked about cocyclic generalised Hadamard matrices.

It is time to put the heavy machinery of this Chapter and Chapter 7 to work, to construct large new classes of generalised Hadamard matrices and from them, new transforms, codes and nonlinear sequences.

Chapter Nine

The Future: Novel Constructions and Applications

This Chapter is a rich storehouse of examples, applications and problems. One third of the open research problems appear here.

Initially we look at several recent uses of cocycles, not necessarily orthogonal, for computation in Galois rings, for cryptography using elliptic curves and for coding over nonbinary alphabets.

In Section 9.2, splitting orthogonal factor pairs are applied to lay the foundations for a general theory of nonlinear functions. These include planar, bent and maximally nonlinear functions, and in Section 9.3, surprising and beautiful connections with finite presemifields are uncovered. A useful technique for forming new orthogonal cocycles as direct sums of orthogonal cocycles is described. These help identify enormous classes of new cocyclic generalised Hadamard matrices.

In turn, in Section 9.4 we create new families of optimal codes, including q-ary codes meeting the Plotkin bound for high distance relative to length, codes defined from planar functions and finite presemifields and extremal self-dual binary codes.

Finally, differential uniformity, an important measure of the resistance of a block encryption cipher to differential attack, is related to well-distributed cocycles. It is extended to 2-D array encryption ciphers, and a class of orthogonal cocycles proposed for testing as array S-box functions.

9.1 NEW APPLICATIONS OF COCYCLES

Cocycles really are everywhere. The emphasis in this book away from cohomology classes and towards cocycles themselves, has led to some 'Eureka' moments of recognition of cocycles in unexpected guises. Some recent appearances follow.

9.1.1 Computation in Galois rings

Here we use a cocycle to describe the Galois ring $GR(p^2, n)$ in Cartesian coordinates from the field $GF(p^n)$, rather than in the usual p-adic coordinates from the Teichmüller set. The advantage of this representation is that the Galois ring operations involve only field arithmetic in $GF(p^n)$. The description results from identifying the additive group $(GR(p^2, n), +)$ as an extension group of $(GF(p^n), +)$ by $(GF(p^n), +)$, defined by a cocycle we term the *Teichmüller* cocycle.

For a prime p, the Galois ring $GR(p^2, n) = \mathcal{R}$ is the Galois extension of degree n of the ring $\mathbb{Z}_{p^2}$. Suppose $f(x)$ is a primitive polynomial of degree n over $GF(p)$ such that the root ω of $f(x)$ is a primitive element of $\mathbb{F}_q = GF(p^n)$. Under the modulo p reduction map $\mathbb{Z}_{p^2}[x] \rightarrow \mathbb{Z}_p[x]$, $f(x)$ has a unique monic preimage $\hat{f}(x)$ in $\mathbb{Z}_{p^2}[x]$ such that a root $\hat{\omega}$ of $\hat{f}(x)$ satisfies $\hat{\omega}^{q-1} = 1$. Then $\mathcal{R} = \mathbb{Z}_{p^2}[\hat{\omega}]$. The additive group of $\mathcal{R}$ is isomorphic to $(\mathbb{Z}_{p^2})^n$ and the radical of $\mathcal{R}$ is the unique maximal ideal $p\mathcal{R}$. The residue class field $\mathcal{R}/p\mathcal{R}$ is isomorphic as a field to $\mathbb{F}_q$ and as an additive subgroup of $\mathcal{R}$, $p\mathcal{R}$ is isomorphic to $(\mathbb{F}_q, +) \cong \mathbb{Z}_p^n$. For a more detailed description, see McDonald [242].

The *Teichmüller set* $\mathcal{T} = \{0 = \hat{0}, 1 = \hat{1}, \hat{\omega}, \ldots, \hat{\omega}^{q-2}\}$ is the set of pth-power elements of $\mathcal{R}$. Every element r of $\mathcal{R}$ has a unique *p-adic representation* in the form $r = r_1 + p r_2$, where $r_1, r_2 \in \mathcal{T}$, so $\mathcal{T}$ is a set of coset representatives of $\mathcal{R}/p\mathcal{R}$. The field isomorphism $\theta : \mathcal{R}/p\mathcal{R} \rightarrow \mathbb{F}_q$ is given by $\theta(\hat{t} + p\mathcal{R}) = t$, $\hat{t} \in \mathcal{T}$. The abelian group isomorphism $\gamma : (p\mathcal{R}, +) \rightarrow (\mathbb{F}_q, +)$ is given by $\gamma(p\hat{t}) = t$, $\hat{t} \in \mathcal{T}$, where we note that for $\hat{t}, \hat{t}_i \in \mathcal{T}, 1 \le i \le j$,

$$\hat{t} \equiv \sum_{i=1}^{j} \hat{t}_i \pmod{p} \implies t = \gamma(p\hat{t}) = \gamma\left(p \sum_{i=1}^{j} \hat{t}_i\right) = \sum_{i=1}^{j} t_i. \qquad (9.1)$$

If $r = r_1 + p r_2$ and $s = s_1 + p s_2$, where $r_1, r_2, s_1, s_2 \in \mathcal{T}$, then naturally

$$rs = r_1 s_1 + p(r_1 s_2 + s_1 r_2), \qquad (9.2)$$

where $r_1 s_1 \in \mathcal{T}$ and $r_1 s_2 + s_1 r_2 \equiv \hat{t} \bmod p$ for some unique $\hat{t} \in \mathcal{T}$.

The inconvenience of using the p-adic representation lies in the difficulty of representing the sum of two elements of $\mathcal{T}$ as an element of $\mathcal{T}$. For prime p and commuting indeterminates X and Y, define $p\, C_p(X, Y) = (X + Y)^p - X^p - Y^p$, that is,

$$C_p(X, Y) = \sum_{k=1}^{p-1} \left[\binom{p}{k} / p \right] X^{(p-k)} Y^k. \qquad (9.3)$$

Mimicking the argument for $p = 2$ of Helleseth and Kumar (cf. [269, p. 205] or [317, Chapter 6]), for $r_1, s_1 \in \mathcal{T}$, write $r_1 + s_1 = t_1 + p t_2$ for $t_1, t_2 \in \mathcal{T}$ and raise both sides of the equation to the qth power. Then $t_1 = (r_1 + s_1)^q = r_1 + s_1 + p\, C_p(r_1^{p^{n-1}}, s_1^{p^{n-1}})$ and $t_2 \equiv -C_p(r_1^{p^{n-1}}, s_1^{p^{n-1}}) \pmod{p}$. Note that $r^{p^{n-1}} = r^{p^{-1}} = \sqrt[p]{r}$. Therefore in $\mathcal{R}$

$$r + s = [r_1 + s_1 + p\, C_p(\sqrt[p]{r_1}, \sqrt[p]{s_1})] + p[r_2 + s_2 - C_p(\sqrt[p]{r_1}, \sqrt[p]{s_1})], \qquad (9.4)$$

where $r_1 + s_1 + p\, C_p(\sqrt[p]{r_1}, \sqrt[p]{s_1}) \in \mathcal{T}$ and $r_2 + s_2 - C_p(\sqrt[p]{r_1}, \sqrt[p]{s_1}) \equiv \hat{t} \bmod p$ for some unique $\hat{t} \in \mathcal{T}$.

Consider the short exact sequence of abelian groups

$$0 \longrightarrow (p\mathcal{R}, +) \overset{\imath}{\longrightarrow} (\mathcal{R}, +) \overset{\pi}{\longrightarrow} (\mathcal{R}/p\mathcal{R}, +) \longrightarrow 0. \qquad (9.5)$$

Since $\mathcal{T}$ is a transversal of $(p\mathcal{R}, +)$ in $(\mathcal{R}, +)$, by Lemma 7.7 applied to (9.5) it determines a cocycle $\psi_{\mathcal{T}}$, namely, for $r_1 + s_1 = t_1 + p t_2, r_1, s_1, t_1, t_2 \in \mathcal{T}$,

$$\psi_{\mathcal{T}}(r_1 + p\mathcal{R}, s_1 + p\mathcal{R}) = \imath^{-1}(r_1 + s_1 - t_1) = -p\, C_p(\sqrt[p]{r_1}, \sqrt[p]{s_1}) \in p\mathcal{R}.$$

For the corresponding short exact sequence of abelian groups

$$0 \longrightarrow (\mathbb{F}_q, +) \overset{\imath \circ \gamma^{-1}}{\longrightarrow} (\mathcal{R}, +) \overset{\theta \circ \pi}{\longrightarrow} (\mathbb{F}_q, +) \longrightarrow 0, \qquad (9.6)$$

$\psi_q = \gamma \circ \psi_{\mathcal{T}} \circ (\theta^{-1} \times \theta^{-1}) : (\mathbb{F}_q, +) \times (\mathbb{F}_q, +) \rightarrow (\mathbb{F}_q, +)$ is a cocycle (cf. Example 6.2.12).

DEFINITION 9.1 *Let* $q = p^n$. *The* Teichmüller *cocycle* $\psi_q : (\mathbb{F}_q, +) \times (\mathbb{F}_q, +) \to$ $(\mathbb{F}_q, +)$ *is the symmetric cocycle*

$$\psi_q(g, h) = -C_p(\sqrt[p]{g}, \sqrt[p]{h})$$

$$= -\sqrt[p]{C_p(g, h)}$$

$$= -\sum_{k=1}^{p-1} \left[\binom{p}{k} / p \right] g^{(p-k)p^{n-1}} h^{kp^{n-1}}. \qquad (9.7)$$

By Corollary 7.10, $(\mathcal{R}, +)$ is isomorphic to the extension group E_{ψ_q} under the mapping $\hat{t_1} + p\hat{t_2} \mapsto (t_2, t_1)$, $\hat{t_1}, \hat{t_2} \in \mathcal{T}$. By (7.6), (9.1) and (9.4), addition in E_{ψ_q} is defined by $(c, g) + (d, h) = (c + d + \psi_q(g, h), g + h)$ for all $c, d, g, h \in (\mathbb{F}_q, +)$. From (9.1) and (9.2), define multiplication on E_{ψ_q} by $(c, g)(d, h) = (ch + dg, gh)$. Proof that E_{ψ_q} is a ring isomorphic to $\mathcal{R}$ is left as an exercise. For consistency with the usual p-adic notation in $\mathcal{R}$, reverse the components in E_{ψ_q} to provide an alternate description of $\mathcal{R}$ using only field arithmetic.

THEOREM 9.2 *Let* $q = p^n$ *and let* ψ_q *be the Teichmüller cocycle (9.7). The* Galois ring $GR(p^2, n)$ *of degree* n *over* $\mathbb{Z}_{p^2}$ *is (isomorphic to) the set* $\mathbb{F}_q \times \mathbb{F}_q$ *with the following addition and multiplication:*

$$(a, b) + (c, d) = (a + c, b + d + \psi_q(a, c)), \qquad (9.8)$$

$$(a, b)(c, d) = (ac, ad + bc). \qquad (9.9)$$

The unique maximal ideal of $GR(p^2, n)$ *is* $M = \{(0, x), x \in \mathbb{F}_q\}$, *the set of* p^{th}-*power elements is* $T = \{(x, 0), x \in \mathbb{F}_q\}$ *and* $GR(p^2, n)/M \cong \mathbb{F}_q$.

This cocyclic definition of addition in a Galois ring is introduced in [227], where arithmetic in the $p = 2$ case is used to great effect, to present the first known family of reversible $(2^{2t}, 4, 2^{2t}, 2^{2t-2})$-RDSs for which the exponent of the forbidden subgroup exceeds 2.

For $p = 2$, another description of $(GR(4, n), +)$ as a multiplicative subgroup of the quotient polynomial ring $\mathbb{F}_{2^n}[x]/(x^3)$ appears in [104]. In the same spirit as the above, it is designed to simplify computing sums of elements of $\mathcal{T}$.

LEMMA 9.3 *When* $p = 2$, *the Teichmüller cocycle is orthogonal. Equivalently, the Teichmüller set* $\mathcal{T}$ *is a* $(2^n, 2^n, 2^n, 1)$-*RDS in* $\mathbb{Z}_4^n$ *relative to* $\mathbb{Z}_2^n$. *Further,* $\mathcal{T} \in$ $\langle 2^n, 2^n, \mathbb{Z}_2^n, \mathbb{Z}_4^n, \mathbb{Z}_2^n, \mu \rangle$.

Proof. When $p = 2$, T (or $\mathcal{T}$) is the set of squares of $GR(4, n)$, and squaring is a bijection on T. The Teichmüller cocycle is $\psi_{2^n}(a, b) = -(ab)^{2^{n-1}} = \sqrt{ab}$. Clearly, for $a \neq 0$ and for each $c \in \mathbb{F}_{2^n}$, $|\{b \in \mathbb{F}_{2^n} : \psi_{2^n}(a, b) = \sqrt{ab} = c\}|$ $= |\{b \in \mathbb{F}_{2^n} : ab = c^2\}| = 1$. That is, by (7.24), the Teichmüller cocycle is orthogonal, yielding another proof (by Corollary 7.31) that the Teichmüller set for $p = 2$ is a $(2^n, 2^n, 2^n, 1)$-RDS in $\mathcal{R}$ relative to $2\mathcal{R}$ (see [31, p. 3, Theorem 1], or it is implied in [139, IIIC P3]). This proof avoids use of the Frobenius ring isomorphism on $\mathcal{R}$. In fact, $\mathcal{T}$ is isomorphic to the canonical RDS R in $E \cong \mathbb{Z}_4^n$ determined by field multiplication μ (see Lemma 9.40). In terms of the classification program of Chapter 8.3.2, $\psi_{2^n} \in \mathcal{B}(\mu)$. □

Note that the Teichmüller cocycle, which is both multiplicative and orthogonal, is used here to define an *addition*, not a multiplication, a fact which becomes significant for $p = 2$ when compared to Theorem 9.32.

When p is odd and $n = 1$, any $(p, p, p, 1)$-RDS must be splitting by [232, Result 2.2], and the Teichmüller set is not a $(p, p, p, 1)$-RDS in $\mathbb{Z}_{p^2}$. When $n = 2$, any abelian $(p^2, p^2, p^2, 1)$-RDS must be in the elementary abelian group by [232, Theorem 3.1], so again, the Teichmüller set is not a $(p^2, p^2, p^2, 1)$-RDS in $(\mathbb{Z}_{p^2})^2$.

Research Problem 61 *If p is an odd prime, does there exist an $n \geq 3$ such that the Teichmüller cocycle on $\mathbb{F}_{p^n}$ is orthogonal?*

9.1.2 Elliptic curve cryptosystems

Many standard cryptosystems (see Chapter 3.5) derive their security level from the perceived difficulty of solving a computational problem in acceptable time. These computational problems are often phrased in the arithmetic of an abelian group, commonly the cyclic group $GF(p^n)^*$.

For example, the perceived difficulty of solving the *Discrete Logarithm Problem* (DLP) in $GF(p^n)^*$ — given x and $y = x^k$, find k — determines the security level of several common algorithms such as DSA for digital signatures and Diffie-Hellman key exchange for high-speed symmetric cryptosystems such as DES.

Even so, p^n has to be very large, because there are subexponential algorithms for solving the DLP.

To avoid this, $GF(p^n)^*$ is replaced by the finite abelian group $E(\mathbb{F}_q)$ of rational points on an elliptic curve E defined over $\mathbb{F}_q = GF(p^n)$. The resulting compression of key alphabets has permitted, for example, implementations of elliptic curve cryptosystems on smart cards. As well, the DLP in elliptic curve groups (ECDLP) is orders-of-magnitude harder to solve than in $\mathbb{F}_q^*$. Refer to [27] for more mathematical background or [244] for information security applications.

Two cryptanalytic attacks using cocycles have surfaced in this area. The first is derived from the anomalous attack on the ECDLP. Well-known attacks on the DLP and ECDLP such as the baby-step/giant-step, Pohlig-Hellman and index-calculus methods are algebraic in nature. Recent attacks on anomalous elliptic curves (where $|E(\mathbb{F}_p)| = p$) have used p-adic methods, by lifting E using Hensel's lemma to an elliptic curve over the p-adic field $\mathbb{Q}_p$. The attack works because associated abelian groups $E_0(\mathbb{Q}_p), E_1(\mathbb{Q}_p)$ and $E_2(\mathbb{Q}_p)$ of points satisfy

$$E_0(\mathbb{Q}_p)/E_1(\mathbb{Q}_p) \cong E_1(\mathbb{Q}_p)/E_2(\mathbb{Q}_p) \cong (\mathbb{F}_p, +).$$

Once the ECDLP in $E(\mathbb{F}_p)$ is lifted to $E_1(\mathbb{Q}_p)$, the p-adic elliptic logarithm in $E_1(\mathbb{Q}_p)$ is applied and the ECDLP can be rewritten as a congruence mod p^2. The congruence is then solved; the p-adic elliptic logarithm in $E_1(\mathbb{Q}_p)$ is easy to compute. For an outline, see [27, Chapter V.3].

Gopalkrishna et al. [128] point out that this attack on the DLP should generalise to any abelian group G for which there is an abelian extension $0 \to C \to E_\psi \to$

$G \to 0$ of an abelian group C, provided a lift of the DLP may be solved more easily in the extension group E_ψ than in G. They illustrate by application to the DLP in $\mathbb{F}_p$, essentially using the Galois ring extension (9.5) to lift the equation $x^k = y$, $x \neq 0$ in $\mathbb{F}_p$ to one in $GR(p^2, 1) \cong \mathbb{Z}_{p^2}$ for which some solutions k are found in polynomial time. Such pairs (x, y) would be weak keys if used in a public key system.

Research Problem 62 *For which abelian extensions $0 \to C \to E_\psi \to G \to 0$ of groups G used in cryptographic protocols is there a more efficient solution to the DLP in E_ψ than in G ?*

The second attack, the *MOV attack*, defined for an integer m with $(m, p) = 1$, uses a cocycle from the group $E[m] \cong \mathbb{Z}_m^2$ of m-torsion points in $E(\mathbb{F}_q)$ to the group of mth roots of unity in the algebraic closure of $\mathbb{F}_q$. This cocycle, called the *Weil pairing*, is multiplicative and alternating [27, III.5]. For m a prime not dividing $q - 1$, its multiplicativity is used to reduce (in polynomial time) the ECDLP to the DLP in $\mathbb{F}_{q^r}$ for r the smallest integer such that $q^r \equiv 1 \bmod m$. Essentially, the MOV attack removes supersingular curves from contention in cryptographic applications, since for them, low values of r exist.

The success of the MOV attack and its generalisations has ushered in *pairing-based cryptography*. This new research area is based on the existence of a nondegenerate bilinear form (that is, a nontrivial multiplicative cocycle) — the *pairing* $\psi : G \times G \to C$, where $G \cong \mathbb{Z}_p$ is written additively, $C \cong \mathbb{Z}_p$ is written multiplicatively and ψ is efficiently computable. Both G and C are assumed to be groups used in cryptographic protocols, and pairing-based cryptography relates the computational complexity of a problem in one group to that of a possibly different problem in the other. Alternatively, schemes are developed which assume that the DLP is hard in both groups, that is, the *Bilinear Diffie-Hellman (BDH) assumption* holds: given $g, k, \ell, m \in G$, the computation of $\psi(g, g)^{k\ell m} \in C$ is hard.

Pairings have made identity-based public key encryption feasible. A survey of pairing-based protocols appears in [103].

It is easy to list the possible pairings theoretically, for p an odd prime. There are $p - 1$ nontrivial multiplicative cocycles in $Z^2(\mathbb{Z}_p, \mathbb{Z}_p)$, and each is orthogonal by Theorem 6.10.2. By Lemma 8.37, each is a coboundary. The corresponding cocyclic matrices are all Vandermonde matrices — essentially matrices of DFTs (4.1). If C is rewritten additively, the pairings are in the bundle $\mathcal{B}(\mu)$ of Example 8.3.1.

LEMMA 9.4 [159, Lemma 4.5, Theorem 4.6.iii] *Let p be an odd prime. The possible pairings $e_i : \mathbb{Z}_p \times \mathbb{Z}_p \to C = \langle x : x^p = 1 \rangle$ for $1 \leq i \leq p - 1$ are given by $e_i(j, k) = x^{ijk}$, $j, k \in \mathbb{Z}_p$. For each $1 \leq i \leq p - 1$, a representative 1-cochain $\phi_i : \mathbb{Z}_p \to C$ for which $e_i = \partial \phi_i$ is given by $\phi_i(0) = \phi_i(1) = 1$, $\phi_i(k) = x^{i k(k-1)/2}$, $2 \leq k \leq p - 1$.*

It remains to be seen whether this information helps when looking for efficient algorithms to compute e_i in specific groups of prime order.

Research Problem 63 *For which groups (of the same prime order p) used in cryptographic protocols and which values i, $1 \leq i \leq p - 1$, do efficient algorithms for computing e_i exist?*

More generally, G can be $E[m]$; for the Tate pairing, efficient algorithms with performance comparable to that of RSA have been found [21].

9.1.3 Cocyclic codes

Many good error-correcting block codes (see Chapter 3.2.1) are derived from $v \times v$ matrices M with entries in a commutative ring R with unity, which have in addition some internal structure.

The rows themselves may form the code. For example, the rows of a generalised Hadamard matrix with entries in $GF(q)$ form codes which meet the q-ary Plotkin bounds (Definition 4.32).

One common construction is to form the $v \times 2v$ matrix $[I_v \ M]$, in order to overcome any linear dependence which may occur between the rows of M. This becomes the generator matrix of a $[2v, v]$ linear block code over R. For example, when M is a binary Hadamard matrix, this method, already encountered in Definition 3.15, is especially useful for construction of binary self-orthogonal and self-dual codes (see, for example, [140, 307]). When M is circulant this construction determines the *double circulant* codes [237, p. 505], which have been discussed extensively in the literature. Examples of these are known which meet the Gilbert-Varshamov bound ([199], [237, pp. 506–507]).

Beth et al. [25] suggest codes constructed this way, where M is either Hadamard or circulant, are likely to have binomially distributed weights for large code length. Hence they should be good candidates for an asymptotically optimal code family on the binary symmetric channel, when decoded by a maximum-likelihood decoder with all codewords having equal prior probabilities.

This construction can be generalised to form $v \times nv$ matrices $[M_1 M_2 \ldots M_n]$, where the $v \times v$ matrices M_i have a common internal structure type. For instance, with M_1 binary normalised Hadamard and M_2 its complement, the rows of $[M_1 M_2]^\top$ form the Hadamard code $\mathcal{C}_v$ of Definition 3.13. If the M_i are all circulant, then, after permuting the coordinates i, $1 \leq i \leq nv$ of the code generated by $[M_1 M_2 \ldots M_n]$ into the order $1, v+1, 2v+1, \ldots, (n-1)v+1, \ldots, v, 2v, \ldots, nv$, we obtain an equivalent quasi-cyclic code with cyclic shift length n [263, p. 60]. More generally, a code is called *quasi-twisted* if an n-fold constacyclic (ω-cyclic) shift of a codeword results in another codeword [131]. If the M_i are all ω-cyclic, then after permuting the coordinates as for the circulant case, we obtain an equivalent quasi-twisted code.

A second common construction takes one or more group developed matrices $M = [\phi(g_i g_j)]$ and treats the first row of each as the coefficients of an element $\sum_{i=0}^{v-1} \phi(g_i) g_i$ in the group ring RG. The corresponding *group ring code* (or *group algebra code* or *G-code* if R is a field) is then defined to be the (one-sided) ideal in RG generated by these elements. All the rows of each M can be regarded as codewords. The cyclic codes ($G = \mathbb{Z}_v$) are the most extensively used of all codes. For example, the Fire codes are used to correct burst errors in transmission. There is a well developed theory of abelian codes, where the underlying group G is abelian. These include the generalised quadratic residue codes [263, Chapter 9]. Sabin [279] has applied these ideas to the (nonabelian) metacyclic group codes.

A third construction applies structured matrices in a completely different manner to obtain codes for the Gaussian (that is, additive white Gaussian noise or AWGN) channel. These are the group codes introduced by Slepian [296] which have many interesting properties from a communications point of view. For example, every word in a group code has the same probability of error on a Gaussian channel, and practical decoding techniques based on the algebraic structure are known. If $\sigma : G \rightarrow GL(n, \mathbb{R})$ is a faithful real representation of G by $n \times n$ orthogonal matrices and $\mathbf{x}$ is a unit vector in $\mathbb{R}^n$, the orbit $\{\sigma(g)\mathbf{x}, \ g \in G\}$ of $\mathbf{x}$ is called an $[m, n]$ *group code* if it contains m distinct vectors which span $\mathbb{R}^n$.

Intrigued by these structured matrix constructions, and by the observation that many good binary Hadamard codes (Definition 3.15) arise from cocyclic Hadamard matrices, the author investigated whether other familiar codes are constructed from cocyclic matrices.

Once the question is asked, it is remarkable how many well-known codes can be seen to derive in some fashion from cocycles. This ubiquity prompted the author to introduce a very general description of *cocyclic* codes [165].

Subsequent research [16, 17, 33, 167, 168, 169, 176, 177, 204, 205, 262, 291] has refined this notion, but the class of cocyclic codes is still far from being fully defined or understood. A preliminary — and not very defensible — first-order classification of cocyclic codes into three categories comes from cocyclic versions of the above three matrix constructions: direct, as ideal and as orbit. These categories are not necessarily disjoint (for instance, there are cyclic codes in both Categories I and II).

Very likely there are other categories. For instance, the convolutional codes identified by the algorithm of Arpasi and Palazzo [12] are cocyclic, since the strongly controllable time-invariant group codes associated to the convolutional encoder have trellis section groups defined in terms of cocycles. Quantum error-correcting codes are cocyclic if they are constructed from cocyclic Hadamard matrices ([24], cf. Chapter 3.2.2).

DEFINITION 9.5 *Categories of cocyclic codes.*

1. *Category I* (cocyclic matrix codes)*: codes derived from the rows of a coupled cocyclic matrix, or from equivalent objects.*

2. *Category II* (twisted group ring codes)*: codes defined as ideals in the twisted group ring $R^\psi G$ determined by the cocycle $\psi \in Z^2(G, R^*)$.*

3. *Category III* (projective group codes)*: codes defined as orbits of a vector in $\mathbb{F}^n$ under the action of a faithful projective representation of G in $GL(n, \mathbb{F})$.*

An outline of Categories II and III follows, before closer scrutiny of Category I.

9.1.3.1 Category II cocyclic codes

Suppose here that R is a commutative ring with unity and group of units R^*. Any group ring code in RE, where E is a central extension of G by an abelian group $C \leq R^*$, is cocyclic, since multiplication in E is defined using a cocycle.

Since the codewords are derived from knowledge of the multiplication in E, or, equivalently, from knowledge of ψ and the multiplication in G, it should be possible to replace a code over E by a corresponding code over G, with immediate gains in coding performance. In particular, the rate of the code improves $|C|$-fold. The coding arithmetic can be performed in the twisted group ring $R^\psi G$, a natural algebraic setting for the study of cocyclic extensions of group ring codes.

DEFINITION 9.6 [198] *If $C \leq R^*$ and $\psi \in Z^2(G,C)$, the* twisted group ring $R^\psi G$ *is defined to be the ring of formal sums* $\{\sum_{g \in G} r_g \, \overline{g} : r_g \in R, \, g \in G\}$, *with addition defined coefficient-wise,*

$$(r_g \, \overline{g}) + (s_g \, \overline{g}) = (r_g + s_g) \, \overline{g}, \; g \in G \; ,$$

and extended by linearity, and multiplication defined distributively using

$$(r_g \, \overline{g})(s_h \, \overline{h}) = \psi(g,h) r_g s_h \, \overline{gh}, \; g, h \in G \; .$$

There is an R-module structure on $R^\psi G$ determined by the action $s(r_g \, \overline{g}) = (sr_g) \, \overline{g}$. When ψ is a coboundary, then $R^\psi G$ and RG are equivalent as twisted rings [198, Lemma 2.2]. The mapping $RE_\psi \rightarrow R^\psi G$ given by $(u, g) \mapsto u \, \overline{g}$ is a ring homomorphism.

DEFINITION 9.7 *A* twisted group ring code *is an ideal in a twisted group ring* $R^\psi G$.

Given ψ, any group ring code in RG determines a cocyclic group ring code in $R^\psi G$, by linear extension of the mapping $g \mapsto \overline{g}$.

Example 9.1.1 *Let $G = \mathbb{Z}_v$ and ψ_ω be as in Example 6.2.2, where $C \leq R^*$. Consider the cyclic code $\langle \sum_{i=0}^{v-1} m_i a^i \rangle$ in RG. The corresponding ideal $\langle \sum_{i=0}^{v-1} m_i a^i \rangle$ in $R^{\psi_\omega} G$ is the span of the codewords $(\sum_{i=0}^{v-1} m_i \overline{a^i}) \overline{a^j} = \sum_{i=0}^{v-1} \omega^{\lfloor (i+j)/v \rfloor} m_i \overline{a^{i+j}}$. This is the quasi-twisted code of a (back) constacyclic matrix.*

Category II includes as a basic case all the group ring codes, in particular the cyclic codes. The simplest nonbasic case includes the *constacyclic* and quasi-constacyclic codes. Hughes [176, 177] uses this point of view to develop a structure theorem for RE when C is a particular kind of subgroup of R^* which he terms a *subtraction* subgroup: for all $a \neq b \in C$, $a - b \in R^*$. All known instances of subtraction subgroups are cyclic.

THEOREM 9.8 [177, Theorem 3.1, Corollary 3.3] *Let $\psi \in Z^2(G,C)$, where C is a subtraction subgroup of R^* of order w. Then $RE_\psi \cong \bigoplus_{\ell=0}^{w-1} R^{\psi^\ell} G$. Furthermore, every group ring code in RE_ψ decomposes into, and may be constructed from, the direct sum of twisted group ring codes in $R^{\psi^\ell} G$, $0 \leq \ell \leq w - 1$.*

In the case E is abelian and R is a suitable finite local ring, the component twisted group ring codes in Theorem 9.8 are multidimensional constacyclic codes. Hughes uses his decomposition to prove, for particular E and R, that no self-dual group ring codes exist, a result then fully generalised by Willems [322].

Sundar Rajan and coauthors [204, 205, 291] have applied these ideas to abelian and dihedral groups G with Galois ring coefficients R (calling the twisted group ring codes *consta-abelian* and *consta-dihedral*, respectively) to develop analogous transforms and decomposition theorems.

9.1.3.2 Category III cocyclic codes

A mapping $\rho : G \to GL(n, \mathbb{F})$ is a *projective (matrix) representation of G over the field* $\mathbb{F}$ if there exists a mapping $\psi : G \times G \to \mathbb{F}^*$ such that $\rho(1) = I_n$ and $\rho(g)\rho(h) = \psi(g, h)\rho(gh)$, for all $g, h \in G$. The associativity of G forces ψ to be a normalised cocycle. Clearly, an ordinary linear representation of G is a projective representation of G for which the cocycle ψ is trivial.

If the centre $\mathbb{F}^* I_n$ (the nonzero scalar multiples of the identity) of $GL(n, \mathbb{F})$ is factored out, we obtain the *projective general linear group* $PGL(n, \mathbb{F})$, and the composition of a projective representation ρ with the canonical epimorphism is a homomorphism $G \to PGL(n, \mathbb{F})$. Conversely, any homomorphism $G \to PGL(n, \mathbb{F})$ determines a projective representation of G. In fact there is also an equivalence between the projective representations ρ of G and the ordinary representations σ of a suitable central extension E of G for which there is a $\tau : E \to \mathbb{F}^*$ such that $\sigma(e) = \rho(\pi(e))\tau(e), \forall e \in E$. The interested reader is referred to Karpilovsky [198] for detailed information on the projective representations of finite groups.

Slepian's group codes can be viewed as a special case of a more general definition of 'projective group codes for the Gaussian channel'. These could be constructed either as the orbit of a unit vector in $\mathbb{R}^n$ under the action of a faithful real projective representation ρ of G in $GL(n, \mathbb{R})$ or, alternatively, as the orbit of a unit vector in $\mathbb{R}^n$ under the action of the image of $\rho(G)$ in $PGL(n, \mathbb{R})$.

The codes resulting from these constructions are cocyclic, because the presence of the cocycle is intrinsic to the definition of the projective representation. Trivially, any group code for the Gaussian channel is cocyclic. We suggest these ideas are worthy of further study.

Research Problem 64 *Let G be a group, $\mathbb{F}$ a field and $\rho : G \to GL(n, \mathbb{F})$ a faithful projective representation. What is the theory of the codes in $\mathbb{F}^n$ which arise as $\rho(G)$-orbits?*

9.1.3.3 Category I cocyclic codes

A cocyclic matrix code, or Category I cocyclic code, is a code derived from some or all of the rows of a coupled cocyclic matrix (Definition 7.18), or from equivalent objects. If the matrix is from a particular family, such as generalised Hadamard or Butson, the code will be correspondingly labelled.

Thus a *cocyclic Hadamard code* will mean a code derived from any of the equivalent objects of the Five-fold Constellation (Theorems 7.29 and 7.40), or from objects equivalent to them.

In Definition 4.34, w-ary Hadamard codes, defined specifically in terms of a $GH(w, v/w)$, are classified by construction technique. By analogy, we obtain a rough second level classification for cocyclic matrix codes. Each class contains very well-known codes.

Class A are codes consisting of the rows of a coupled cocyclic matrix M or its translates — these may be linear or nonlinear (but perhaps additive). For instance, the simplex, first-order Reed-Muller and punctured first-order Reed-Muller codes are examples of binary Class A cocyclic Hadamard codes, by Chapters 3.2.2 and 6.4.1.

Class B are linear codes which are derived from some rows of M or from objects equivalent to M. For instance, the duals of the simplex, first-order Reed-Muller and punctured first-order Reed-Muller codes are defined from some rows of a Sylvester Hadamard matrix, so they are Class B cocyclic Hadamard codes. Since the Paley Type I Hadamard matrices are Ito Hadamard matrices, they are dihedral-cocyclic (Chapter 6.4.4), so the resulting extended quadratic residue codes are binary Class B cocyclic Hadamard codes.

Class C are linear codes with generator matrix $[I \ A]$ for some matrix A associated with M. For instance, A could be the incidence matrix of an associated design or could be M itself. By Example 3.2.2 and Chapter 6.4.4 the $[24, 12, 8]$ extended Golay code is a binary Class C D_{12}-cocyclic Hadamard code.

The fact that the Teichmüller set is a central semiregular RDS in $\mathbb{Z}_4^n$ relative to $\mathbb{Z}_2^n$ (Lemma 9.3) allows us to identify some quaternary codes in Chapter 4.4.3 as Class B cocyclic Hadamard codes. By the Five-fold Constellation (Corollary 7.31) the Teichmüller set is equivalent to a $\mathbb{Z}_2^n$-cocyclic generalised Hadamard matrix over $\mathbb{Z}_2^n$, namely $M_{\psi_{2^n}}$, where ψ_{2^n} is the Teichmüller cocycle (Definition 9.1).

Example 9.1.2   *The $\mathbb{Z}_4$-linear Kerdock code is a $\mathbb{Z}_2^n$-cocyclic Hadamard code, since it is generated by the all-1s vector and the Teichmüller set vector [139]. Similarly, its dual (the $\mathbb{Z}_4$-linear Preparata code) and the $\mathbb{Z}_4$-linear Goethals code are cocyclic Hadamard codes, since their parity check matrices are derived from the Teichmüller set.*

Other good q-ary codes, for q a prime power, are also Class B cocyclic matrix codes.

Example 9.1.3   *The rows of a $k \times v$ generating matrix for a generalised Reed-Solomon code $GR_k(\mathbf{c}, \mathbf{1})$, where $\mathbf{c} = (1, c, \dots, c^{v-1})$ for some $c \in GF(q)$ of order v (see [263, (14) p. 73]), are rows of a cocyclic matrix (Example 6.2.3), so these Reed-Solomon codes are Class B cocyclic matrix codes.*

In particular, for $v = p$, an odd prime, the DFT generalised Hadamard matrix of Example 4.1.1 is $\mathbb{Z}_p$-cocyclic over $\mathbb{Z}_p$. The resulting Reed-Solomon codes are p-ary Class B cocyclic Hadamard codes.

Example 9.1.4   *[168, 4.3] For $G = C = (\mathbb{F}_{q^r}, +)$, $q = p^n$, the field multiplication cocycle μ of Example 6.2.7 is orthogonal. The corresponding canonical RDS*

$\{(1, g) : g \in \mathbb{F}_{q^r}\}$ *of Theorem 7.14 is isomorphic to the quadratic relative difference set* $Q = \{(g^2, g) : g \in \mathbb{F}_{q^r}\}$. *The orthogonal cocycle corresponding to* Q *is* $-\mu$. *Since for* $q^r = p$, Q *identifies the quadratic residues of* $\mathbb{F}_p$, *the binary quadratic residue codes are Class B cocyclic Hadamard codes.*

New cocyclic Butson codes are described in the next subsection and new cocyclic Hadamard codes are presented in Section 9.4.

9.1.4 Cocyclic Butson matrices and codes

By Definitions 4.35 and 7.18, a Generalised Butson Hadamard matrix B of order $v \geq 2$, indexed by G with entries from $N \leq R^*$, is coupled cocyclic if there is a factor pair $(\psi, \varepsilon) \in F^2(G, N)$ such that $B \sim M_{(\psi,\varepsilon)}$.

For instance the Fourier Transform matrix $\mathcal{F}_G$ for an abelian group G of order v and exponent m is a cocyclic $BH(m, v)$, by Example 6.2.5.

Not all GBH matrices are cocyclic, however. The $GH(3, 2)$ of Example 4.3.1 is a $BH(3, 6)$ by Example 4.3.3 but is not cocyclic by Example 7.4.2.

A particular advantage in restricting to coupled cocyclic matrices when searching for GBH matrices is computational. The invertibility condition defining a GBH reduces to preservation of the necessary zero-row-sum condition under all coupling actions.

Compare this with the computational cutdown achieved when searching for generalised Hadamard matrices amongst coupled cocyclic matrices (Theorem 7.22), where the Hadamard condition reduces to row balance of the decoupled matrix.

LEMMA 9.9 *If* $(\psi, \varepsilon) \in F^2(G, N)$, *denote the coupled cocyclic matrix* $M_{(\psi,\varepsilon)}$ *of (7.28) by* $[m(g, h)]_{g,h \in G}$. *Suppose* $N \leq R^*$ *as in Definition 4.35. Then* $M_{(\psi,\varepsilon)}$ *is a GBH matrix if and only if, for all* $g \neq 1$ *and all* $k \in G$,

$$\sum_{h \in G} m(g, h)^{\varepsilon(k)} = 0.$$

In particular, if $\varepsilon \equiv 1$, $M_{(\psi,1)}$ *is a GBH matrix if and only if, in each row of the decoupled matrix* M_ψ *apart from the row indexed by 1, the elements sum to 0 in R.*

Proof. Work with the equivalent matrix $\overline{M}_{(\psi,\varepsilon)}$ of Corollary 7.19. For $h \neq k \in G$, let $S(h, k) = \sum_{g \in G} \psi(h, h^{-1}g)\psi(k, k^{-1}g)^{-1}$. Let $x = k^{-1}g$, so by (7.2)

$$S(h, k) = \sum_{x \in G} \psi(h, h^{-1}kx)\psi(k, x)^{-1} = \sum_{x \in G} \psi^{-1}(h^{-1}k, x)^{\varepsilon(h)} \, \psi(h, h^{-1}k).$$

Thus $S(h, k) = 0$ if and only if $\sum_{x \in G} \psi^{-1}(h^{-1}k, x)^{\varepsilon(h)} = 0$ if and only if (by (7.1) and Lemma 7.3.4) $\sum_{h \in G}(\psi^{-1}(g, h)^{\varepsilon(g)^{-1}})^{\varepsilon(k)} = 0$. If $\varepsilon \equiv 1$, the row sum condition on M_ψ follows by adapting the row quotient argument in the proof of Theorem 7.22. $\square$

To illustrate, the cocyclic matrix $M_\psi = \begin{bmatrix} 1 & 1 & 1 \\ 1 & \alpha & \gamma \\ 1 & \gamma & \alpha^{-1}\gamma \end{bmatrix}$ of Example 6.3.1 is GBH if and only if $1 + \alpha + \gamma = \alpha + \gamma + \alpha\gamma = 0$ in R, only if $\gamma = \alpha^{-1}$ in R^*.

Therefore, with $\alpha = \beta, \gamma = \beta^2$ and $1 + \beta + \beta^2 = 0$, the $GBH(3,3)$ B_3 of Example 4.5.4 is cocyclic. (When $\beta = e^{2\pi/3}$, B_3 is the 3-point DFT matrix $\mathcal{F}_3$.)

9.1.4.1 Cocyclic jacket GBH matrices

The examples of primary GBH matrices with jacket weight 1 listed in Chapter 4.5.1 are all cocyclic. When $v = 2$, we get $\mathcal{S}_1 = \mathcal{F}_2$. When $v = 2n \geq 4$ and α is a complex primitive v^{th} root of unity, equation (4.20) shows that there is an indexing of $\mathcal{F}_v$ which equals $\mathcal{K}_n(\alpha)$.

The matrix $K_2(r)$ is cocyclic, where $G = \mathbb{Z}_2 \times \mathbb{Z}_2$ and C is the abelian subgroup of R^* generated by r and -1, on setting $\alpha = \beta = -r$, $\gamma = r^{-1}$ and $\kappa = 1$ in Example 6.3.2.

The matrix $K_4(i)$ is cocyclic where $G = \mathbb{Z}_8$ and $C = \{\pm 1, \pm i\} \leq \mathbb{C}^*$. The mapping $\phi : G \to C$ with $\phi(0) = \phi(1) = 1, \phi(2) = i, \phi(3) = \phi(4) = 1, \phi(5) = -1, \phi(6) = i, \phi(7) = -1$ determined by the perfect quaternary sequence (4.5) defines a coboundary $\partial\phi(k,l) = \phi(k+l) - \phi(k) - \phi(l)$. Switching to mixed radix indices (with $n = 4$) and permuting row and column indices as in equation (4.20) shows that there is an indexing of $M_{\partial\phi}$ which equals $K_4(i)$.

Application of Example 6.2.14 to Corollary 4.41 gives the remaining examples.

9.1.4.2 New cocyclic Butson matrices and codes

Pinnawala and Rao [262] have constructed cocyclic Butson matrices by applying the trace map to the Galois ring $GR(2^s, n)$, the Galois extension of degree n of the ring $\mathbb{Z}_{2^s}$.

Their construction parallels the Singer difference set construction of cocyclic Hadamard matrices equivalent to $\mathcal{S}_n$ which is shown in Lemma 2.14, and to which it specialises if $s = 1$.

In brief, $GR(2^s, n) = \mathbb{Z}_{2^s}[\zeta]$, where ζ is a root of a polynomial of degree n, irreducible over $\mathbb{Z}_{2^s}$, and ζ has order $2^n - 1$ in $GR(2^s, n)$. The abelian group $(GR(2^s, n), +)$ is generated by $\{\zeta^j, 0 \leq j \leq n - 1\}$ and is isomorphic to $(\mathbb{Z}_{2^s})^n$. The Teichmüller set is $\mathcal{T} = \{0, \zeta^j, 0 \leq j \leq 2^n - 2\}$, and every element of $GR(2^s, n)$ has a unique 2-adic representation $r = \sum_{j=0}^{s-1} 2^j r_j$, where $r_j \in \mathcal{T}$. The element r is invertible if and only if $r_0 \neq 0$. (Compare with the description of $GR(p^2, n)$ given in Section 9.1.1.)

The Frobenius automorphism of $GR(2^s, n)$ is $\sigma(r) = \sum_{j=0}^{s-1} 2^j r_j^2$. The trace map Tr: $GR(2^s, n) \to \mathbb{Z}_{2^s}$, given by $\mathrm{Tr}(r) = \sum_{j=0}^{n-1} \sigma^j(r)$, is an abelian group homomorphism.

THEOREM 9.10 [262, Theorem 5.1] *Let* $\omega = e^{2\pi i/2^s}$ *and let* $\psi : (GR(2^s, n), +) \times (GR(2^s, n), +) \to \langle\omega\rangle$ *be*

$$\psi(x, y) = \omega^{\mathrm{Tr}(xy)}, \quad x, y \in (GR(2^s, n), +).$$

Then $\psi \in Z^2((\mathbb{Z}_{2^s})^n, \mathbb{Z}_{2^s})$ *and* M_ψ *is a cocyclic* $BH(2^s, 2^{sn})$.

Proof. That ψ is a cocycle follows directly, or on applying Example 6.2.12 with $\gamma = \vartheta \circ \mathrm{Tr}$ to multiplication in $GR(2^s, n)$ (Example 6.2.7), where ϑ is the exponentiation isomorphism $\mathbb{Z}_{2^s} \cong \langle \omega \rangle$ given by $1 \mapsto \omega$. That M_ψ is a Butson matrix follows from Lemma 9.9 and the distribution properties of the trace map. $\qquad \square$

They then show the rows of the cocyclic matrix $[\text{Tr}(xy)]_{x,y \in GR(2^s,n)}$ form a $\mathbb{Z}_{2^s}$-linear code with generator matrix consisting of the n linearly independent rows

$$[\text{Tr}(\zeta^j y), \; y \in GR(2^s, n)], 0 \leq j \leq n - 1,$$

and consequently its columns consist of all the distinct vectors in $(\mathbb{Z}_{2^s})^n$. Hence it is equivalent to the simplex code of type α over $\mathbb{Z}_{2^s}$ (introduced by Gupta [132], following Carlet [45]).

COROLLARY 9.11 [262, Theorem 5.2] *The rows of* $[\text{Tr}(xy)]_{x,y \in GR(2^s,n)}$ *form a simplex code of type* α *over* $\mathbb{Z}_{2^s}$ *which is a Class A cocyclic Butson code.*

9.2 NEW GROUP DEVELOPED GENERALISED HADAMARD MATRICES

This Section develops one of the most accessible and important constructions of generalised Hadamard matrices: that of coupled group developed matrices from splitting factor pairs. There are two objectives for the Section. The first is to use the relationship of PN functions (particularly planar functions) and orthogonal splitting factor pairs to construct coupled group developed GH matrices. The second is to use splitting factor pairs and the literature on abelian PN functions as the basis for a general theory of highly nonlinear functions (slightly improved from the earlier version [161]).

9.2.1 Group developed GH matrices and PN functions

We resume study of coupled G-developed $GH(w, v/w)$ over N, which are completely determined by their top rows $(\phi(x), \; x \in G)$ and homomorphism $\varrho : G \to \text{Aut}(N)^{\text{op}}$. By Definition 7.34, such a top row describes a perfect nonlinear (PN) function ϕ relative to ϱ, and vice versa.

Perfect nonlinear functions are optimal with respect to a specific measure of nonlinearity: uniformity of their difference distribution (that is, balance of their directional derivatives) perhaps with an additional twist (as in Definition 7.34). Nyberg's original PN functions [251, Definition 3.1] for $G = \mathbb{Z}_r^n$ and $N = \mathbb{Z}_r^m$, $n \geq m$, were designed as S-box functions with maximum resistance to differential attack. The binary case $r = 2$ appears briefly in Chapter 3.5.2. However, there is no requirement in cryptographic applications to restrict G and N to the elementary abelian 2-groups. The cryptosystem GOST (a Russian analogue of the DES) uses S-box functions with $G = \mathbb{Z}_{16}$ and $N = \mathbb{Z}_2^4$ (cited in [268]).

If we know one PN function, we can generate many others. Obviously, if ϕ is PN relative to ϱ, then $\mathcal{D}(\phi, \varrho) = v/w$ by (7.33). Thus every function in its bundle $\text{b}(\phi, \varrho)$ is PN relative to some ϱ', by Theorem 8.24.

A less obvious example is its *dual* PN function ϕ^*, produced in the splitting case of Lemma 7.26.

LEMMA 9.12 *Let G and N be finite groups of order v and w, respectively, where $w|v$, let $\phi \in C^1(G, N)$ and let $\varrho : G \to \mathrm{Aut}(N)^{\mathrm{op}}$ be a homomorphism. Define the dual $\phi^* \in C^1(G, N)$ of ϕ relative to ϱ to be $\phi^*(x) = \phi^{-1}(x^{-1})^{\varrho(x)}$.*

1. *If $H = [\phi(xy)^{\varrho(x^{-1})}]_{x,y \in G}$ then $H^* \sim [\phi^*(xy)^{\varrho(x^{-1})}]_{x,y \in G}$.*

2. *ϕ is PN relative to ϱ if and only if ϕ^* is PN relative to ϱ .*

Proof. For part 1, let $(\psi, \varepsilon) = (\partial^{-1}\phi, \overline{\phi}\varrho) \sim_\phi (1, \varrho)$ as in Corollary 7.21. By Lemma 7.3.4, $\varepsilon^*(x) = \overline{\psi^{-1}(x, x^{-1})}\varepsilon(x) = \overline{\phi^*(x)}\varrho(x) = \varrho(x)\overline{\phi^{-1}(x^{-1})}$. Thus by (7.32) and (7.16), $(\psi^*, \varepsilon^*) \sim_{\phi^*} (1, \varrho)$. Part 2 follows from Lemma 7.26.   □

Recall that if ϕ is PN it may still be inequivalent to ϕ^* — cf. Example 7.3.1, although it is not known whether the inequivalent transpose pair of cocyclic Hadamard matrices in that example are group developed. By Corollary 8.13, it is also possible that $\phi^* \notin \mathbf{b}(\phi, \varrho)$ but ϕ and ϕ^* generate equivalent $GH(w, v/w)$.

Research Problem 65 *Find a dual pair ϕ, ϕ^* of PN functions which are the top rows of inequivalent coupled G-developed $GH(w, v/w)$ over N. Find a dual pair ϕ, ϕ^* of PN functions for which $\mathbf{b}(\phi, \varrho) \neq \mathbf{b}(\phi^*, \varrho)$ but which are the top rows of equivalent coupled G-developed $GH(w, v/w)$ over N.*

Recognition of new PN functions may be made easier because of their numerous identities in the Splitting Five-fold Constellation (Theorems 7.30 and 7.35).

For clarity and ease of reference, the 'vanilla' case $\varrho \equiv 1$ of the Splitting Five-fold Constellation is recorded next, using (7.33) and Corollaries 7.6, 7.15 and 7.16. The last three equivalences have already been encountered as Corollary 4.23.

THEOREM 9.13 *Let G and N be finite groups of order v and w, respectively, where $w|v$ and let $\phi \in C^1(G, N)$. Then the following are equivalent:*

1. *the function ϕ is PN;*

2. *in the group ring $\mathbb{Z}N$,*

$$\forall\, x \neq 1 \in G, \quad \sum_{y \in G} \phi(xy)\phi(y)^{-1} = (v/w)\, N; \qquad (9.10)$$

3. *the splitting factor pair $(\partial^{-1}\phi, \overline{\phi})$ is orthogonal;*

4. *the G-developed matrix $[\phi(xy)]_{x,y \in G}$ is a $GH(w, v/w)$ over N;*

5. *the transversal $R_\phi = \{(\phi(x), x) : x \in G\}$ is a splitting $(v, w, v, v/w)$-RDS in $N \times G$ relative to $N \times \{1\}$, isomorphic to the canonical RDS $R_{(\partial^{-1}\phi, \overline{\phi})} = \{(1, x) : x \in G\}$ in $E_{(\partial^{-1}\phi, \overline{\phi})}$ relative to $\overline{N}$;*

6. *the design $\mathrm{dev}(R_\phi)$ is a $(v, w, v, v/w)$-divisible design with regular group $N \times G$, class regular with respect to $N \times \{1\}$.*

The situation for *abelian* PN functions (between abelian groups G and C) is surveyed in [48] and in [268]. Carlet and Ding [48] also record the equivalence of abelian PN functions with G-developed generalised Hadamard matrices over C [48, Theorem 12], a result going back at least to 1992 [81]. Pott [268] records the equivalence of abelian PN functions and abelian splitting semiregular relative difference sets. This equivalence had been the author's introduction to PN functions, through the (then) startling realisation that orthogonal coboundaries $\partial\phi$ were equivalent to PN functions ϕ [154, Corollary 2].

For completeness, the case $\varrho \equiv 1$ of Lemma 9.12 is recorded next. The second equivalence is (9.10) for ϕ^*, rewritten in terms of ϕ.

LEMMA 9.14 *Let G and N be finite groups of order v and w, respectively, where $w|v$, and let $\phi \in C^1(G, N)$. Then ϕ is PN*

1. *if and only if the function ϕ^* is PN, where $\phi^*(x) = \phi(x^{-1})^{-1}$;*

2. *if and only if in the group ring $\mathbb{Z}N$,*

$$\forall\, y \neq 1 \in G, \quad \sum_{x \in G} \phi^{-1}(xy)\phi(x) = (v/w)\, N. \qquad (9.11)$$

9.2.1.1 Example: Planar functions — the case $v = w$, odd

If $v = w$ (and $\varrho \equiv 1$), we do know rather more about PN functions. They are called *planar* functions, and were introduced by Dembowski and Ostrom [94] to describe affine planes with certain properties (cf. [60, p. 21]). It is known that for a planar function to exist, v must be odd. When G is abelian, v must be an odd prime power [30], and it is conjectured that this must be true for all G. When $G = C$ is cyclic, v must be odd and square-free (see [222]), so v must be an odd prime.

Examples of planar functions when v is an odd prime power $q = p^n$ do exist, but very few are known to describe nonisomorphic planes. They are energetically sought, both in the hope of discovering new planes and for cryptographic applications. Most effort focusses on the elementary abelian G of odd order, and concentrates on polynomial power functions in $GF(p^n)$ (see, for example, [63, 101, 146]).

For $G = (GF(q), +)$, every $\phi : G \to G$ may be obtained as the evaluation map of some polynomial $\phi(x) \in GF(q)[x]$ of degree less than q. The homomorphisms $\mathrm{Hom}(G, G)$ are precisely the linearised polynomials (8.33). The elements of $C^1(G, G)/\mathrm{Hom}(G, G) \cong B^2(G, G)$ — see (8.27) — are represented by the polynomials $\phi(x) \in GF(q)[x]$ with $\phi(0) = 0$ and with no linearised summand, so we make the splitting identifications

$$C^1(G, G) = C_0^1(G, G) \oplus \mathrm{Hom}(G, G); \qquad (9.12)$$

$$C_0^1(G, G) = \left\{ \phi(x) = \sum_{i=1}^{q-1} \lambda_i x^i, \lambda_i \in GF(q) : i \neq p^j, 0 \leq j \leq n - 1 \right\}. \qquad (9.13)$$

Planarity of a polynomial is preserved by affine transformations: if $\phi(x)$ is planar, so is $\alpha\,\phi(\lambda x + \mu) + \beta$, where $\alpha \neq 0, \lambda \neq 0, \beta, \mu \in G$. For instance, if ϕ is planar and $a \in G$, then the shift $\phi \cdot a$ is also planar, because by (8.21) it is the affine transformation $(\phi \cdot a)(x) = \phi(x + a) - \phi(a)$.

More generally, many constructions which preserve planarity of a polynomial: addition of a linearised polynomial of G, shift by an element of G, or pre- or post-composition with an LPP, are equivalences according to Definition 8.22. This underlines the naturalness of Definition 8.22, because planar functions in the same affine bundle will determine isomorphic planes (see [63, p. 169]). For planar functions, Corollaries 8.20 and 8.21 translate as follows.

COROLLARY 9.15 *Let $|G| = |N|$ be odd and $\varrho \equiv 1$. The map $\widehat{\partial}^{-1} : \mathbf{b}(\phi) \mapsto \mathcal{B}(\partial^{-1}(\phi))$ is a set isomorphism from the set of bundles of planar functions in $C^1(G, N)$ to the set of bundles of orthogonal splitting factor pairs. When N is abelian, $\widehat{\partial}^{-1} = \widehat{\partial} : \mathbf{b}(\phi) \mapsto \mathcal{B}(\partial(\phi))$, the coboundary operator.*

We will see in Section 9.3.2 that if $G = N = (GF(p^n), +)$, planar functions from different bundles might still determine isomorphic planes. However, this cannot occur if n is odd: in this case, by Theorem 9.33 and Corollary 9.42, different bundles of planar functions determine different planes.

A list of known families of planar functions for which $\partial\phi$ is multiplicative may be derived from [197, §5]. The total number of corresponding known pairwise nonisomorphic planes of order p^n is less than $\log_2 p^n$. If n is even, there are 5 construction methods known (including the first two in the next Example 9.2.1). If n is odd, it is possible (cf. [63, p. 183]) that every planar function on $G = (GF(p^n), +)$ determines a plane isomorphic to one determined by the following four families. Three are power mappings. The first three determine multiplicative $\partial\phi$, but for $b \not\equiv \pm 1 \pmod{2n}$ the fourth does not.

Example 9.2.1 *Let $q = p^n$, for p an odd prime, and $G = (GF(q), +)$. The following four families of normalised functions $G \to G$ are planar (PN):*

1. $\phi_1(x) = x^2$, $x \in G$;

2. $\phi_2(x) = x^{p^b + 1}$, $x \in G$, where $n/(b, n)$ is odd;

3. (Ding and Yuan [95], Coulter and Henderson [62, Theorem 1])
   $\phi_3(x) = x^{10} - ux^6 - u^2 x^2$, $x \in G$, where $p = 3$, n is odd and $u \neq 0$, or $n = 2$ and $u = \pm 1$;

4. (Coulter and Matthews [63, Theorem 4.1, Theorem 6.2], see also [147])
   $\phi_{(n,b)}(x) = x^{(3^b + 1)/2}$, $x \in G$, where $p = 3$, b is odd and $(b, n) = 1$.

It seems reasonable to suppose that in the wider context of functions which are PN relative to ϱ, it will be easier to find examples with $v = w$. As far as the author is aware, this is virgin territory for research.

DEFINITION 9.16 *Let G and N be finite groups of the same order v and let $\varrho :$ $G \to \mathrm{Aut}(N)^{\mathrm{op}}$ be a homomorphism. A function $\phi \in C^1(G, N)$ which is PN* *relative to ϱ will be called* planar relative to ϱ. *That is, by (7.33), φ is planar relative to ϱ if and only if, for every $x \neq 1 \in G$ and $u \in N$,*

$$|\{y \in G : \phi(xy)(\phi(y)^{-1})^{\varrho(x)} = u\}| = 1. \qquad (9.14)$$

Research Problem 66 *Find new equivalence classes of planar functions relative to ϱ.*

Research Problem 67 *What is the geometric significance of a planar function relative to ϱ when $\varrho \neq 1$?*

9.2.2 PN functions and a theory of highly nonlinear functions

In the binary case, when PN functions exist, they are also bent, that is, optimal for another measure of nonlinearity: they are maximally distant (in a specific sense) from linear functions. The measuring instrument is the Walsh-Hadamard Transform. The function $\phi : V(n,2) \to V(m,2)$, with even $n \geq 2m$, is PN if and only if for each $c \neq 0 \in V(m,2)$ the component ϕ_c is bent, that is, if and only if for each $c \neq 0 \in V(m,2)$ the WHT of $(-1)^{\phi_c}$ takes only the values $\pm 2^{n/2}$ (see Definition 3.31.1 and Corollary 3.33).

The analogue of this result holds for abelian PN functions $\phi : G \to C$, if the rôle of the Walsh-Hadamard Transform is taken by the Fourier Transform for the abelian group C (Definition 4.3), and for each $c \in C$, the *component* ϕ_c of ϕ is defined to be $\phi_c = \chi_c \circ \phi$. The requisite definition of a *bent* function comes from Logachev et al. [228] (cited in [48]).

DEFINITION 9.17 *Let C be a finite abelian group and suppose $\varphi : C \to \mathbb{C}$ takes values in the complex unit circle. Then φ is bent if its FT $\widehat{\varphi}$ has constant magnitude $|\widehat{\varphi}(a)| = \sqrt{|C|}$ for every $a \in C$.*

For example, if p is prime and $C = \mathbb{Z}_p^n$, we know from Example 4.1.3 that $\mathcal{F}_C = \otimes^n \mathcal{F}_p$. Matsufuji and Suehiro [240] state some of the following p-ary generalisations of Lemma 3.29, for which proof is analogous.

Example 9.2.2 *For p prime, let $\omega = \exp(-2\pi i/p)$ and order $\mathbb{Z}_p^n$ lexicographically. For a function $f : \mathbb{Z}_p^n \to \mathbb{Z}_p$, define $F : \mathbb{Z}_p^n \to \mathbb{C}$ by $F(\mathbf{v}) = \omega^{f(\mathbf{v})}$. The following are equivalent:*

1. *F is bent;*

2. *$|\widehat{F}(\mathbf{u})| = p^{n/2}$ for all $\mathbf{u} \in \mathbb{Z}_p^n$;*

3. *$[p^{-n/2}\widehat{F}(\mathbf{u} + \mathbf{v})]$ is a Butson matrix;*

4. *$[F(\mathbf{u} + \mathbf{v})]$ is a Butson matrix.*

Pott [268] extends the definition of *maximal nonlinearity* from the binary case (Definition 3.31.2) to the abelian case. As for bentness, this is a character-theoretic definition, which Pott gives in terms of the characters of a transversal of C in $C \times G$.

DEFINITION 9.18 *Let G and C be finite abelian groups, let $\widehat{C \times G}$ be the character group of $C \times G$, let $\phi : G \to C$ and let $R_\phi = \{(\phi(g), g) : g \in G\} \subset C \times G$. The* maximum nonlinearity of ϕ *is $\mathcal{L}(\phi) = \max\{|\chi(R_\phi)| : \chi \neq \chi_0 \in \widehat{C \times G}\}$ and ϕ is* maximally nonlinear *if it attains the minimum possible value for $\mathcal{L}(\phi)$ for functions from G to C.*

Pott shows that $\mathcal{L}(\phi) \geq \sqrt{|G|}$. When $|C|$ divides $|G|$, he shows that functions with maximum nonlinearity coincide with PN functions by proving the transversal R_ϕ is a splitting abelian RDS. His proof invokes the dual definition, in terms of its characters, of an abelian (v, w, k, λ)-RDS [24, Vol. 1, Lemma 10.9]. (This is found by taking the FT of (4.12), with the converse following by Fourier inversion.)

He suggests that the transversal $R_\phi = \{(\phi(g), g) : g \in G\}$ is the correct instrument for measuring the nonlinear behaviour of any $\phi : G \to C$.

THEOREM 9.19 *Let G and C be abelian groups of orders v and w, respectively, where $w|v$, and let $\phi \in C^1(G, C)$. Then the following are equivalent:*

1. *ϕ is PN;*

2. *[48, Theorem 16] for every $c \neq 1 \in C$ the component $\phi_c = \chi_c \circ \phi$ is bent, that is, its FT $\widehat{\phi_c}$ has magnitude $\widehat{\phi_c}(g) = \sqrt{v}$ for every $g \in G$;*

3. *[268, Theorem 8] ϕ is maximally nonlinear with maximal nonlinearity $\sqrt{v}$.*

These two extra characterisations of abelian PN functions (additional to Theorem 9.13) should still somehow hold true for our most general form of PN function, although obviously we will have to adapt the idea of 'linearity' appropriately. The crossed homomorphisms take this rôle quite naturally.

From (7.33) it is clear that ϕ is PN relative to ϱ if and only if, for every $x \neq 1 \in G$ and $u \in N$,

$$|\{y \in G : \phi(xy)(\phi(y)^{-1})^{\varrho(x)} = u\}| = v/w. \tag{9.15}$$

However, if ϕ is a ϱ-crossed homomorphism (Definition 8.16), then the left-hand side of (9.15) takes only two values: 0 (if $\phi(x) \neq u$) and v (if $\phi(x) = u$), so that the frequency distributions, as $x \neq 1$ runs through G, are at opposite extremes: a sequence of delta-functions for crossed homomorphisms but of uniform distributions for PN functions.

How are we to capture this optimal difference of PN functions relative to ϱ from ϱ-crossed homomorphisms?

When N is abelian but G is arbitrary, a character-theoretic formulation of PN functions is developed in Section 9.3.1 (see Theorem 9.27.2), which might serve as a definition of bentness in this case.

Character theoretic techniques begin to falter when N is nonabelian. It is not known how best to apply the Fourier Transform of a complex-valued function on N — even though it is defined for any N and for any set of complex matrix representations of N [239, Definition 3.1] — to extend Theorem 9.19. One possibility would be to work only with the linear characters of N, that is, the homomorphisms from N to $\mathbb{C}^*$. However, there are only $|N/N'|$ of these: the lifts of the linear characters of N/N' [185, Theorem 17.11].

Research Problem 68 *Extend Theorem 9.19 to nonabelian N or $\varrho \not\equiv 1$.*

Alternatively, the notions of bentness and maximal nonlinearity for a function $\phi : G \to N$ with N nonabelian could be extended using the theory we have developed.

In the abelian case, the rows of the Fourier Transform $\mathcal{F}_c$, a cocyclic GHT, form a finite set of mutually orthogonal basis functions. Perhaps we can move away altogether from complex-valued functions by testing function $\phi : G \to N$ directly against all the ϱ-crossed homomorphisms $\chi : G \to N$, using the coupled G-developed matrices $[\chi(xy)^{\varrho(x^{-1})}]_{x,y \in G}$ as transform matrices. However, the advantages of Fourier inversion and transform may be lost if there is not a set of mutually orthogonal ϱ-crossed homomorphisms to work with.

DEFINITION 9.20 *Let G and N be finite groups of order v and w, respectively, where $w|v$, and let $\varrho : G \to \mathrm{Aut}(N)^{\mathrm{op}}$ be a homomorphism. For $\phi \in C^1(G, N)$ and $\chi \in \mathrm{Hom}_\varrho(G, N)$, define $\langle \chi, \phi \rangle : G \to \mathbb{Z}N$ by*
$$\langle \chi, \phi \rangle(x) = \sum_{y \in G} \chi(xy)^{\varrho(x^{-1})} \phi^{-1}(y), \quad x \in G.$$
Then ϕ is bent relative to ϱ if, for all $x \neq 1 \in G$ and $\chi \in \mathrm{Hom}_\varrho(G, N)$,
$$\langle \chi, \phi \rangle(x) = (v/w)N.$$

Research Problem 69 *Develop the linear approximation (LA) theory of functions $\phi \in C^1(G, N)$ relative to ϱ, with bentness defined in Definition 9.20. How consistent is it with other approaches to this problem?*

Pott's approach suggests that maximal nonlinearity could reasonably be defined by existence of a splitting RDS, with the set to be measured for nonlinearity being the transversal $\{(\phi(x), x) : x \in G\}$ of N in an appropriate split extension of N by G. Then the maximal cases are given by Theorem 7.14 and Corollary 7.15.

DEFINITION 9.21 *Let G and N be finite groups of order v and w, respectively, and let $\phi \in C^1(G, N)$. Let $\varrho : G \to \mathrm{Aut}(N)^{\mathrm{op}}$ be a homomorphism.*

Then ϕ is maximally nonlinear relative to ϱ if for some $k > 1$ there exists a k-subset D of G such that $R_\phi = \{(\phi(x), x) : x \in D\} \subset N \rtimes_\varrho G$ is a splitting (v, w, k, λ)-RDS relative to $N \times \{1\}$ lifting D.

Research Problem 70 *Develop the difference distribution (DD) theory of functions $\phi \in C^1(G, N)$ relative to ϱ, with maximality defined in Definition 9.21. How consistent is it with other approaches to this problem?*

What do equivalence classes of functions look like when $\varrho \equiv 1$? This is the 'vanilla' case of Theorem 8.5 and Definition 8.18: the only bundles which are likely to be of practical interest for some time. The affine bundles are similarly found from Definition 8.22.

THEOREM 9.22 *(The case $\varrho \equiv 1$) The following statements are equivalent:*

  1. The functions $\phi, \varphi \in C^1(G, N)$ are equivalent;

2. $\mathbf{b}(\phi) = \mathbf{b}(\varphi)$;

3. *there exist* $s \in G$, $\theta \in \mathrm{Aut}(G)$, $\gamma \in \mathrm{Aut}(N)$, $f \in \mathrm{Hom}(G, N)$ *such that*

$$\varphi = (\gamma \circ (\phi \cdot s) \circ \theta) \, f, \quad where$$

$$(\phi \cdot s)(x) = \phi(s)^{-1} \, \phi(sx), \; x \in G;$$

4. *the transversals* $T_\varphi = \{(\varphi(x), x) : x \in G\}$ *and* $T_\phi = \{(\phi(x), x) : x \in G\}$
   of $N \times \{1\}$ *in* $N \times G$ *are equivalent, that is, there exist* $\alpha \in \mathrm{Aut}(N \times G)$
   and $s \in G$ *such that*

$$\alpha(N \times \{1\}) = N \times \{1\} \quad and$$

$$\alpha(T_\varphi) = (\phi^{-1}(s), s^{-1}) \, T_\phi.$$

We illustrate Theorem 9.22 with an application when $G = N$.

COROLLARY 9.23 *Let* $\varrho \equiv 1$, $G = N$ *and suppose* $\phi \in C^1(G, G)$ *is a permutation with inverse* $\mathrm{inv}(\phi)$. *Then* ϕ *and* $\mathrm{inv}(\phi)$ *are equivalent, that is,* $\mathbf{b}(\phi) = \mathbf{b}(\mathrm{inv}(\phi))$.

Proof. Clearly, $T_\phi = \{(\phi(x), x) : x \in G\}$ is a normalised transversal of $G \times \{1\}$ in $G \times G$ and $T' = \{(x, \phi(x)) : x \in G\}$ is a normalised transversal of $\{1\} \times G$ in $G \times G$. The splitting factor pair determined by T_ϕ is $(\partial^{-1}\phi, \overline{\phi})$. Let $\tau(x, y) = (y, x)$ for all $x, y \in G$. Then $\tau \in \mathrm{Aut}(G \times G)$, $\tau(\{1\} \times G) = G \times \{1\}$ and $\tau(T') = T_\phi$, so T' and T_ϕ are equivalent. By Theorem 8.5 and Definition 8.10, the corresponding splitting factor pairs lie in the same bundle $\mathcal{B}((\partial^{-1}\phi, \overline{\phi}))$. But as a transversal of $G \times \{1\}$ in $G \times G$, $T' = T_{\mathrm{inv}\phi} = \{(\mathrm{inv}(\phi)(x), x) : x \in G\}$, so it determines the splitting factor pair $(\partial^{-1}\mathrm{inv}(\phi), \overline{\mathrm{inv}(\phi)})$. By Theorem 9.22, $\mathbf{b}(\phi) = \mathbf{b}(\mathrm{inv}(\phi))$. □

Bundle equivalence takes a familiar form when $G = (GF(p^n), +)$ and $N = (GF(p^m), +)$, written additively, and $\varrho \equiv 1$.

When $p = 2$ and $m = 1$, affine bundle equivalence includes affine equivalence of Boolean functions [46, §4.1].

When $G = N$, it includes equivalence of planar functions, as shown in Section 9.2.1. For projective planes coordinatised by presemifields, it coincides with *strong isotopism* of presemifields, in a way to be made precise in Section 9.3.2.

Bundle equivalence also includes the linear equivalence used in cryptography and probably in other contexts as well. Two functions $\phi, \varphi \in C^1(\mathbb{Z}_p^n, \mathbb{Z}_p^m)$ are *linearly equivalent* [35, p. 80] if there exist invertible linear transformations β of G and γ of N and $\chi \in \mathrm{Hom}(G, N)$ such that

$$\varphi(x) = (\gamma \circ \phi \circ \beta)(x) + \chi(x), \tag{9.16}$$

in which case, by Theorem 9.22, $\mathbf{b}(\phi) = \mathbf{b}(\varphi)$.

The nature of equivalence for cryptographic functions has attracted considerable attention recently, and competing definitions have been proposed [47, 35, 39]. These have been prompted by the observation that if ϕ is invertible, then $\mathrm{inv}(\phi)$ has the same cryptographic robustness as ϕ, so the inverse of a function is also quoted as being equivalent to it.

Both [47, Proposition 3] (as cited in [39]) and [35] appear to have arrived independently at the same weakening of linear equivalence when $G = N = \mathbb{Z}_2^n$ which will include permutations and their inverses in the same equivalence class. The weakening in [47] is called *CCZ equivalence* in [39] and the weakening in [35] is called *generalised linear equivalence*. Both equivalences are the case $E = G \times G$ of equivalence of un-normalised transversals (cf. Definition 8.1), so may be replaced in the resulting theory by the corresponding normalised functions and transversals, without loss of generality. In [39] the transversal $T = \{(\phi(x), x) : x \in G\}$ is called the *graph* of ϕ and translation is on the right. In [35] it is called the *implicit embedding* and no translation is included. By Corollary 9.23, a permutation and its inverse lie in the same bundle, so bundle equivalence explains and unifies these ideas.

In [39] two functions $f, f' : \mathbb{Z}_2^n \to \mathbb{Z}_2^n$ are called *extended affine equivalent*, if β, γ and χ in (9.16) are allowed to be affine rather than linear functions (and *affine equivalent* if $\chi \equiv 0$). Clearly f and $f \cdot 1$ are affine equivalent since we may set $\beta(x) = x$ and $\gamma(x) = x - f(1)$ so that $\gamma \circ f \circ \beta(x) = (f \cdot 1)(x)$. By [39, Proposition 3], such extended affine equivalent functions $f, f' : \mathbb{Z}_2^n \to \mathbb{Z}_2^n$ are CCZ equivalent.

THEOREM 9.24 *If $f, f' : \mathbb{Z}_2^n \to \mathbb{Z}_2^n$ are extended affine equivalent functions, or CCZ equivalent functions, or generalised linear equivalent functions, they are in the same affine bundle.*

In [39], families of maximally nonlinear and of APN functions are found, which are extended affine inequivalent to any power function, but they are in the affine bundle of the APN function f_1.

9.3 NEW COCYCLIC GENERALISED HADAMARD MATRICES

This Section presents further constructions of the most accessible orthogonal factor pairs: from direct sums of orthogonal cocycles and from multiplicative cocycles. I believe they will prove important.

9.3.1 Direct sum constructions

As remarked in Chapter 6.2.4, the characterisation of orthogonality for a cocycle in terms of orthogonality of its direct summands is a subtle problem. The only published solutions are for $C = \mathbb{Z}_p^n$ and require all nontrivial $\mathbb{Z}_p$-linear combinations of the summands to be orthogonal.

Let C be a finite abelian group of order w on which the finite group G of order v acts trivially, and assume $C \not\cong \mathbb{Z}_{p^k}$ for any prime power p^k. Then C has at least one decomposition as a direct sum of proper subgroups. Fix one such, and write $C = C_1 \times \cdots \times C_n, n \geq 2$, using the standard isomorphism between a finite internal direct sum and external direct product. For $\psi \in Z^2(G, C)$, set $\psi_j = \pi_j \circ \psi$, where $\pi_j : C \to C_j$ is the jth projection epimorphism, so that (cf. Example 6.2.13)

$\psi(g,h) = (\psi_1(g,h), \ldots, \psi_n(g,h))$. We write $\psi = (\psi_1, \ldots, \psi_n)$. Similarly, for $\phi \in C^1(G,C)$, write $\phi = (\phi_1, \ldots, \phi_n)$.

By Corollary 6.8, if ψ is orthogonal, so is each factor ψ_j, but the converse does not hold. For instance, if ψ_j is orthogonal, the cocycle $(\psi_j, \psi_j) : G \times G \to C_j \times C_j$ is not even surjective. Thus an orthogonal cocycle cannot have any repeated direct factors. We record some straightforward consequences when ψ is orthogonal.

LEMMA 9.25 *Assume $\psi = (\psi_1, \ldots, \psi_n) \in Z^2(G, C_1 \times \cdots \times C_n)$, $n \geq 2$, is orthogonal. Then each $\psi_j \in Z^2(G, C_j)$ is orthogonal. Further,*

1. *If $i \neq j$ but there is an isomorphism $\alpha : C_i \cong C_j$, then $\alpha \circ \psi_i \neq \psi_j$.*

2. *If $C_j = \mathbb{Z}_r$ and $k \in \mathbb{Z}_r$, then the scalar multiple $k\psi_j$ is orthogonal if and only if $(k,r) = 1$.*

3. *If p is prime and $C_j = \mathbb{Z}_p$, $1 \leq j \leq n$, then every nontrivial $\mathbb{Z}_p$-linear combination $\sum_{j=1}^n c_j \psi_j$ is an orthogonal cocycle in $Z^2(G, \mathbb{Z}_p)$.*

Proof. For part 1, suppose to the contrary that $\alpha \circ \psi_i = \psi_j$. Compose ψ with the epimorphism $\gamma : C_1 \times \cdots \times C_n \to C_j \times C_j$ which sends factors C_k, for $k \neq i, j$, to the identity and is α on C_i. By Corollary 6.8, $\gamma \circ \psi = (\psi_j, \psi_j)$ is orthogonal, a contradiction. Part 2 follows from the definitions. Part 3 also follows from Corollary 6.8, since every nontrivial $\mathbb{Z}_p$-linear combination $\sum_{j=1}^n c_j \psi_j$ is a composition $c \circ (\psi_1, \ldots, \psi_n)$ of ψ with the epimorphism $c : \mathbb{Z}_p^n \to \mathbb{Z}_p$, where c takes the jth unit vector of $\mathbb{Z}_p^n$ to c_j, with at least one $c_j \neq 0$ and, vice versa, every epimorphism is of this form. $\square$

The converse of Lemma 9.25.3 is proved by LeBel [217, Theorem 6.2]. His result is generalised by LeBel and Horadam [218], who adapt the character-theoretic formulation of balance derived by Carlet and Ding [48, Theorem 14]. Recall that, following Nyberg [251, p. 381], we call a surjective function $f : G \twoheadrightarrow C$ *balanced* if $w|v$ and

$$\forall c \in C, \ |\{g \in G : f(g) = c\}| = v/w. \tag{9.17}$$

By the fundamental theorem for finite abelian groups, we may assume that the factors of C are all cyclic, say $C_j = \mathbb{Z}_{m_j}$, $1 \leq j \leq n$. In this case, the character group isomorphism $\chi : C \to \widehat{C}$ of Definition 4.3 may be chosen as follows. Suppose C has exponent m and $\omega = e^{2i\pi/m} \in \mathbb{C}$. Select $\omega_j = \omega^{m/m_j} = e^{2i\pi/m_j}$ as the m_j^{th} root of unity used to define the character group $\widehat{C_j}$. Then, for all $\mathbf{c} = (c_1, c_2, \ldots, c_n)$, $\mathbf{d} = (d_1, d_2, \ldots, d_n) \in C$,

$$\chi_{\mathbf{c}}(\mathbf{d}) = \omega^{\mathbf{c} * \mathbf{d}}, \quad \text{where} \quad \mathbf{c} * \mathbf{d} = \sum_{j=1}^n c_j d_j m/m_j.$$

In particular, when $m_j = m$ for all $1 \leq j \leq n$, that is, $C = \mathbb{Z}_m^n$, $\mathbf{c} * \mathbf{d} = \mathbf{c} \cdot \mathbf{d} = \sum_{j=1}^n c_j d_j$. When m is a prime p, this is the exponential sum of the Fourier Transform, Example 4.1.3.

More generally (see [166]), if C_j is the additive group of a Galois ring of exponent m_j with trace map $T_j : C_j \to \mathbb{Z}_{m_j}$, $1 \le j \le n$, then the *weighted trace map* is $T : C \to \mathbb{Z}_m$ defined by

$$T(\mathbf{d}) = \sum_{j=1}^{n} T_j(d_j) m/m_j, \ \mathbf{d} \in C, \qquad (9.18)$$

and we may realise the additive characters of C as $\chi_{\mathbf{c}}(\mathbf{d}) = \omega^{T(\mathbf{cd})}$, $\mathbf{c} \in C$.

LEMMA 9.26 [218, Corollary 2.4] *Let* $C = \mathbb{Z}_{m_1} \times \cdots \times \mathbb{Z}_{m_n}$ *be abelian of order* w *and exponent* m, *and let* $\omega = e^{2i\pi/m}$. *Then* $\phi \in C^1(G, C)$ *is balanced if and only if, in* $\mathbb{C}$, *for every* $\mathbf{c} \ne \mathbf{0} \in C$,

$$\sum_{g \in G} \omega^{\mathbf{c} * \phi(g)} = 0.$$

By Definition 6.7 and (8.2), $\psi \in Z^2(G, C)$ is orthogonal if and only if, for each $d \ne 1 \in G$, the mapping $\psi_d : G \to C$ given by $\psi_d(g) = \psi(d, g)$ is balanced. By (9.10) it is clear that $\phi : G \to C$ is PN if and only if, for each $d \ne 1 \in G$, the *directional derivative* $(\Delta\phi)_d$ of ϕ in direction d, given by

$$(\Delta\phi)_d(g) = \phi(dg) - \phi(g), \ g \in G \qquad (9.19)$$

is balanced. Lemma 9.26 applies in each case.

THEOREM 9.27 [218, Theorem 3.3] *Let* G *be a group of order* v *and let* $C = \mathbb{Z}_{m_1} \times \cdots \times \mathbb{Z}_{m_n}$, $m_j \ge 2$, $1 \le j \le n$, *be an abelian group of exponent* m *and order* w, *where* $w|v$. *Let* $\psi = (\psi_1, \ldots, \psi_n) \in Z^2(G, C)$ *and for every* $\mathbf{c} = (c_1, \ldots, c_n) \in C$, *define the cocycle* $\mathbf{c} * \psi \in Z^2(G, \mathbb{Z}_m)$ *to be*

$$(\mathbf{c} * \psi)(g, h) = \sum_{j=1}^{n} c_j \, \psi_j(g, h) \, m/m_j, \ g, h \in G. \qquad (9.20)$$

1. *Then* ψ *is orthogonal if and only if, for each* $\mathbf{c} \ne \mathbf{0} \in C$, *the cocycle* $\mathbf{c} * \psi$ *satisfies*

$$\sum_{g \in G} \omega^{(\mathbf{c}*\psi)_d(g)} = 0, \ \forall d \ne 1 \in G. \qquad (9.21)$$

2. *If* $\psi = \partial\phi = (\partial\phi_1, \ldots, \partial\phi_n) \in B^2(G, C)$, *then* $\phi : G \to C$ *is PN if and only if, for each* $\mathbf{c} \ne \mathbf{0} \in C$,

$$\sum_{g \in G} \omega^{\mathbf{c}*(\Delta\phi)_d(g)} = 0, \ \forall d \ne 1 \in G. \qquad (9.22)$$

When each m_j is the prime p, so C is elementary abelian, (9.21) is equivalent to orthogonality for the linear combination cocycle $\mathbf{c} \cdot \psi = \sum_{j=1}^{n} c_j \psi_j$. That is because, for any $k \ne 0 \in \mathbb{Z}_p$ and $\mathbf{c} \in C$, $k\mathbf{c} = \mathbf{0} \Leftrightarrow \mathbf{c} = \mathbf{0}$. Since $k(\mathbf{c} \cdot \psi) = (k\mathbf{c}) \cdot \psi$, Lemma 9.26 applies.

THEOREM 9.28 [217, Theorem 6.2] *Let G be a group of order v and let $C = \mathbb{Z}_p^n$, where $p^n | v$. Then $\psi \in Z^2(G, C)$ is orthogonal if and only if, for each $\mathbf{c} \neq \mathbf{0} \in \hat{C}$, the cocycle $\mathbf{c} \cdot \psi = \sum_{j=1}^n c_j \psi_j$ in $Z^2(G, \mathbb{Z}_p)$ is orthogonal.*

For a multiplicative $\psi \in Z^2(G, \mathbb{Z}_p^n)$, Theorem 9.27 may be proved (cf. Mac-Donald [233]) using the matrix representations of the factors ψ_j (proof of Theorem 6.10.2), since necessarily $G \cong \mathbb{Z}_p^t$. For a direct proof see Chen et al. [54, Lemma 2.3].

COROLLARY 9.29 [233, 54] *Let $t \geq n$, let $\psi = (\psi_1, \ldots, \psi_n)$ in $Z^2(\mathbb{Z}_p^t, \mathbb{Z}_p^n)$ be multiplicative, and represent the bilinear form ψ_j by matrix M_j, $1 \leq j \leq n$. Then ψ is orthogonal if and only if every nontrivial $\mathbb{Z}_p$-linear combination of the M_j is nonsingular, that is, if and only if for any $(c_1, c_2, \ldots, c_n) \neq 0 \in \mathbb{Z}_p^n$, $\sum_{j=1}^n c_j M_j \in GL(t, p)$.*

For any abelian C which is not elementary abelian, nonorthogonal nontrivial linear combinations of the direct factors of an orthogonal cocycle exist, by Lemma 9.25.2.

Research Problem 71 *For $\psi = (\psi_1, \ldots, \psi_n) \in Z^2(G, C = C_1 \times \cdots \times C_n)$, $n \geq 2$, where C is not elementary abelian, how does condition (9.21) for orthogonality of ψ relate to orthogonality of the weighted sum cocycles $\mathbf{c} * \psi$, $\mathbf{c} \neq \mathbf{0} \in \hat{C}$?*

Clearly, we have a new technique for constructing PN functions and orthogonal cocycles by direct sums.

Research Problem 72 *Use Theorem 9.27 to find new bundles of PN functions and orthogonal cocycles in $Z^2(G, \mathbb{Z}_{m_1} \times \cdots \times \mathbb{Z}_{m_n})$ from orthogonal cocycles in $Z^2(G, \mathbb{Z}_{m_j})$, $j = 1, \ldots, n$.*

Implementation of Theorem 9.28 for small elementary abelian 2-groups shows it gives a faster algorithm for finding all orthogonal cocycles than does direct exhaustive search.

9.3.1.1 Computational results

From Table 6.1, all orthogonal cocycles in $Z^2(\mathbb{Z}_2^t, \mathbb{Z}_2)$ with $1 \leq t \leq 3$ are multiplicative. Hence all orthogonal cocycles in $Z^2(\mathbb{Z}_2^t, \mathbb{Z}_2^n)$ with $1 < n \leq t \leq 3$ are multiplicative; otherwise projection onto one factor would give a contradiction.

LeBel [217] applies Theorem 9.28 to determine all the orthogonal cocycles in $Z^2(\mathbb{Z}_2^t, \mathbb{Z}_2^n)$ with $1 < n \leq t \leq 4$, using the orthogonal cocycles in $Z^2(\mathbb{Z}_2^t, \mathbb{Z}_2)$ he found in computation of Table 6.1. He found no nonmultiplicative orthogonal cocycles in $Z^2(\mathbb{Z}_2^4, \mathbb{Z}_2^2)$. This implies (by projection, again) that all orthogonal cocycles in $Z^2(\mathbb{Z}_2^4, \mathbb{Z}_2^3)$ and $Z^2(\mathbb{Z}_2^4, \mathbb{Z}_2^4)$ are multiplicative.

LEMMA 9.30 (LeBel [217]) *When $2 \leq n \leq t \leq 4$, all orthogonal cocycles in $Z^2(\mathbb{Z}_2^t, \mathbb{Z}_2^n)$ are multiplicative.*

All the orthogonal cocycles for $t = n$ are then computed by applying Theorem 9.28 to direct sums of distinct multiplicative orthogonal cocycles.

Example 9.3.1 *(Compare with Table 6.1) The total number o of orthogonal cocycles in $Z^2(\mathbb{Z}_2^n, \mathbb{Z}_2^n)$, $1 \le n \le 4$, is tabulated. In each case, they are all multiplicative.*

n	1	2	3	4
o	1	12	96,768	2,160,666,869,760
		[156]	[217]	$\approx 2.2 \times 10^{12}$ [217]

LeBel [217] conjectures that if $|C| > 2$, all orthogonal cocycles between elementary abelian 2-groups must be multiplicative.

Research Problem 73 *(LeBel) For $2 \le n \le t < \infty$, are all orthogonal cocycles in $Z^2(\mathbb{Z}_2^t, \mathbb{Z}_2^n)$ multiplicative?*

For odd primes, this is not true, even for $G = \mathbb{Z}_3^4$. When $p = 3$, the Coulter-Matthews planar mapping of Example 9.2.1.4 determines orthogonal coboundaries in $Z^2(\mathbb{Z}_3^{2k}, \mathbb{Z}_3^{2k})$ which are not multiplicative.

Nevertheless, multiplicative cocycles are a very significant source of orthogonal cocycles, which we study next.

9.3.2 Multiplicative orthogonal cocycles and presemifields

By Theorem 6.10, if a multiplicative cocycle $\psi \in Z^2(G, C)$ is orthogonal, there is a prime p such that both G and C are elementary abelian p-groups.

The multiplicative orthogonal cocycles with $G = C$ may be characterised in terms of *presemifields*, a class of algebraic systems which includes the semifields (which coordinatise certain projective planes) as well as the fields.

Many constructions for finite presemifields which are not fields are known.

DEFINITION 9.31 *A presemifield $F = (F, +, *)$ consists of a set F with two binary operations $+$ and $*$ such that*

1. *$(F, +)$ is an abelian group (with additive identity 0);*

2. *$(F \backslash \{0\}, *)$ is a quasigroup (that is, for any g, h in $F^* = F \backslash \{0\}$, there are unique solutions in F^* to both equations $g * x = h$ and $y * g = h$); and*

3. *both distributive laws hold.*

A *semifield* is a presemifield with a multiplicative identity ([58, VI.8.4] and [172, p. 116]). If F is a finite commutative semifield which is not a field, the only field property which does not hold is associativity of multiplication. Semifields are also called *planar ternary rings* or 'nonassociative division rings', but use of the term 'ring' is confusing, as they need not satisfy the usual ring axiom of associativity of multiplication. Conversely, finite rings which are not fields need not satisfy some semifield axioms. For example, by (9.9) a Galois ring $GR(p^2, n)$ is not a semifield because not every nontrivial linear equation has a unique solution. The term semifield is preferred. Refer to the texts [58, 134, 172, 151], Knuth [206] or the survey by Cordero and Wene [59] for further information.

Multiplication $*$ in a finite presemifield $(F, +, *)$ is a multiplicative cocycle on its additive group $(F, +)$, since by the distributive laws it is homomorphic in each coordinate, so $g * h + (g + h) * k = g * h + g * k + h * k = g * (h + k) + h * k$ for all $g, h, k \in G$. Furthermore, $g * 0 = 0 * h = 0$ so multiplication is normalised.

THEOREM 9.32 *Suppose G is an additively written finite abelian group and let $\psi \in Z^2(G, G)$. Then ψ is multiplicative and orthogonal if and only if $(G, +, \psi)$ is a presemifield.*

Proof. If $(G, +, *)$ is a presemifield with multiplication ψ (that is, for all $g, h \in G$, $\psi(g, h) = g * h$), then ψ is a multiplicative cocycle, and given any $g \neq 0$ and k in G, $|\{h \in G : g * h = k\}| = |\{h \in G : \psi(g, h) = k\}| = 1$ and ψ is orthogonal. Conversely, if ψ is orthogonal and multiplicative, then $G = C = \mathbb{Z}_p^n$ for some prime p, and for any $g, h \neq 0 \in G$, there are unique solutions in G to $\psi(g, x) = h$ and $\psi(y, g) = h$, because M_ψ is row balanced and column balanced. Finally, both distributive laws hold by multiplicativity. $\square$

So another corollary of Theorem 6.10 is a different proof (cf. Knuth [206, p. 185]) that the additive group of a finite presemifield is isomorphic to $\mathbb{Z}_p^n$ for some prime p. A semifield which is not a field necessarily has $p^n \geq 16$ and $n \geq 3$ [206, Theorem 6.1].

For convenience, Lemmas 7.3.6 and 7.42 are re-recorded as they apply to presemifield multiplication.

THEOREM 9.33 *Set $G = (GF(p^n), +)$, let $\psi \in Z^2(G, G)$ be multiplicative and orthogonal, let E_ψ be its extension group (7.6) and let F be the presemifield $F = (G, +, \psi)$. Then*

1. *E_ψ is abelian if and only if ψ is symmetric if and only if F is commutative;*

2. *E_ψ has exponent p if $p > 2$ and exponent 4 with $2^n(2^n - 1)$ elements of order 4 if $p = 2$;*

3. *if p is odd and ψ is symmetric then $\psi = \partial\phi \in B^2(G, G)$ and ϕ is planar.*

As a first application of Theorem 9.32, all the LP cocycles of Definition 8.31 determine presemifields.

Example 9.3.2   *Set $G = (GF(p^n), +)$. For each LPP λ of $GF(p^n)$, $(G, +, \mu_\lambda)$ is a presemifield.*

As a second application of Theorem 9.32, each bundle of multiplicative orthogonal cocycles determines a unique set of presemifields of order p^n.

DEFINITION 9.34 *Let $G = (GF(p^n), +)$ and $F = (G, +, \psi)$ be a presemifield. The presemifield bundle $\mathcal{PB}(\psi)$ of F is the set of presemifields $\mathcal{PB}(\psi) = \{(G, +, \diamond) : \diamond \in \mathcal{B}(\psi)\}$.*

In fact, bundle action on presemifield multiplications translates to a stronger form of *isotopism*, a natural equivalence relation on presemifields [206, (4.12), p. 201].

DEFINITION 9.35 *Set* $G = (GF(p^n), +)$. *Two presemifields* $F = (G, +, *)$ *and* $F' = (G, +, \diamond)$ *are* isotopic *if there exist* $\tau, \theta, \delta \in \mathrm{Aut}(G)$, *such that*

$$\delta(g \diamond h) = \tau(g) * \theta(h), \quad g, h \in G,$$

and (τ, θ, δ) *is called an* isotopism *from* F *to* F' . *If* (θ, θ, θ) *is an isotopism from* F *to* F' *then* F *and* F' *are* isomorphic, *and* $\theta : F \to F'$ *is a presemifield isomorphism.*

Obviously, (θ, θ, δ) is an isotopism from $F = (G, +, *)$ to $F' = (G, +, \diamond)$ for some δ, θ in $\mathrm{Aut}(G)$ if and only if $\diamond = \delta^{-1} \circ * \circ (\theta \times \theta)$ for some δ, θ in $\mathrm{Aut}(G)$ if and only if $\mathcal{B}(\diamond) = \mathcal{B}(*)$ if and only if $\mathcal{PB}(\diamond) = \mathcal{PB}(*)$.

We should think of bundle action on presemifields as being 'halfway' between isotopism and isomorphism, with each isotopism class of presemifields partitioned into bundles and each bundle of presemifields partitioned into isomorphism classes. In fact Coulter and Henderson [61] call an isotopism (θ, θ, δ) a *strong isotopism* or *weak isomorphism* of presemifields. Thus bundles and strong isotopism classes of presemifields are identical concepts.

The relationship between the techniques in Theorem 9.33 and Example 9.3.2 for generating presemifields will become apparent once we extend Example 9.3.2 to presemifield multiplication, as in [168, Lemma 3.1].

DEFINITION 9.36 *Set* $G = (GF(p^n), +)$ *and* $F = (G, +, *)$. *For each LPP* λ *of* $GF(p^n)$ *the* linearized permutation (LP) *cocycle* $\mu_\lambda^* : G \times G \to G$ *is*

$$\mu_\lambda^*(g, h) = \lambda(g) * h, \quad g, h \in G.$$

The cases with monomial λ *are termed* power *cocycles and denoted* $\mu_i^*(g, h) = g^{p^i} * h$, $i = 0, \ldots, n-1$. *When* $i = 0$, *write* $\mu_0^* = \mu^*$. *When* $*$ *is field multiplication, write* $\mu_\lambda^* = \mu_\lambda$. *When* $\lambda = 1$, *write* $\mu_\lambda^* = \mu^*$.

It is important to remember here that λ is defined in terms of addition and multiplication in the field $GF(p^n)$ but that μ_λ^* multiplies elements from G using the presemifield multiplication $*$.

Since the LP cocycles are orthogonal, $(\lambda, 1, 1) : (G, +, *) \to (G, +, \mu_\lambda^*)$ is an isotopism for any λ in $\mathrm{Aut}(G)$. Because any isotopism $(\tau, \theta, \delta) : (G, +, *) \to (G, +, \diamond)$ may be factored as $(\tau, \theta, \delta) = (\theta, \theta, \delta) \circ (\lambda, 1, 1) = (\kappa, 1, 1) \circ (\theta, \theta, \delta)$, where $\lambda = \theta^{-1} \circ \tau$ and $\kappa = \tau \circ \theta^{-1}$, we have the induced commuting diagram on bundles

$$
\begin{array}{ccc}
\mathcal{PB}(*) & \xrightarrow[(\theta,\theta,\delta)]{=} & \mathcal{PB}(*) = \mathcal{PB}(\mu_{\kappa^{-1}}^\diamond) \\
{\scriptstyle (\lambda,1,1)}\downarrow & & \downarrow{\scriptstyle (\kappa,1,1)} \\
\mathcal{PB}(\mu_\lambda^*) & \xrightarrow[(\theta,\theta,\delta)]{=} & \mathcal{PB}(\mu_\lambda^*) = \mathcal{PB}(\diamond)
\end{array}
\qquad , \qquad (9.23)
$$

and the isotopism class of $(G, +, *)$ partitions into bundles $\mathcal{PB}(\mu_\lambda^*)$, $\lambda \in \mathrm{Aut}(G)$.

DEFINITION 9.37 *Set* $G = (GF(p^n), +)$. *Let* $F = (G, +, *)$ *be a presemifield. The* LP-*orbit of* $*$ *is the set of bundles* $\{\mathcal{B}(\mu_\lambda^*), \ \lambda \in \mathrm{Aut}(G)\}$. *The* LP-*orbit of* F *is the corresponding set of presemifield bundles* $\{\mathcal{PB}(\mu_\lambda^*), \ \lambda \in \mathrm{Aut}(G)\}$.

Every presemifield is isotopic to at least one semifield as follows [206, Theorem 4.5.4]: if $F = (G, +, *)$ has no identity, and $a \neq 0 \in G$, then a semifield $S = (G, +, \diamond)$ with identity $a * a$ may be defined from F by

$$(x * a) \diamond (a * y) = x * y, \quad x, y \in G. \tag{9.24}$$

So, at least one of the bundles in the LP-orbit of F contains a semifield (and any semifield isomorphic to it). Therefore the set of bundles of presemifields of order p^n collects, by isotopism class, into LP-orbits, each containing a bundle containing a semifield.

THEOREM 9.38 *Set* $G = (GF(p^n), +)$ *and let* $\mathcal{M}(G)$ *denote the set of bundles of orthogonal cocycles in* $M^2(G, G)$. *Then* $\mathcal{M}(G)$ *consists of a finite set of disjoint LP-orbits of semifield multiplications. Each isotopism class of presemifields of order* p^n *partitions as exactly one LP-orbit of presemifields.*

The complete listing of bundles for $p^n < 16$ is known, by Example 8.3.1 and Table 8.1. In each case, the bundles form a single LP-orbit, partitioning the sole isotopism class, that of the Galois field $GF(p^n)$. This will also be true for any p whenever $n = 1$ or 2, since proper semifields can only exist for $n \geq 3$.

LEMMA 9.39 *If* $n \leq 2$ *or* $p^n < 16$, *any presemifield of order* p^n *is isotopic to* $GF(p^n)$. *For each such order, the bundles form a single LP-orbit.*

In this case, by Theorem 8.32 there are at least n bundles in the single LP-orbit, one containing the Galois field and $n - 1$ bundles of noncommutative presemifields. For $n = 1$, this bound is exact by Example 8.3.1: the LP-orbit of $GF(p)$ consists of a single bundle; but for $n = 2$, Table 8.1 immediately shows us this lower bound is not tight. From this tiny sample we might wonder if, for order p^2, there are p bundles in the LP-orbit.

Research Problem 74 *How many distinct bundles are there in the LP-orbit of* $GF(p^2)$ *? Are there always at least* p?

The smallest order for which there exist proper semifields (that is, which are not fields) is 16, and this example is treated in Section 9.3.2.1 below. It is known that there are 3 isotopism classes of presemifields of order 16, 2 isotopism classes of presemifields of order 27 and 6 isotopism classes of presemifields of order 32 [59].

By a construction of Knuth [206, Section 4.4] any semifield determines 6 potentially nonisotopic semifields. If p is odd, the power n is even and the semifield is proper, Ball and Brown [19] show that Knuth's process yields at least 3 isotopism classes and often 6.

Quite remarkably, Kantor [197, Theorem 1.1] has combined their work with constructions of noncommutative semifields derived from symplectic spreads to show that the number of isotopism classes of *commutative* semifields of order 2^n is not bounded above by any polynomial in the order. To round out our general theory, we will consider this simplest case: the commutative presemifields.

First, recall that a bundle of commutative semifields of order p^n equates to a bundle of symmetric multiplicative orthogonal cocycles in $Z^2(\mathbb{Z}_p^n, \mathbb{Z}_p^n)$ and thus by Corollary 8.12 to an equivalence class of abelian $(p^n, p^n, p^n, 1)$-RDSs in an extension group E of $\mathbb{Z}_p^n$ by $\mathbb{Z}_p^n$.

All the constructions of abelian $(p^n, p^n, p^n, 1)$-RDSs in the literature [31, 75, 189, 266] are in equivalence classes of RDSs of this kind. Effectively, this is the only type of construction yet known. For odd p, all these RDSs are splitting (as we know from Theorem 9.33 must be the case). For $p = 2$, all these RDSs correspond to $\mathcal{B}(\mu)$, the bundle of the Galois field $GF(2^n)$, but by Kantor's result there must in fact be many other RDS equivalence classes.

LEMMA 9.40 [168, §4] *If $p = 2$, all abelian $(p^n, p^n, p^n, 1)$-RDS constructions known (Teichmüller, diagonal and quadratic RDS) are isomorphic to the canonical RDS defined by field multiplication μ.*

If p is odd, all abelian $(p^n, p^n, p^n, 1)$-RDS constructions known are isomorphic to the canonical RDS defined either by field multiplication μ (diagonal and quadratic RDS) or by multiplication in a commutative proper semifield.

Theorem 9.38 and Lemma 9.40 prove the statement made in [168] that it is finite presemifields, not semifields, which hold one key to the classification problem for equivalence classes of central $(p^n, p^n, p^n, 1)$-RDSs — and most apparently, for nonabelian central RDSs.

If $F = (G, +, *)$ is a commutative presemifield, any semifield S defined from it by (9.24) must be commutative. Even more, the isotopism $F \rightarrow S$ is a strong isotopism $(\theta, \theta, 1)$ by the simple argument [61] that the mapping $\theta(x) = a * x = x * a$ is an LPP.

How many distinct bundles of commutative semifields can exist in a single LP-orbit, that is, in a single isotopism class? Coulter and Henderson [61] have shown that unless a fairly restrictive condition holds, there can be at most one.

THEOREM 9.41 [61, Theorem 2.6] *Let $F = (G, +, *)$ and $F' = (G, +, \diamond)$ be isotopic commutative presemifields of order p^n. Suppose the middle nuclei of corresponding commutative semifields S and S' in (9.24) have orders p^m and p^k, respectively. Then $\mathcal{PB}(*) = \mathcal{PB}(\diamond)$, unless m/k is even and the only isotopisms from S to S' are of the form $(\alpha * \theta, \theta, \delta)$, where α is a nonsquare element of the middle nucleus of S.*

COROLLARY 9.42 [61, Corollaries 2.7, 2.8] *If $F = (G, +, *)$ is a commutative presemifield of even order, or of odd order p^n and n is odd, then the LP-orbit of F contains exactly one bundle of commutative presemifields.*

Example 8.3.1, Table 8.1 and Table 8.2 demonstrate this, with one bundle of commutative presemifields in each LP-orbit (the bundle numbered 1, indexed by the identity LPP and containing $GF(p^n)$).

Examples of commutative semifields of odd order are treated in more detail in Section 9.3.2.2.

9.3.2.1 Example: Presemifields of order 16

Let us return to the set of bundles, derived from field multiplication in $GF(16)$, in Table 8.2. By Definition 9.37 this set is a single LP-orbit in $\mathcal{M}(\mathbb{Z}_2^4)$. The first bundle contains the field multiplication cocycle μ. The remaining 31 bundles all contain nonsymmetric multiplications and, by Theorem 9.33 (or Corollary 9.42), correspond to bundles of noncommutative presemifields, and to equivalence classes of nonabelian central $(16, 16, 16, 1)$-RDSs.

There are 23 nonisomorphic semifields of order 16 which are not isotopic to $GF(16)$, but they lie in only 2 isotopism classes. Consequently $\mathcal{M}(\mathbb{Z}_2^4)$ consists of 3 LP-orbits, with the LP-orbit of $GF(16)$ containing 32 bundles.

One of these isotopism classes, containing 18 nonisomorphic semifields, is represented by a noncommutative semifield denoted V and the other, containing 5, is represented by a noncommutative semifield denoted W ([206, 2.2], [58, Examples VI.8.42, 43]).

The LP-orbits of both semifield V and semifield W are unknown. Nor is it known how many bundles are in each. At least by Example 9.3.1, we do know exactly how many distinct presemifields of order 16 comprise the three isotopy classes, in terms of a fixed generating set for $\mathbb{Z}_2^4$: $2, 160, 666, 869, 760$. This is only a little more than one-quarter of the total number $|\mathrm{Aut}(\mathbb{Z}_2^4)|^3$ of isotopisms of any element within each isotopy class.

Research Problem 75 *Determine the LP-orbit in semifield V and in semifield W of the bundle defined by multiplication.*

Construction of the noncommutative semifields V and W is now detailed.

First, we represent the nonzero elements of $\mathbb{Z}_2^4$ as powers of a primitive element α of $GF(16)$ satisfying $\alpha^4 = \alpha + 1$. Setting $\omega = \alpha^5$, we know $\omega^2 = \omega + 1$, and, since $GF(16)$ is a quadratic extension of $GF(4) \cong \{0, 1, \omega, 1+\omega = \omega^2\}$, every element of $\mathbb{Z}_2^4$ also has a unique representation $u + \alpha v = (a + b\omega) + \alpha(c + d\omega)$, $a, b, c, d \in GF(2)$: $u, v \in GF(4)$.

Semifield V: The elements of the semifield V have the form $u + \alpha v, u, v \in GF(4)$. Addition is defined component-wise, and multiplication is defined as follows:

$$(u_1 + \alpha v_1)(u_2 + \alpha v_2) = (u_1 u_2 + v_1^2 v_2) + \alpha(v_1 u_2 + u_1^2 v_2 + v_1^2 v_2^2).$$

Semifield W: Semifield W has the same elements and the same addition as V, but multiplication is defined as follows:

$$(u_1 + \alpha v_1)(u_2 + \alpha v_2) = (u_1 u_2 + \omega v_1^2 v_2) + \alpha(v_1 u_2 + u_1^2 v_2).$$

The multiplication tables of semifield V and semifield W are displayed next. Nonzero entries are represented as powers of α. Rows and columns are in the reverse lexicographical order of $\mathbb{Z}_2^4 \cong (GF(16), +)$ (with the least significant bit at the left-hand end — cf. Definition 3.2).

In each case, the second row and column are symmetric but the matrices are asymmetric, confirming that they cannot be obtained from any LP cocycle derived from field multiplication in $GF(16)$.

The $\mathbb{Z}_2^4$-cocyclic matrix given by multiplication in semifield V is

$$
\begin{bmatrix}
0 & 0 & 0 & 0 & 0 & 0 & 0 & 0 & 0 & 0 & 0 & 0 & 0 & 0 & 0 & 0 \\
0 & 1 & \alpha & \alpha^4 & \alpha^2 & \alpha^8 & \alpha^5 & \alpha^{10} & \alpha^3 & \alpha^{14} & \alpha^9 & \alpha^7 & \alpha^6 & \alpha^{13} & \alpha^{11} & \alpha^{12} \\
0 & \alpha & \alpha^4 & 1 & \alpha^{12} & \alpha^{13} & \alpha^6 & \alpha^{11} & \alpha^{10} & \alpha^8 & \alpha^2 & \alpha^5 & \alpha^3 & \alpha^9 & \alpha^7 & \alpha^{14} \\
0 & \alpha^4 & 1 & \alpha & \alpha^7 & \alpha^3 & \alpha^9 & \alpha^{14} & \alpha^{12} & \alpha^6 & \alpha^{11} & \alpha^{13} & \alpha^2 & \alpha^{10} & \alpha^8 & \alpha^5 \\
0 & \alpha^2 & \alpha^{13} & \alpha^{14} & \alpha^5 & \alpha & \alpha^7 & \alpha^{12} & \alpha^6 & \alpha^3 & 1 & \alpha^8 & \alpha^9 & \alpha^{11} & \alpha^{10} & \alpha^4 \\
0 & \alpha^8 & \alpha^{12} & \alpha^9 & \alpha & \alpha^{10} & \alpha^{13} & \alpha^3 & \alpha^2 & 1 & \alpha^7 & \alpha^{11} & \alpha^5 & \alpha^4 & \alpha^{14} & \alpha^6 \\
0 & \alpha^5 & \alpha^{11} & \alpha^3 & \alpha^{14} & \alpha^{12} & \alpha^{10} & 1 & \alpha^7 & \alpha^{13} & \alpha^8 & \alpha^4 & \alpha & \alpha^2 & \alpha^6 & \alpha^9 \\
0 & \alpha^{10} & \alpha^6 & \alpha^7 & \alpha^{13} & \alpha^9 & 1 & \alpha^5 & \alpha^4 & \alpha^2 & \alpha^{12} & \alpha^3 & \alpha^{11} & \alpha^{14} & \alpha & \alpha^8 \\
0 & \alpha^3 & \alpha^2 & \alpha^6 & 1 & \alpha^{14} & \alpha^8 & \alpha^{13} & \alpha^5 & \alpha^{11} & \alpha & \alpha^9 & \alpha^{10} & \alpha^{12} & \alpha^4 & \alpha^7 \\
0 & \alpha^{14} & \alpha^5 & \alpha^{12} & \alpha^8 & \alpha^6 & \alpha^4 & \alpha^9 & \alpha^{11} & \alpha^{10} & \alpha^3 & 1 & \alpha^7 & \alpha & \alpha^{13} & \alpha^2 \\
0 & \alpha^9 & \alpha^{10} & \alpha^{13} & \alpha^{11} & \alpha^2 & \alpha^{14} & \alpha^4 & 1 & \alpha^7 & \alpha^5 & \alpha^6 & \alpha^{12} & \alpha^8 & \alpha^3 & \alpha \\
0 & \alpha^7 & \alpha^8 & \alpha^{11} & \alpha^9 & 1 & \alpha^{12} & \alpha^2 & \alpha^{14} & \alpha & \alpha^6 & \alpha^{10} & \alpha^4 & \alpha^3 & \alpha^5 & \alpha^{13} \\
0 & \alpha^6 & \alpha^{14} & \alpha^8 & \alpha^{10} & \alpha^7 & \alpha^{11} & \alpha & \alpha^9 & \alpha^5 & \alpha^4 & \alpha^{12} & \alpha^{13} & 1 & \alpha^2 & \alpha^3 \\
0 & \alpha^{13} & \alpha^7 & \alpha^5 & \alpha^4 & \alpha^{11} & \alpha^3 & \alpha^8 & \alpha & \alpha^{12} & \alpha^{14} & \alpha^2 & 1 & \alpha^6 & \alpha^9 & \alpha^{10} \\
0 & \alpha^{11} & \alpha^9 & \alpha^2 & \alpha^3 & \alpha^5 & \alpha & \alpha^6 & \alpha^{13} & \alpha^4 & \alpha^{10} & \alpha^{14} & \alpha^8 & \alpha^7 & \alpha^{12} & 1 \\
0 & \alpha^{12} & \alpha^3 & \alpha^{10} & \alpha^6 & \alpha^4 & \alpha^2 & \alpha^7 & \alpha^8 & \alpha^9 & \alpha^{13} & \alpha & \alpha^{14} & \alpha^5 & 1 & \alpha^{11}
\end{bmatrix}
$$

and the $\mathbb{Z}_2^4$-cocyclic matrix given by multiplication in semifield W is

$$
\begin{bmatrix}
0 & 0 & 0 & 0 & 0 & 0 & 0 & 0 & 0 & 0 & 0 & 0 & 0 & 0 & 0 & 0 \\
0 & 1 & \alpha & \alpha^4 & \alpha^2 & \alpha^8 & \alpha^5 & \alpha^{10} & \alpha^3 & \alpha^{14} & \alpha^9 & \alpha^7 & \alpha^6 & \alpha^{13} & \alpha^{11} & \alpha^{12} \\
0 & \alpha & \alpha^5 & \alpha^2 & \alpha^9 & \alpha^3 & \alpha^6 & \alpha^{11} & \alpha^{13} & \alpha^{12} & \alpha^7 & \alpha^{14} & \alpha^{10} & \alpha^8 & 1 & \alpha^4 \\
0 & \alpha^4 & \alpha^2 & \alpha^{10} & \alpha^{11} & \alpha^{13} & \alpha^9 & \alpha^{14} & \alpha^8 & \alpha^5 & 1 & \alpha & \alpha^7 & \alpha^3 & \alpha^{12} & \alpha^6 \\
0 & \alpha^2 & \alpha^3 & \alpha^6 & \alpha^4 & \alpha^{10} & \alpha^7 & \alpha^{12} & \alpha^5 & \alpha & \alpha^{11} & \alpha^9 & \alpha^8 & 1 & \alpha^{13} & \alpha^{14} \\
0 & \alpha^8 & \alpha^9 & \alpha^{12} & \alpha^{10} & \alpha & \alpha^{13} & \alpha^3 & \alpha^{11} & \alpha^7 & \alpha^2 & 1 & \alpha^{14} & \alpha^6 & \alpha^4 & \alpha^5 \\
0 & \alpha^5 & \alpha^{11} & \alpha^3 & \alpha^{14} & \alpha^{12} & \alpha^{10} & 1 & \alpha^7 & \alpha^{13} & \alpha^8 & \alpha^4 & \alpha & \alpha^2 & \alpha^6 & \alpha^9 \\
0 & \alpha^{10} & \alpha^6 & \alpha^7 & \alpha^{13} & \alpha^9 & 1 & \alpha^5 & \alpha^4 & \alpha^2 & \alpha^{12} & \alpha^3 & \alpha^{11} & \alpha^{14} & \alpha & \alpha^8 \\
0 & \alpha^3 & \alpha^{14} & 1 & \alpha^6 & \alpha^2 & \alpha^8 & \alpha^{13} & \alpha^{12} & \alpha^{10} & \alpha^5 & \alpha^{11} & \alpha^4 & \alpha^7 & \alpha^9 & \alpha \\
0 & \alpha^{14} & \alpha^7 & \alpha & \alpha^3 & 1 & \alpha^4 & \alpha^9 & \alpha^{10} & \alpha^{11} & \alpha^6 & \alpha^8 & \alpha^{12} & \alpha^5 & \alpha^2 & \alpha^{13} \\
0 & \alpha^9 & \alpha^{12} & \alpha^8 & \alpha^5 & \alpha^6 & \alpha^{14} & \alpha^4 & \alpha & \alpha^3 & \alpha^{13} & \alpha^{10} & \alpha^2 & \alpha^{11} & \alpha^7 & 1 \\
0 & \alpha^7 & \alpha^{13} & \alpha^5 & \alpha & \alpha^{14} & \alpha^{12} & \alpha^2 & \alpha^9 & 1 & \alpha^{10} & \alpha^6 & \alpha^3 & \alpha^4 & \alpha^8 & \alpha^{11} \\
0 & \alpha^6 & 1 & \alpha^{13} & \alpha^{12} & \alpha^4 & \alpha^{11} & \alpha & \alpha^{14} & \alpha^8 & \alpha^3 & \alpha^2 & \alpha^5 & \alpha^9 & \alpha^{10} & \alpha^7 \\
0 & \alpha^{13} & \alpha^4 & \alpha^{11} & \alpha^7 & \alpha^5 & \alpha^3 & \alpha^8 & 1 & \alpha^6 & \alpha & \alpha^{12} & \alpha^9 & \alpha^{10} & \alpha^{14} & \alpha^2 \\
0 & \alpha^{11} & \alpha^{10} & \alpha^{14} & \alpha^8 & \alpha^7 & \alpha & \alpha^6 & \alpha^2 & \alpha^9 & \alpha^4 & \alpha^{13} & 1 & \alpha^{12} & \alpha^5 & \alpha^3 \\
0 & \alpha^{12} & \alpha^8 & \alpha^9 & 1 & \alpha^{11} & \alpha^2 & \alpha^7 & \alpha^6 & \alpha^4 & \alpha^{14} & \alpha^5 & \alpha^{13} & \alpha & \alpha^3 & \alpha^{10}
\end{bmatrix}
$$

9.3.2.2 Example: Commutative presemifields from planar functions

Theorem 9.13 says that no presemifield of order 2^n can have multiplication equal to a coboundary, since there are no planar functions between groups of even order.

On the other hand, if p is odd, the same Theorem, together with Theorem 9.32, says that any planar function $\phi : \mathbb{Z}_p^n \to \mathbb{Z}_p^n$ for which $\partial\phi$ is multiplicative determines a commutative presemifield multiplication $\partial\phi$, while Theorem 9.33 ensures that any commutative presemifield multiplication μ^* is of this form.

Thus the bundle of a commutative presemifield of odd order is defined by the image under the coboundary operator of a bundle of planar functions (Corollary 9.15).

The simplest illustration is the quadratic planar function $\phi_1(g) = g^2$ on $GF(p^n)$ for p odd, of Example 9.2.1, which gives $\partial\phi_1 = 2\mu$, so $\mu = \partial(2^{-1}\phi_1)$. Consequently, the bundle $\partial(\mathbf{b}(\phi_1)) = \mathcal{B}(\mu)$ defines the bundle of the Galois field $GF(p^n)$.

Conversely, for each symmetric μ^*, a planar map ϕ such that $\mu^* = \partial\phi$ may be derived computationally by solving the simultaneous linear equations (6.7) over $GF(p^n)$ in the $p^n - 1$ unknowns $\phi(g)$, $g \neq 0 \in G$. Once the map ϕ is determined, it may be expressed as a polynomial using the Mattson-Solomon transform. Alternatively (see Corollary 9.54 below), by mimicking Example 6.2.6 a planar map ϕ can be found very quickly as half the *diagonal* of ψ; that is $\phi = 2^{-1}D\psi$, where $D\psi(x) = \psi(x, x)$.

To illustrate this latter technique, suppose p is odd and μ_i is a power cocycle (Definition 8.31 or 9.36) defined on $GF(p^n)$. Though μ_i itself is not symmetric if $i \neq 0$, its symmetrisation μ_i^+, with $\mu_i^+(x, y) = x^{p^i}y + x y^{p^i}$ (Example 6.2.17) is symmetric and multiplicative. Then $\phi(x) = 2^{-1}D\mu_i^+(x) = x^{p^i+1}$ is planar if and only if $n/(i, n)$ is odd. In this case $\phi(x) = \phi_2(x)$ in Example 9.2.1, $\partial\phi_2 = \mu_i^+$ and $(G, +, \mu_i^+)$ is a commutative presemifield. Under (9.24), it defines a commutative semifield in the same bundle, which is an Albert's twisted field ([58, Example VI.8.45], see also [197, 5.1]). So we have identified the commutative semifields corresponding to the first two planar functions in Example 9.2.1.

Example 9.3.3 *Let $G = (GF(p^n), +)$ for p odd.*

1. *Let $\phi_1(x) = x^2$. Then $\partial(\mathbf{b}(\phi_1)) = \mathcal{B}(\mu)$ defines the bundle $\mathcal{PB}(\mu)$ of the Galois field $GF(p^n)$.*

2. *Let $\phi_2(x) = x^{p^i+1}$, where $n/(i, n)$ is odd. Then $\partial(\mathbf{b}(\phi_2)) = \mathcal{B}(\mu_i^+)$ defines the bundle $\mathcal{PB}(\mu_i^+)$ of a commutative Albert's twisted field.*

   There are $[(n-1)/2]$ distinct LP-orbits of this kind [197, p. 107].

The nonmonomial ϕ_3 of Example 9.2.1 also define multiplicative coboundaries. For $n \geq 5$ odd, the corresponding commutative presemifields define planes in two more distinct isotopism classes, see [62].

A solution ϕ of $\mu^* = \partial\phi$ will not be unique, but by (9.12) any two solutions will differ by a linearised polynomial and one of the form (9.13) may be found.

As illustration, applying the simultaneous linear equations technique to the multiplication of the Dickson commutative semifield of order $81 = 3^4$ produces a planar map which is also not a simple power function. It is a *Dembowski-Ostrom (DO) polynomial* [63, Definition 3.1], that is, is a mapping $\phi: GF(p^n)[x] \to GF(p^n)[x]$ which, when reduced modulo $x^{p^n} - x$, is of the form

$$\phi(x) = \sum_{j=0}^{n-1} \sum_{i=0}^{j} \lambda_{ij} \, x^{p^i + p^j}, \quad \lambda_{ij} \in GF(p^n). \tag{9.25}$$

Example 9.3.4   [169, Example 3.2] *Let $(G, +, \mu^*)$ be the Dickson commutative semifield of order 81 and let α be a primitive element of $GF(81)$. A planar function f for which $\mu^* = \partial f$ is given by $f(g) = g^{54} + g^{30} + \alpha^{55}g^{27} + \alpha^{12}g^{18} + \alpha^{12}g^{10} + \alpha^{42}g^9 + g^6 + \alpha^4 g^3 + \alpha^9 g^2 + \alpha g$. Its linearised summand is $\lambda(g) = \alpha^{55}g^{27} + \alpha^{42}g^9 + \alpha^4 g^3 + \alpha g$, with $\partial\lambda = 0$. Their difference $\phi = f - \lambda$ is a DO polynomial*

$$\phi(g) = g^{54} + g^{30} + \alpha^{12}g^{18} + \alpha^{12}g^{10} + g^6 + \alpha^9 g^2$$

and is a planar function of simpler form which also satisfies $\mu^* = \partial\phi$.

The first three planar functions, including the nonmonomial ϕ_3, of Example 9.2.1 — all of which determine multiplicative coboundaries — are also DO polynomials. This is no coincidence.

THEOREM 9.43 *Let $G = (GF(p^n), +)$, p odd, let $f \in C^1(G, G)$ have linearised summand $\lambda \in \mathrm{Hom}(G, G)$ and set $\phi = f - \lambda$. Then*

1. *$\partial\phi = \partial f$ is multiplicative if and only if ϕ is a DO polynomial;*

2. *$\mathcal{B}(\partial\phi) = \partial(\mathbf{b}(\phi))$ is a bundle of commutative presemifields if and only if ϕ is a planar DO polynomial.*

Proof. For part 1, coboundary ∂f is multiplicative if and only if, for all $x, y, a \in G$, $\partial f(x + a, y) = \partial f(x, y) + \partial f(a, y)$, if and only if, for all $x, y, a \in G$, $f(x + y + a) - f(x + y) = f(x + a) + f(y + a) - f(x) - f(y) - f(a)$, if and only if, for all $x, a \neq 0 \in G$, $L_a(x) = f(x + a) - f(x) - f(a) = (\Delta f)_a(x) - f(a)$ is linearised, where $(\Delta f)_a$ is the directional derivative of f in direction a. By [63, Theorem 3.2] the last condition holds if and only if ϕ is DO. Part 2 follows immediately. $\square$

Research Problem 76 *Let p be an odd prime and $G = (GF(p^n), +)$. Which DO polynomials $\phi : G \to G$ are planar (modulo their bundle)?*

It is thought that, up to isomorphism of the corresponding planes, the planar functions for $G = C = \mathbb{Z}_p^n$ where n is odd have all been listed in Example 9.2.1. By Corollary 9.42 each of these planar functions represents the sole bundle of commutative presemifields in its LP-orbit.

Research Problem 77 *Let $G = (GF(p^n), +)$, let ϕ_1, ϕ_2 and ϕ_3 be the planar DO polynomials of Example 9.2.1 and let n be odd. Must a planar DO polynomial on G lie in the same bundle as one of ϕ_1, ϕ_2 or ϕ_3?*

9.3.3 Swing action

Does the LP-action of Section 9.3.2 have a more widespread counterpart? For the sake of opening the area to investigation, define an action, the *swing* action, of $\mathrm{Aut}(G)$ on $C^2(G, C)$ as follows: for each $\tau \in \mathrm{Aut}(G)$ and $\Phi \in C^2(G, C)$, define $\Phi \cdot \tau \in C^2(G, C)$ to be $(\Phi \cdot \tau)(g, h) = \Phi(\tau(g), h)$, $g, h \in G$. For multiplicative orthogonal cocycles and $G = C = (GF(p^n), +)$ this clearly specialises to the LP action. But in general it does not preserve cocycles, even multiplicative cocycles.

Research Problem 78 *Investigate the properties of swing action on $C^2(G,C)$.*
When is it an action on $Z^2(G,C)$? When is it an action on factor pairs $F^2(G,C)$?
What is the action on coboundaries?

9.4 NEW HADAMARD CODES

A multitude of new and optimal or near optimal codes and sequences awaits our
discovery. Some early and exciting applications of all the theory behind us can
now be presented. First look at the best-understood cases: the cocyclic codes obtained from cocyclic Hadamard matrices arising from multiplicative cocycles and
coboundaries in Sections 9.2 and 9.3.

9.4.1 Class A cocyclic Hadamard codes

The following construction of optimal Class A cocyclic Hadamard codes follows
immediately from Definition 4.32 and Theorem 7.29.

LEMMA 9.44 *Let G and N be finite groups of order v and w, respectively, where*
w divides v. Let $(\psi, \varepsilon) : G \times G \to N$ be an orthogonal factor pair.
  The rows of $M_{(\psi,\varepsilon)}$ without the first column form a $(v-1, v, v(w-1)/w)$ w-ary
code $\mathcal{A}_{(\psi,\varepsilon)}$ meeting the Plotkin bound $A_w(n,d) = wd/(w-1)$.
  The rows of the translates $uM_{(\psi,\varepsilon)}$, $u \in N$, of $M_{(\psi,\varepsilon)}$ form a $(v, vw, v(w-1)/w)$ w-ary code $\mathcal{C}_{(\psi,\varepsilon)}$ meeting the Plotkin bound $A_w(n,d) = wn$.

Now apply Lemma 9.44 with $\varepsilon \equiv 1$, $G = N = (GF(q), +)$, $q = p^n$, to
the presemifield multiplications of Section 9.3.2. Suppose $F = (G, +, *)$ is a
presemifield.

Because the multiplication μ^* is additive in the first coordinate, the code $\mathcal{A}_{\mu^*}$ in
Lemma 9.44 is in fact the linear span over $\mathbb{F}_p$ of n rows $[g_i * h, \ h \in G, h \neq 0]$,
where $\{g_i, \ 1 \leq i \leq n\}$ is some minimal generating set of G. These n rows are
linearly independent so the code has $\mathbb{F}_p$-dimension n. A similar argument applies
to $\mathcal{C}_{\mu^*}$. Since $v = w$, the optimal code $\mathcal{D}_{\mu^*}$ of Definition 4.32 is defined here too.

THEOREM 9.45 *Let $F = (G, +, *)$ be a presemifield of order $q = p^n$, with addi-*
tive group $G = (GF(q), +)$. Then
  $\mathcal{A}_{\mu^*}$ is a q-ary $(q-1, q, q-1)$ cocyclic Hadamard code with $\mathbb{F}_p$-dimension n,
which meets the Plotkin bound and
  $\mathcal{C}_{\mu^*}$ is a q-ary $(q, q^2, q-1)$ cocyclic Hadamard code with $\mathbb{F}_p$-dimension $2n$,
which meets the Plotkin bound and
  $\mathcal{D}_{\mu^*}$ is a q-ary $(q+1, q^2, q)$ cocyclic Hadamard code with $\mathbb{F}_p$-dimension $2n$,
which meets the Plotkin bound.

The isotopism class of F is $\{(G, +, \ \delta^{-1} \circ \mu^* \circ (\tau \times \theta)), \ \delta, \tau, \theta \in \mathrm{Aut}(G)\}$. Fix
isotopism (τ, θ, δ) and set $\mu^\diamond = \delta^{-1} \circ \mu^* \circ (\tau \times \theta)$. Then $M_{\mu^\diamond} = \delta^{-1}(M_{\mu^* \circ (\tau \times \theta)})$
and the Hadamard codes resulting from $M_{\mu^\diamond}$ and $M_{\mu^* \circ (\tau \times \theta)}$ are equivalent. The
rows of the latter are just a reordering of the rows of $M_{\mu^* \circ (1 \times \theta)}$ and therefore

p	n	$|\mathrm{Aut}(\mathbb{Z}_p^n)|$	Presemifield	$\dim_q M_{\mu^*}$
2	1	1	$\mathbb{F}_2$	1
2	2	6	$\mathbb{F}_4$	1
2	3	168	$\mathbb{F}_8$	1
2	4	$20,160$	$\mathbb{F}_{16}$	1
2	4	$20,160$	Semifield W	2
2	4	$20,160$	Semifield V	3
3	1	2	$\mathbb{F}_3$	1
3	2	48	$\mathbb{F}_9$	1
3	3	$11,232$	$\mathbb{F}_{27}$	1
3	3	$11,232$	Albert presemifield	2
3	4	$24,261,120$	$\mathbb{F}_{81}$	1
3	4	$24,261,120$	Albert presemifield	2
3	4	$24,261,120$	Dickson semifield	3

Table 9.1 [167, Table II] q-ary rank of rows in presemifield multiplication tables

determine the same Hadamard codes as those of $M_{\mu^* \circ (1 \times \theta)}$. Finally, permuting columns of $M_{\mu^* \circ (1 \times \theta)}$ according to θ^{-1} gives M_{μ^*} and the resulting Hadamard codes are equivalent. Recall that every presemifield is isotopic to some semifield.

LEMMA 9.46 *The isotopism class of a semifield determines an equivalence class of each type of Class A Hadamard code.*

The optimal Class A codes determined by the orthogonal cocycles of Example 9.1.4, and Lemma 9.40 for $p = 2$, are in the code equivalence classes determined by multiplication in the fields $GF(q)$. New code classes with the same parameters (n, k, d) will be found by using the isotopism classes of other (pre)semifields $(G, +, \mu^*)$, for example, those in [19, 197].

Research Problem 79 *How do the optimal Class A Hadamard semifield codes determined by different isotopism classes of semifields of the same order differ, for example in their weight enumerators, additivity and performance characteristics?*

Table 9.1, from [167], gives some preliminary results. It lists the q-ary dimension of the code spanned by the rows of M_{μ^*} for several well-known presemifields of small order. Notation follows [58, VI.8.4]. All isotopism classes of semifields for $p = 2$, $n \le 4$ and $p = 3$, $n \le 3$ are listed, but for $p = 3$, $n = 4$ at least 4 isotopism classes containing commutative semifields are known [59, 197].

The last column of Table 9.1 is a quick indicator that different semifield isotopism classes give inequivalent Class A Hadamard codes, according to their additive properties, but all are optimal with respect to the Plotkin bound. It seems likely that distinct isotopism classes will determine distinct code equivalence classes, but as yet it is not proved.

Research Problem 80 *Do distinct isotopism classes of semifields of order p^n always determine distinct equivalence classes of Class A Hadamard codes?*

If $n = rm$, the relative trace mapping $\mathrm{tr}_r : (GF(p^n), +) \to (GF(p^m), +)$ is an epimorphism, so by Corollary 6.8 the composition $\mathrm{tr}_r \circ \mu^*$ is orthogonal and Lemma 9.44 applies, generalising Theorem 9.45. If $\mu^* = \mu$ is the field multiplication of $GF(q)$, then the resulting codes are linear, and an argument similar to that giving Corollary 9.11 may be applied.

LEMMA 9.47 (see [169, Lemma 2.2]) *Suppose $n = rm$, and let tr_r denote the relative trace function from $G = (GF(p^n), +)$ to $C = (GF(p^m), +)$.*

1. *The rows $\{\mathrm{tr}_r \circ \mu^*(g, h),\ h \in G\}$, $g \in G$, form a p^m-ary $(q, q, q - p^{n-m})$ code.*

2. *These rows together with their p^m-ary translates, form a p^m-ary $(q, qp^m, q - p^{n-m})$ code.*

3. *If $\mu^* = \mu$ is the field multiplication of $GF(p^n)$, these codes are linear, and the code of part 2 has parameters $[(p^m)^r,\ r + 1,\ (p^m)^r - (p^m)^{r-1}]$ and is equivalent to the generalised first-order Reed-Muller code. In particular, the rows of M_μ form a q-ary $[q, 1, q-1]$ linear code, $\mathcal{A}_\mu$ is a q-ary (trivial) MDS code and $\mathcal{C}_\mu$ is a q-ary MDS code.*

All the Class A cocyclic Hadamard codes listed in Lemma 9.47 arise from multiplicative orthogonal cocycles on G. It is conjectured that when $p = 2$ all orthogonal cocycles on G are multiplicative (Research Problem 73), but this is definitely not true for odd p.

THEOREM 9.48 *Let p be odd and $G = (GF(p^n), +)$. Every planar function $\phi : G \to G$ determines a generalised Hadamard matrix $M_{\partial \phi}$ and, consequently, Hadamard codes of all the types constructed in Definition 4.34.*

When $p = 3$, for the class of planar power functions $\phi_{(n,b)}$ over $GF(3^n)$ in Example 9.2.1.4, if $b \not\equiv \pm 1 \pmod{2n}$, then the symmetric orthogonal coboundary $\partial \phi_{(n,b)}$ cannot be multiplicative, so it is *not* a presemifield multiplication.

In particular, the resulting ternary Hadamard codes are not linear 3^n-ary codes. The linear Class B codes they determine are studied in [169].

9.4.2 Class B cocyclic Hadamard codes

In Table 9.2, adapted from [169], the authors also calculate, for $n \leq 6$, the 3^n-ary dimension of the Class B code generated by the rows of $M = M_{\partial \phi_{(n,b)}}$ and the 3-ary dimension of the Class B code generated by the rows of the matrix resulting from taking the absolute trace of M. The same is done for the codes of the ternary Galois fields, the Albert presemifields of [58, Example VI.8.45] and Dickson commutative semifields. The author does not know whether the Dickson commutative semifields are isotopic to any commutative presemifields with multiplication $\partial \phi$ where ϕ is a power function (see Example 9.3.4).

n	Construction	Cocycle	d	$k = \dim_{3^n} M$	$\dim_3 \mathrm{tr}(M)$
1	Galois field	$\partial\phi_1$	2	1	1
2	Galois field	$\partial\phi_1$	2	1	2
3	Galois field	$\partial\phi_1$	2	1	3
3	Albert presemifield	$\partial\phi_2$	4	2	3
4	Galois field	$\partial\phi_1$	2	1	4
4	Albert presemifield	$\partial\phi_2$	4	2	4
4	Dickson semifield	$\partial\phi, \phi\,\mathrm{DO}$	?	3	4
4	Coulter-Matthews	$\partial\phi_{(4,3)}$	14	10	16
5	Galois field	$\partial\phi_1$	2	1	5
5	Albert presemifield	μ^*	-	2	5
5	Coulter-Matthews	$\partial\phi_{(5,3)}$	14	10	20
6	Galois field	$\partial\phi_1$	2	1	6
6	Albert presemifield	μ^*	-	2	6
6	Dickson semifield	$\partial\phi, \phi\,\mathrm{DO}$	?	3	6
6	Coulter-Matthews	$\partial\phi_{(6,5)}$	122	46	90

Table 9.2 [169, Table 2] Ternary cocyclic Hadamard codes; if applicable, $d := x^d = \phi(x)$

If an Albert presemifield is noncommutative, its multiplication cannot be a co-boundary, and hence cannot be derived from a planar map. However, all the Albert presemifields for $n = 4$ determine codes with the same parameters and their $\mathcal{A}_{\mu^*}$ code is almost-MDS, with parameters $[80, 2, 78]$. The Dickson semifield linear 3^n-ary codes generated by the rows of M_{μ^*} have parameters $[81, 3, 54(= 3^4 - 3^3)]$ for $n = 4$ and $[729, 3, 648(= 3^6 - 3^4)]$ for $n = 6$.

From Lemma 9.47.3 we know that for field multiplication on G, the rows of M_μ form a p^n-ary linear $[p^n, 1, p^n - 1]$ code and the rows of its absolute trace $\mathrm{tr}(M_\mu)$ form a p-ary linear $[p^n, n, p^n - p^{n-1}]$ code. For both Albert presemifield multiplication and Dickson semifield multiplication on G, the rows of the absolute trace matrix $\mathrm{tr}(M_{\mu^*})$ similarly form a p-ary linear code with parameters $[p^n, n]$.

Research Problem 81 *Let* $G = (GF(p^n), +)$. *Is it always true that, for Albert presemifield multiplication on* G, *the Class B code generated by* M_{μ^*} *has parameters* $[p^n, 2]$, *while for Dickson semifield multiplication on* G, *the Class B code generated by* M_{μ^*} *has parameters* $[p^n, 3]$?

Although every known example of a generalised Hadamard matrix has w equal to a prime power, the only restriction on v is that it is a multiple of w. If G is an abelian p-group of rank n, a modified linearity argument may be applied.

LEMMA 9.49 [167, Lemma 1] *If* G *is an abelian p-group of rank n and exponent* p^m, *and* $\psi : G \times G \to C$ *is a multiplicative orthogonal cocycle, then* M_ψ *has* $\mathbb{Z}_{p^m}$-*rank at most n. The rows of* M_ψ *generate a linear code of at most dimension n over* $\mathbb{F}_p$.

The basic problems for Class B codes are still to determine the dimension and distance parameters of these linear codes. In addition, the question of whether the code $C_p(H)$ of a cocyclic generalised Hadamard matrix H, defined as the linear span over $\mathbb{F}_p$ of the rows of H together with those of its translates aH, $a \in C$, is self-orthogonal has still to be answered.

Research Problem 82 *Let $\psi \in Z^2(G,C)$ be orthogonal. For which G, C, ψ and primes p is the code $C_p(M_\psi)$ self-orthogonal?*

9.4.3 Class C cocyclic Hadamard codes

Binary Class C codes are defined by their generator matrix $[I_{4t} \ A]$, where A is the binary version of a Hadamard matrix H. The dimension of these linear codes is known ($= 4t$), and the class contains self-dual codes, so the main problems here are to determine weight enumerators and equivalence classes, and to find extremal self-dual codes. The self-dual codes with all codewords having weights divisible by 4 are called *doubly-even*. *Extremal* doubly-even self-dual codes are those of length n which meet the distance bound $d \leq 4\lfloor \frac{n}{24} \rfloor + 4$.

Rao (= Baliga) [16, 17] has used the orthogonal cocycles over $\mathbb{Z}_2^2 \times \mathbb{Z}_t$ and D_{4t} listed in Table 6.2 to obtain a ready source of Hadamard matrices for use in this construction. Her aim is to identify new doubly- and singly-even self-dual binary cocyclic Hadamard codes and, just possibly, to find the 'Holy Grail' in this area, an extremal doubly-even self-dual $[72, 36, 16]$ code.

This approach to computation has been successful; for example, for $t = 5$ she identifies 27 distinct equivalence classes of extremal binary doubly-even self-dual $[40, 20, 12]$ codes, giving a total of over 30,000 such codes, which are extremal Class C cocyclic Hadamard codes.

By (8.17) we know that cocycles ψ and φ in the same bundle in $Z^2(G, \{\pm 1\})$ determine equivalent cocyclic matrices M_ψ and M_φ. Then it is readily checked that they generate equivalent binary Class C codes.

LEMMA 9.50 *Suppose $\psi, \varphi \in Z^2(G, \{\pm 1\})$. If $\mathcal{B}(\psi) = \mathcal{B}(\varphi)$, then $[I_{4t} \ A_\psi]$ and $[I_{4t} \ A_\varphi]$ generate equivalent binary Class C codes.*

In [274] Rao investigates shift action on $\psi \in [(1, -1, -1)] \in H^2(D_{4t}, \mathbb{Z}_2)$ (see Chapter 6.4.4). She shows that for $\psi \cdot a$, $a \in D_{4t}$, the Class C codes generated by the equivalent un-normalised matrices of (6.19) are all equivalent codes. (This result is not necessarily true if the binary version of an arbitrary matrix equivalent to M_ψ is substituted.)

For $t = 5$ Rao then identifies 35 distinct shift orbits of the matrices of form (6.19) — and some equivalent matrices — which she used to derive the 21 equivalence classes of extremal doubly-even self-dual $[40, 20, 12]$ codes found in [17]. Each orbit contained 160 matrices.

Research Problem 83 *Does an extremal binary doubly-even self-dual $[72, 36, 16]$ code exist? If so, is there a cocyclic Class C Hadamard example?*

9.5 NEW HIGHLY NONLINEAR FUNCTIONS

For the last topic of this book, we take up again the question of nonlinearity for functions intended to provide confusion in cryptographic algorithms. S-box functions were introduced for binary-based cryptosystems in Chapter 3.5 and for functions of arbitrary finite groups in Section 9.2.1. There, the properties of perfect nonlinearity, bentness and maximal nonlinearity are described, and their interdependencies and characterisations in terms of generalised Hadamard matrices and the Fourier Transform are laid out. Some of the results below are peppered through [154, 157, 158] but some are new.

9.5.1 1-D differential uniformity

PN functions provide maximum resistance to differential attack. However, Nyberg [252, p. 58] also developed a more general measure of resistance to differential attack on 'DES-like' ciphers where the S-box functions are between finite abelian groups, but PN functions may be few, or nonexistent. She defined a function $\phi : G \to C$ of abelian groups to be *differentially m-uniform* when $m = \max_{x \neq 1 \in G, c \in C} |\{y \in G : \phi(xy)\phi(y)^{-1} = c\}|$. If ϕ is an S-box function, its susceptibility to differential cryptanalysis is minimised if m is as small as possible. For contributions to the more general theory of differential uniformity for block ciphers see, for example, [49, 143, 42, 44, 264].

When $|G| = |C|$, differentially 1-uniform, PN and planar functions all coincide (although none exist when $|G|$ is even).

When $|G| = |C|$ is even, Coulter and Henderson [60, §2] call certain differentially 2-uniform functions *semi-planar*. When $G = C = \mathbb{Z}_p^n$, a differentially 2-uniform function is also termed *almost perfect nonlinear* (APN), and when $p = 2$ — the case first encountered in Chapter 3.5.2 — it will also be semi-planar.

Now we extend Nyberg's definition to arbitrary G and to $C^2(G, C)$, using Definitions 8.14 and 8.23 with $\varrho \equiv 1$, in order to relate differential uniformity of ϕ to that for $\partial \phi$ and cocycles in general.

DEFINITION 9.51 *Let G and C be finite groups with C abelian, written multiplicatively, let $\phi \in C^1(G, C)$ and let $\Phi \in C^2(G, C)$. Set $n_\phi = n_{(\phi,1)}$ in Definition 8.23.*

1. *Define ϕ to be differentially Δ_ϕ-uniform if*

$$\max\{n_\phi(g, c) : g \in G, c \in C, g \neq 1\} = \Delta_\phi.$$

2. *Define Φ to be differentially Δ_Φ-row uniform if*

$$\max\{N_\Phi(g, c) : g \in G, c \in C, g \neq 1\} = \Delta_\Phi.$$

Example 9.5.1   *For the LP cocycles μ_λ^* in Definition 9.36, $\Delta_{\mu_\lambda^*} = 1$. For the mappings Φ in Example 8.1.1, $\Delta_\Phi = 2$ for the $\mathbb{Z}_7$ mapping and $\Delta_\Phi = 4$ for the $|G| = 8$ mapping.*

These maxima are invariant under the coboundary operator and bundle actions, by Theorems 8.15 and 8.24.

COROLLARY 9.52 *If $\phi \in C^1(G, C)$ and $\psi \in Z^2(G, C)$, then*

1. $\Delta_\phi = \Delta_\varphi$ *for all $\varphi \in \mathbf{b}(\phi)$;*

2. $\Delta_\phi = \Delta_{\partial\phi}$;

3. $\Delta_\psi = \Delta_{\psi'}$ *for all $\psi' \in \mathcal{B}(\psi)$.*

There also exists a *diagonal operator* $D : C^2(G, C) \rightarrow C^1(G, C)$, which is a homomorphism of abelian groups in the reverse direction to that of ∂. Here, the *diagonal* $D\Phi : G \rightarrow C$ of Φ is

$$D\Phi(g) = \Phi(g, g), \ g \in G. \tag{9.26}$$

The diagonal operator may be thought of as generalising the relationship between bilinear and quadratic forms (Example 6.2.6). We have seen it used in Section 9.3.2.2. In general, D does not preserve distributions, but, for a multiplicative cocycle, the distributions of its diagonal and its symmetrisation (Example 6.2.17) are the same.

LEMMA 9.53 *Suppose $\psi \in Z^2(G, C)$ is multiplicative. Then*

1. *ψ^+ is a cocycle and $\partial D(\psi) = \psi^+$; when ψ is symmetric, $\partial D(\psi) = \psi^2$;*

2. *when $\psi = \partial\phi$ is symmetric, $D\psi = \phi^2 f$ for some $f \in \mathrm{Hom}(G, C)$;*

3. *$\mathcal{D}(D\psi) = \mathcal{D}(\psi^+)$ and $\Delta_{D\psi} = \Delta_{\psi^+}$.*

Proof. Part 1 follows by definition. Part 2 follows from it since $\partial(D(\partial\phi)) = \partial(\phi^2)$ so $D(\partial\phi)(\phi^2)^{-1} \in \ker \partial$. Since ψ is multiplicative, $D\psi(gh)D\psi(h)^{-1} = \psi(gh, gh)\psi(h, h)^{-1} = \psi(g, g)\psi^+(g, h)$, so $n_{D\psi}(g, c) = N_{\psi^+}(g, c\psi(g, g)^{-1})$, giving part 3. $\qquad\square$

If square roots may be extracted in C (or, if C is written additively, if values in C can be halved), Lemma 9.53.2 permits us to compute from any multiplicative symmetric ψ known to be a coboundary, a suitable (but not unique) 1-cochain ϕ such that $\psi = \partial\phi$. This technique for calculating ϕ from ψ is clearly faster that given by solution of the simultaneous linear equations (6.7), mentioned in Section 9.3.2.2.

COROLLARY 9.54 *Suppose $\psi \in B^2(G, C)$ is a multiplicative and symmetric coboundary. If square roots may be extracted in C, define $(D\psi)^{\frac{1}{2}}$ by $(D\psi)^{\frac{1}{2}}(g) = (D\psi(g))^{\frac{1}{2}}$, $g \in G$. Then $\psi = \partial(D\psi)^{\frac{1}{2}}$, and ψ and $(D\psi)^{\frac{1}{2}}$ have the same differential uniformity.*

Obviously, by Section 9.3.2.2, the DO polynomials are good candidates to test for differential uniformity. We start with the DO monomials.

THEOREM 9.55 [158, Theorem 2, Lemma 4] *Let $G = C = (GF(p^n), +)$ and let $\phi : G \rightarrow G$ be a DO monomial, $\phi(g) = g^{p^i + p^j}$, $g \in G$, where $i \leq j$.*

1. *There exists $0 \leq b \leq n$ such that $N_{\partial\phi}(g,c) = p^b$ or 0 for every $g \neq 0 \in G$ and every $c \in G$. Consequently $\partial\phi$ is differentially p^b-row uniform and $\partial\phi : G \times G \to \mathrm{Im}(\partial\phi)$ is orthogonal.*

2. *When $p = 2$, $\Delta_{\partial\phi} = 2$ when $i < j$ and $(n, j-i) = 1$; $\Delta_{\partial\phi} = 2^{(j-i)}$ when $i < j$ and $(n, j-i) > 1$ and $\Delta_{\partial\phi} = 2^n$ when $i = j$.*

3. *When p is odd, $\Delta_{\partial\phi} = 1$ when $i = j$ or when $i < j$ and $n/(n,j-i)$ is odd; otherwise $\Delta_{\partial\phi} = p^{(n,j-i)}$.*

This simple construction (with $i = 0$) accounts for the two planar functions ϕ_1 and ϕ_2 of Example 9.2.1 and the binary-based APN power function f_1 of Table 3.1.

Can it be adapted to account for other functions with low differential uniformity? The next result is a simple consequence of Lemma 9.53 and the definitions.

COROLLARY 9.56 *Let $G = C = (GF(p^n), +)$ and let $\phi : G \to G$ be a DO polynomial $\phi(g) = \sum_{j=0}^{n-1} \sum_{i=0}^{j} \lambda_{ij} g^{p^i + p^j}$, $\lambda_{ij} \in GF(p^n)$, $g \in G$.*

Define $\varphi \in Z^2(G,G)$ from ϕ to be $\varphi(g,h) = \sum_{j=0}^{n-1} \sum_{i=0}^{j} \lambda_{ij} g^{p^i} h^{p^j}$. Then φ is multiplicative, $D\varphi = \phi$ and so $\varphi^+ = \partial\phi$.

Hence $\Delta_{D\varphi} = \Delta_\phi = \Delta_{\varphi^+} = \Delta_{\partial\phi}$.

Research Problem 84 *Let ϕ be a DO polynomial on $GF(p^n)$ and let φ^+ be as in Corollary 9.56. Determine the value of $\Delta_\phi = \Delta_{\varphi^+}$.*

Differential m-row uniformity for cocycles is easy to characterise using the shift action: no more than m entries can equal 1 in any noninitial row of each matrix corresponding to a shift of the cocycle.

To the best of the author's knowledge, this is the first application of shift action in a cryptographic context.

THEOREM 9.57 [158, Theorem 1] *Let $\psi \in Z^2(G,C)$. Then*
$$\Delta_\psi \leq m \iff N_{\psi \cdot k}(g,1) \leq m, \quad \forall\, g \neq 1,\ k \in G.$$
If ψ is multiplicative, $\Delta_\psi \leq m \iff N_\psi(g,1) \leq m$, $\forall\, g \neq 1 \in G$.

Proof. If $\Delta_\psi \leq m$ then by Corollary 9.52.3, $\Delta_{\psi \cdot k} \leq m$ for any $k \in G$ and the row condition holds. Conversely, if $\Delta_\psi \geq m+1$ then for some $g' \neq 1 \in G$, there exist $m+1$ distinct elements $x, y_1, y_2, \ldots, y_m \in G$ such that $\psi(g', x) = \psi(g', y_1) = \psi(g', y_2) = \cdots = \psi(g', y_m)$. Write $y_i = xu_i$, $i = 1, \ldots, m$, and $g = x^{-1}g'x$. By (6.6), $\psi(g', xu_i)\psi(g', x)^{-1} = (\psi \cdot x)(g, u_i) = 1$, $i = 1, \ldots, m$, giving m distinct values $h \neq 1$ for which $(\psi \cdot x)(g, h) = 1$. If ψ is multiplicative, it is fixed by every shift (Lemma 8.30). $\square$

Special cases of this result give known characterisations of PN and APN functions. Dembowski and Ostrom ([94], cited in [63, (2)]) gave a symmetric characterisation of planar DO polynomials, which is a particular instance of the case $m = 1$, $\psi = \partial\phi$ multiplicative, of Theorem 9.57. Canteaut's characterisation [42] of binary-based APN functions in terms of 'second derivatives', namely, ϕ is APN if and only if, for all nonzero $g \neq h \in G$ and all $k \in G$, $\Delta((\Delta\phi)_h)_g(k) \neq 0$, is the case $m = 2$, $G = C = (GF(2^a), +)$, $\psi = \partial\phi$, of Theorem 9.57.

COROLLARY 9.58 *Let $G = (GF(p^n), +)$ and $\phi \in C^1(G, G)$.*

1. *(Dembowski and Ostrom) Let p be odd and let ϕ be the DO polynomial $\phi(g) = \sum_{j=0}^{n-1} \sum_{i=0}^{j} \lambda_{ij} \, g^{p^i + p^j}$. Then ϕ is planar if and only if*

$$\sum_{j=0}^{n-1} \sum_{i=0}^{j} \lambda_{ij} \left(g^{p^i} h^{p^j} + g^{p^j} h^{p^i} \right) = 0 \Leftrightarrow g = 0 \text{ or } h = 0.$$

2. *(Canteaut) Let $p = 2$. Then ϕ is APN if and only if, for all $k \in G$ and $g \neq h$, $g \neq 0$, $h \neq 0 \in G$, $\left((\partial \phi) \cdot k \right)(g, h) \neq 0$.*

Proof. By Corollary 9.56, $\Delta_{D\varphi} = \Delta_{\phi} = \Delta_{\varphi^+}$ and, since φ^+ is multiplicative, $\Delta_{\phi} = 1$ if and only if $N_{\varphi^+}(g, 0) = 1$, for all $g \neq 0 \in G$, giving part 1. For part 2, note that $\Delta((\Delta\phi)_h)_g(k) = \phi(k + g + h) - \phi(k + g) - \phi(k + h) + \phi(k) = (\partial(\phi \cdot k))(g, h) = ((\partial\phi) \cdot k)(g, h)$. □

We now give a construction of $m\ell$-row uniform functions from m-row uniform functions, which may be a more promising technique for discovering new functions with low differential uniformity, since it is not restricted to the mere translation of properties for ϕ into properties for $\partial\phi$.

THEOREM 9.59 *Let $\Phi \in C^2(G, C)$ and let $\alpha \in \mathrm{Hom}(C, C)$. If $\Delta_{\Phi} = m$ and $\alpha(C)$ has index ℓ in C, then $\Delta_{\alpha \circ \Phi} \leq m\ell$. In particular, if $|G| = |C|$ and $\Delta_{\Phi} = 1$, then $\Delta_{\alpha \circ \Phi} = \ell$.*

Proof. If $c' \notin \mathrm{Im}\,\alpha$, then $N_{\alpha \circ \Phi}(g, c') = 0$. If $c' \in \mathrm{Im}\,\alpha$ and $\alpha(c) = c'$, then $N_{\alpha \circ \Phi}(g, c') = \sum_{d \in \mathrm{Ker}\,\alpha} N_{\Phi}(g, c + d) \leq m\ell$. □

9.5.2 Differential 2-row uniformity and APN functions

The most robust resistance to differential attack in the absence of a PN function is provided by a differentially 2-uniform function. In this subsection (see [157]) two new techniques for deriving differentially 2-row uniform functions from differentially 1-row uniform functions are presented.

The first, for $G = (GF(2^n), +)$, is nothing more than Theorem 9.59 for $\ell = 2$, applied to the enormous collection — implicit in Section 9.3.2 — of binary-based differentially 1-row uniform cocycles that are presemifield multiplications. Of course none of these multiplications are themselves coboundaries, since there are no binary-based planar functions. However, their image under a homomorphism with index 2 image may be a coboundary. A necessary condition for this to happen is also stated.

COROLLARY 9.60 *Let $G = (GF(2^n), +)$ and let $\psi \in Z^2(G, G)$ have $\Delta_{\psi} = 1$. Suppose $\alpha \in \mathrm{Hom}(G, G)$, where $\alpha(G)$ has index 2 in G, so there exists an isomorphism $\beta : \alpha(G) \cong \mathbb{Z}_2^{n-1}$.*

1. *If ψ is multiplicative, $(G, +, \psi)$ is a presemifield.*

2. *$\Delta_{\alpha \circ \psi} = 2$ and $\beta \circ (\alpha \circ \psi) \in Z^2(\mathbb{Z}_2^n, \mathbb{Z}_2^{n-1})$ is PN.*

3. *Suppose $\alpha \circ \psi = \partial\phi$, so ϕ is APN. Then there is a unique $x \neq 0 \in G$ such that $D\psi(g) = x$ for all $g \neq 0 \in G$.*

Proof. Parts 1 and 2 follow by definition and Theorem 9.59. For part 3, for any $g \in G$, $\partial\phi(g,g) = 0 = \alpha \circ D\psi(g)$. But α maps only one nonzero element x of G to 0. If $D\psi(g) = 0$ for some $g \neq 0$, then $\psi(g,0) = \psi(g,g) = 0$, contradicting $\Delta_\psi = 1$, so $D\psi(g) = x$ for all $g \neq 0$.   □

Although the binary-based APN power function f_1 is DO, none of the representatives $f_2, \ldots, f_6$ of the other known families are DO (see Table 3.1 in Chapter 3.5.2). Perhaps they, and new families, can be constructed using Corollary 9.60.

Research Problem 85 *Let $G = (GF(2^n), +)$. For which presemifields $(G, +, \psi)$, and which $\alpha \in \mathrm{Hom}(G, G)$ with $\alpha(G)$ of index 2, is $\alpha \circ \psi$ a coboundary? Do any of these coboundaries determine the families represented by the APN power functions $f_2, \ldots, f_6$ of Table 3.1? Which of these coboundaries determine new bundles of APN?*

On the other hand, this construction does not apply to multiplications in presemifields of odd order, since they have no subgroups of index 2.

Our second technique, for the odd case, just relies on the quadratic residues in $GF(p^n)$.

LEMMA 9.61 *Let $G = (GF(p^n), +)$, p odd, and $\psi \in Z^2(G, G)$, where $\Delta_\psi = 1$. Then $\Delta_{\psi^2} = 2$.*

Proof. Since $N_\psi(g, c) = 1$, for all $g \neq 0$, $c \in G$, $N_{\psi^2}(g, d) = 2$, for all $g \neq 0$, $d = c^2 \neq 0 \in G$; $N_{\psi^2}(g, d) = 0$, for all $g \neq 0$, $d \neq c^2 \in G$; and $N_{\psi^2}(g, 0) = 1$ for all $g \neq 0 \in G$.   □

In the instances of Lemma 9.61 where ψ is multiplicative, ψ is a presemifield multiplication, but ψ^2 is not, because it is not multiplicative. By Lemma 9.53, if ψ is also symmetric, it must be a coboundary $\psi = \partial\phi$, in which case ϕ is PN; in fact, ϕ is a planar DO polynomial (up to a linearised summand) by Theorem 9.43.

COROLLARY 9.62 *Let $G = (GF(p^n), +)$, p odd, let $\psi \in Z^2(G, G)$ and let $(G, +, \psi)$ be a presemifield. Then ψ^2 is differentially 2-row uniform.*

A list of families of APN power functions on $GF(p^n)$, p odd, has been established by Helleseth and colleagues [147, 146] and appears in [48]. Perhaps these, and new families, can be constructed using Corollary 9.62.

Research Problem 86 *Let $G = (GF(p^n), +)$, p odd. For which presemifields $(G, +, \psi)$ is ψ^2 a coboundary? Do any of these coboundaries determine the families of APN power functions described in [147, 146]? Which of these coboundaries determine new bundles of APN?*

There is some evidence to suggest APN power functions for odd p may arise more generally from symmetrisations of differentially 1-row uniform functions which are not multiplicative. For example, if ϕ is a DO monomial and $\Phi(g, h) = \phi(g)h$, then $\Delta_\Phi = 1$ (though Φ may not be a cocycle). If the symmetrisation Φ^+ is a coboundary, does $\Delta_{\Phi^+} = 2$?

Research Problem 87 *Generalise Corollary 9.62 to $\psi \in Z^2(G, G)$ which are not multiplicative.*

Perhaps the constructions of Corollary 9.60 and Lemma 9.61 are reversible. This would deliver us potentially new differentially 1-row uniform and PN functions.

Research Problem 88 *Let $G = (GF(p^n), +)$ and $\psi \in Z^2(G, G)$. If $p = 2$, let $\alpha \in \mathrm{Hom}(G, G)$ where $\alpha(G)$ has index 2 in G, and suppose $\Delta_{\alpha \circ \Phi} = 2$. Otherwise, suppose $\Delta_{\psi^2} = 2$. Under what circumstances is $\Delta_\psi = 1$?*

The final topic covered in this volume, albeit briefly, is that promised in Chapter 7.4: a different approach from that of the Five-fold Constellation to finding a property of 2-D functions which extends perfect nonlinearity. Rather than extracting a 1-D representative (which will be PN in the splitting case) from an orthogonal cocycle, the idea is to extend perfect nonlinearity from 1-D to 2-D functions.

9.5.3 2-D total differential uniformity

Recent interest in encryption algorithms involving arrays, sparked by the choice of *Rijndael* as the AES algorithm, raises the problem of differential cryptanalysis of ciphertext which is genuinely array-encrypted (rather than, for example, encrypted by a set of key-dependent S-boxes each encrypting an input block).

Highly nonlinear functions of arrays may themselves also provide a potential source of key-dependent S-boxes for block inputs, or of mixer functions for iterative block ciphers, or of hash-based MACs.

We extend the ideas of Section 9.5.1 to two dimensions.

A differential attack on an S-box function $\Phi \in C^2(G, C)$ with 2-D array inputs would involve fixing an input pair $(a, b) \neq (1, 1) \in G \times G$ and looking for bias in the frequencies of output differences $\Phi(ag, hb) - \Phi(g, h)$, as (g, h) runs through $G \times G$. Consequently, the susceptibility of such a function to differential attack is minimised if the maximum of these frequencies is as small as possible.

DEFINITION 9.63 *For $\Phi \in C^2(G, C)$ and for each $(a, b) \in G \times G$ and $c \in C$, set $n_\Phi(a, b; c) = |\{(g, h) \in G \times G : \Phi(ag, hb) - \Phi(g, h) = c\}|$.*

Define Φ to be totally differentially m-uniform *if*

$$\max\{n_\Phi(a, b; c) : (a, b) \neq (1, 1) \in G \times G, c \in C\} = m.$$

Optimal total differential uniformity will occur only when every element of C is a total differential equally often, that is, when $|C|$ divides $|G|^2$ and $m = |G|^2/|C|$.

For $\Phi \in C^2(G, C)$, and for $k \in G$, the *left first partial derivative* $(\Delta_1 \Phi)_k : G \times G \to C$ of Φ in direction k is $(\Delta_1 \Phi)_k(g, h) = \Phi(kg, h) - \Phi(g, h)$ and the *right first partial derivative* $(\nabla_1 \Phi)_k : G \times G \to C$ of Φ in direction k is $(\nabla_1 \Phi)_k(g, h) = \Phi(gk, h) - \Phi(g, h)$. Corresponding definitions apply for the second partial derivatives $(\Delta_2 \Phi)_k$ and $(\nabla_2 \Phi)_k$.

LEMMA 9.64 [158, Lemmas 1, 5] *Let $\psi \in Z^2(G, C)$ and $L_\psi(h, c) = |\{g \in G : \psi(g, h) = c\}|$, $h \in G, c \in C$. Then*

$$n_\psi(a, b; c) = |\{(g, h) \in G \times G : (\Delta_1\psi)_{ag}(h, b) + (\nabla_2\psi)_h(a, g) = c\}|.$$

If $\psi \in Z^2(G, C)$ is multiplicative and $(a, b) \neq (1, 1) \in G \times G$, then

$$n_\psi(a, b; c) = \sum_{e \in C} N_\psi(a, e)L_\psi(b, c - e).$$

Consequently, if ψ is multiplicative and orthogonal, then ψ has optimal total differential uniformity.

Multiplicative cocycles are wholly determined by their values on pairs of elements from a minimal generating set for G. As any other values are found by additions only, they are fast to compute by comparison with nonmultiplicative cocycles, for which a generating set may be difficult to determine and from which other entries must be computed from the cocycle equation (6.6) by an algorithm such as one of those in Chapter 6.3.

Since the multiplicative cocycles are so highly structured, we always have to balance their potential utility as array-input S-box functions against the ease of recovering them. However, in the most likely case that $G = \mathbb{Z}_p^n$, there are $|C|^{n^2}$ multiplicative cocycles, so, for example, in the binary case we only need, say, $n = 32$ for a search space of size 2^{1024}, prohibitively large for exhaustion using present computing power.

Apart from Lemma 9.64 the relationship, if any, between orthogonality and total differential uniformity has not been explored.

Research Problem 89 *Let $\psi \in Z^2(G, C)$. When does orthogonality imply total differential uniformity for ψ? When does total differential uniformity imply orthogonality for ψ?*

Finally we emphasise that in the binary case, a 2-D point of view allows us to construct cryptographic functions which are better than their 1-D counterparts. The following instance, the simplest demonstration of this statement, will be very familiar to the reader.

Let $G = C = (GF(p^n), +)$. The quadratic function $\phi(g) = g^2, g \in G$ and the corresponding coboundary $\partial\phi(g, h) = 2gh$, $g, h \in G$ have the same differential row uniformity, by Corollary 9.52.2.

When $p = 2$, ϕ has worst-possible differential uniformity $\Delta_\phi = 2^n$ by Theorem 9.55.1, and $\partial\phi = 0$. Nonetheless, the field multiplication $\psi(g, h) = gh$ on $GF(2^n)$ is optimal; in fact it has optimal total differential uniformity.

Apart from their optimal resistance to differential attack on array inputs, nothing is known about total differentially uniform functions. They may not be resistant to standard 1-D attacks. A basic research program is proposed (cf. Research Problem 31 for the theory of linear array codes).

Research Problem 90 *Create a uniform framework for the theory of total differentially uniform cryptographic functions.*

1. *What are the advantages and disadvantages of an optimal total differentially uniform $v \times v$ array S-box function when compared with a simultaneous standard differential attack on each of the rows or columns, or on a length v^2 vector S-box function read out from it?*

2. Investigate the susceptibility of a total differentially uniform array S-box function to a linear array attack.

Over 2 trillion presemifield multiplications of order 16 exist (Section 9.3.2.1). They provide an attractive experimental space of total differentially uniform functions with which to investigate this problem.

We have seen, especially in this final Chapter, something of the richness and interconnectedness of the cocyclic approach to generalised Hadamard matrices. The subject is still in its infancy, with open questions ranging in difficulty from simple to profound.

Good fortune to those hunting for solutions.

Bibliography

[1] S. S. Agaian, *Hadamard Matrices and Their Applications*, LNM 1168, Springer, Berlin, 1985.

[2] N. Ahmed and K. R. Rao, *Orthogonal Transforms for Digital Signal Processing*, Springer, Berlin, 1975.

[3] J. L. Alperin and R. B. Bell, *Groups and Representations*, Springer, New York, 1995.

[4] V. Alvarez, J. A. Armario, M. D. Frau and P. Real, An algorithm for computing cocyclic matrices developed over some semidirect products, *AAECC-14*, S. Boztaş, I. Shparlinski, eds., LNCS 2227, Springer, Berlin, 2001, 287–296.

[5] V. Alvarez, J. A. Armario, M. D. Frau and P. Real, A genetic algorithm for cocyclic Hadamard matrices, *AAECC-16*, M. Fossorier et al., eds., LNCS 3857, Springer, Berlin, 2006, 144–153.

[6] V. Alvarez, J. A. Armario, M. D. Frau and P. Real, Calculating cocyclic Hadamard matrices in *Mathematica*: exhaustive and heuristic searches, *ICMS 2006*, A. Iglesias, N. Takayama, eds., LNCS 4151, Springer, Berlin 2006, 419–422.

[7] F. W. Anderson and K. R. Fuller, *Rings and Categories of Modules*, 2nd ed., GTM 13, Springer, New York, 1992.

[8] K. T. Arasu and W. de Launey, Two-dimensional perfect quaternary arrays, *IEEE Trans. Inform. Theory* 47 (2001) 1482–1493.

[9] K. T. Arasu, W. de Launey and S.-L. Ma, On circulant complex Hadamard matrices, *Des. Codes Cryptogr.* 25 (2002) 123–142.

[10] K. T. Arasu and J. F. Dillon, Perfect ternary arrays, *Difference Sets, Sequences and Their Correlation Properties*, A. Pott, P. V. Kumar, T. Helleseth and D. Jungnickel, eds., NATO Science Series C, 542, Kluwer, Dordrecht, 1999, 1–15.

[11] K. T. Arasu, D. Jungnickel, S.-L. Ma and A. Pott, Relative difference sets with $n = 2$, *Discr. Math.* 147 (1995) 1–17.

[12] J. P. Arpasi and R. Palazzo Jr., An algorithm to construct strongly controllable group codes from an abstract group, *Proc. 1998 ISIT*, IEEE (1998), 154.

[13] M. Aschbacher, *Finite Group Theory*, 2nd ed., CUP, Cambridge, 2000.

[14] E. F. Assmus Jr and J. D. Key, *Designs and Their Codes*, CUP, Cambridge, 1992.

[15] R. Baer, Erweiterung von Gruppen und ihre Automorphismen, *Math. Z.* 38 (1934) 375–416.

[16] A. Baliga, New self-dual codes from cocyclic Hadamard matrices, *J. Combin. Math. Combin. Comput.* 28 (1998) 7–14.

[17] A. Baliga, Cocyclic codes of length 40, *Des. Codes Cryptogr.* 24 (2001) 171–179.

[18] A. Baliga and K. J. Horadam, Cocyclic Hadamard matrices over $\mathbb{Z}_t \times \mathbb{Z}_2^2$, *Australas. J. Combin.* 11 (1995) 123–134.

[19] S. Ball and M. R. Brown, The six semifield planes associated with a semifield flock, *Adv. Math.* 189 (2004) 68–87.

[20] R. H. Barker, Group synchronisation of binary digital systems, *Communication Theory* (Proc. Second London SIT), Butterworth, London, 1953, 273–287.

[21] P. S. L. M. Barreto, H. K. Kim, B. Lynn and M. Scott, Efficient algorithms for pairing-based cryptosystems, *CRYPTO 2002*, LNCS 2442, Springer, Berlin, 2002, 354–368.

[22] N. Bauer, P. Carmany and K. W. Smith, All $(8, 8, 8, 1)$ relative difference sets, draft manuscript, February 2006.

[23] K. G. Beauchamp, *Walsh Functions and Their Applications*, Academic Press, London, 1975.

[24] T. Beth, D. Jungnickel and H. Lenz, *Design Theory*, 2nd ed., CUP, Cambridge, 1999.

[25] T. Beth, H. Kalouti and D. E. Lazic, Which families of long binary linear codes have a binomial weight distribution?, *AAECC-11*, G. Cohen, M. Giusti, T. Mora, eds., LNCS 948, Springer, Berlin, 1995, 120–130.

[26] E. Biham and A. Shamir, Differential cryptanalysis of DES-like cryptosystems, *J. Cryptology* 4 (1991) 3–72.

[27] I. Blake, G. Seroussi and N. Smart, *Elliptic Curves in Cryptography*, LMS Lecture Note Series 265, CUP, Cambridge, 1999.

[28] M. Blaum, J. Bruck and A. Vardy, Interleaving schemes for multidimensional cluster errors, *IEEE Trans. Inform. Theory* 44 (1998) 730–743.

[29] M. Blaum, P. G. Farrell and H. C. A. van Tilborg, Array codes, Chapter 22, *Handbook of Coding Theory*, V. S. Pless and W. C. Huffman, eds., North-Holland, Amsterdam, 1998.

[30] A. Blokhuis, D. Jungnickel and B. Schmidt, Proof of the prime power conjecture for projective planes of order n with abelian collineation groups of order n^2, *Proc. Amer. Math. Soc.* 130 (2002) 1473–1476.

[31] A. Bonnecaze and I. M. Duursma, Translates of linear codes over $\mathbb{Z}_4$, *IEEE Trans. Inform. Theory* 43 (1997) 1–13.

[32] W. Bosma, J. Cannon and C. Playoust, The MAGMA algebra system I: the user language, *J. Symbol. Comp.* 24 (1997) 235–265.

[33] S. Boztaş, Constacyclic codes and constacyclic DFTs, *Proc. 1998 ISIT*, IEEE (1998) 235.

[34] S. Boztaş, R. Hammons and P. V. Kumar, 4-phase sequences with near-optimum correlation properties, *IEEE Trans. Inform. Theory* 38 (1992) 1101–1113.

[35] L. Breveglieri, A. Cherubini and M. Macchetti, On the generalized linear equivalence of functions over finite fields, *ASIACRYPT 2004*, P. J. Lee, ed., LNCS 3329, Springer, Berlin, 2004, 79–91.

[36] K. Brincat, F. C. Piper and P. R. Wild, Stream ciphers and correlation, *Difference Sets, Sequences and Their Correlation Properties*, A. Pott et al., eds., NATO Science Series C, 542, Kluwer, Dordrecht, 1999, 17–44.

[37] B. W. Brock, Hermitian congruence and the existence and completion of generalised Hadamard matrices, *J. Combin. Theory A* 49 (1988) 233–261.

[38] K. S. Brown, *Cohomology of Groups*, GTM 87, Springer, New York, 1982.

[39] L. Budaghyan, C. Carlet and A. Pott, New classes of almost bent and almost perfect nonlinear polynomials, *IEEE Trans. Inform. Theory* 52 (2006) 1141–1152.

[40] A. T. Butson, Generalised Hadamard matrices, *Proc. Amer. Math. Soc.* 13 (1962) 894–898.

[41] D. Calabro and J. K. Wolf, On the synthesis of two-dimensional arrays with desirable correlation properties, *Inform. Control* 11 (1968) 537–560.

[42] A. Canteaut, Cryptographic functions and design criteria for block ciphers, *INDOCRYPT 2001*, C. Pandu Rangan and C. Ding, eds., LNCS 2247, Springer, Berlin, 2001, 1–16.

[43] A. Canteaut, P. Charpin and H. Dobbertin, Binary m-sequences with three-values crosscorrelation: a proof of Welch's conjecture, *IEEE Trans. Inform. Theory* 46 (2000) 4–8.

[44] A. Canteaut and M. Videau, Degree of composition of highly nonlinear functions and applications to higher order differential cryptanalysis, *EUROCRYPT-02*, LNCS 2332, Springer, Berlin, 2002, 518–533.

[45] C. Carlet, $\mathbb{Z}_{2^k}$-linear codes, *IEEE Trans. Inform. Theory* 44 (1998) 1543–1547.

[46] C. Carlet, Boolean functions for cryptography and error-correcting codes; and, Vectorial Boolean functions for cryptography, *Boolean Methods and Models*, P. Hammer and Y. Crama, eds., CUP, Cambridge, 2006, to appear.

[47] C. Carlet, P. Charpin and V. Zinoviev, Codes, bent functions and permutations suitable for DES-like cryptosystems, *Des. Codes Cryptogr.* 15 (1998) 125–156.

[48] C. Carlet and C. Ding, Highly nonlinear mappings, *J. Complexity* 20 (2004) 205–244.

[49] F. Chabaud and S. Vaudenay, Links between linear and differential cryptanalysis, *EUROCRYPT-94*, LNCS 950, Springer, New York, 1995, 356–365.

[50] W. K. Chan and M. K. Siu, Summary of perfect $s \times t$ arrays, $1 \leq s \leq t \leq 100$, *Electron. Lett.* 27 (1991) 709–710, Errata, same volume 1112.

[51] W. K. Chan, M. K. Siu and P. Tong, Two-dimensional binary arrays with good autocorrelation, *Inform. Control* 42 (1979) 125-130.

[52] C. Charnes, M. Rötteler and T. Beth, Homogeneous bent functions, invariants, and designs, *Des. Codes Cryptogr.* 26 (2002) 139–154.

[53] Y. Q. Chen, On the existence of abelian Hadamard difference sets and a new family of difference sets, *Finite Fields Appl.* 3 (1997) 234–256.

[54] Y. Q. Chen, K. J. Horadam and W. H. Liu, Relative difference sets fixed by inversion (III) — Cocycle theoretical approach, *Discr. Math.* (2007), to appear.

[55] W. Chu and C. J. Colbourn, Optimal $(n, 4, 2)$-OOC of small orders, *Discr. Math.* 279 (2004) 163–172.

[56] J. J. Chua and A. Rao, An image-guided heuristic for planning an exhaustive enumeration, *Proc. 2004 HIS*, IEEE (2005) 136–141; and personal communication May 2005.

[57] C. J. Colbourn and W. de Launey, Difference matrices, Chapter IV.11, *The CRC Handbook of Combinatorial Designs*, CRC Press, Boca Raton, 1996.

[58] C. J. Colbourn and J. H. Dinitz, eds., *The CRC Handbook of Combinatorial Designs*, CRC Press, Boca Raton, 1996.

[59] M. Cordero and G. P. Wene, A survey of finite semifields, *Discr. Math.* 208 (1999) 125–137.

[60] R. S. Coulter and M. Henderson, A class of functions and their application in constructing semi-biplanes and association schemes, *Discr. Math.* 202 (1999) 21–31.

[61] R. S. Coulter and M. Henderson, Commutative presemifields and semifields, preprint 2004.

[62] R. S. Coulter and M. Henderson, A new class of commutative presemifields of odd order, preprint 2005.

[63] R. S. Coulter and R. W. Matthews, Planar functions and planes of Lenz-Barlotti Class II, *Des. Codes Cryptogr.* 10 (1997) 167–184.

[64] R. Craigen, Complex Golay sequences, *J. Combin. Math. Combin. Comput.* 15 (1994) 161–169.

[65] R. Craigen, Hadamard matrices and designs, Chapter IV.24, *The CRC Handbook of Combinatorial Designs*, C. J. Colbourn and J. H. Dinitz, eds., CRC Press, Boca Raton, 1996.

[66] R. Craigen, Signed groups, sequences, and the asymptotic existence of Hadamard matrices, *J. Combin. Theory A* 71 (1995) 241–254.

[67] R. Craigen and H. Kharaghani, Hadamard matrices from weighing matrices via signed groups, *Des. Codes Cryptogr.* 12 (1997) 49–58.

[68] R. Craigen and H. Kharaghani, Hadamard matrices and designs, Part V.1, *The CRC Handbook of Combinatorial Designs*, 2nd ed., C. J. Colbourn and J. H. Dinitz, eds., CRC Press, Boca Raton, 2006.

[69] R. Craigen and W. D. Wallis, Hadamard matrices: 1893–1993, *Congr. Numer.* 97 (1993) 99–129.

[70] R. Craigen and R. Woodford, Power Hadamard matrices, *Discr. Math.* (2007), to appear.

[71] C. W. Curtis and I. Reiner, *Representation Theory of Finite Groups and Associative Algebras*, Wiley-Interscience, New York, 1962.

[72] J. Daemen, L. R. Knudsen and V. Rijmen, The block cipher Square, *Fast Software Encryption '97*, E. Biham, ed., LNCS 1267, Springer, Berlin, 1997, 149–165.

[73] J. Daemen and V. Rijmen, *The Design of Rijndael: AES — The Advanced Encryption Standard*, Springer, Berlin, 2002.

[74] J. A. Davis and J. Jedwab, A survey of Hadamard difference sets, *Groups, Difference Sets and the Monster*, K. T. Arasu et al., eds., de Gruyter, Berlin, 1996, 145–156.

[75] J. A. Davis and J. Jedwab, A unifying construction for difference sets, *J. Combin. Theory A* 80 (1997) 13–78.

[76] W. de Launey, $(0, G)$-*Designs with applications*, Ph.D. Thesis, University of Sydney, Sydney, Australia, 1987.

[77] W. de Launey, A survey of generalised Hadamard matrices and difference matrices $D(k, \lambda; G)$ with large k, *Utilitas Math.* 30 (1986) 5–29.

[78] W. de Launey, Square GBRDs over non-abelian groups, *Ars Combin.* 27 (1989) 40-49.

[79] W. de Launey, On the construction of n-dimensional designs from 2-dimensional designs, *Australas. J. Combin.* 1 (1990) 67-81.

[80] W. de Launey, A note on N-dimensional Hadamard matrices of order 2^t and Reed-Muller codes, *IEEE Trans. Inform. Theory* 37 (1991) 664–667.

[81] W. de Launey, Generalised Hadamard matrices which are developed modulo a group, *Discr. Math.* 104 (1992) 49–65.

[82] W. de Launey, Cocyclic Hadamard matrices and relative difference sets, Hadamard Centenary Conference, U. Wollongong, Wollongong, Australia, December 1993, unpublished.

[83] W. de Launey, On the asymptotic existence of partial complex Hadamard matrices and related combinatorial objects, *Discr. Appl. Math.* 102 (2000) 37–45.

[84] W. de Launey, On a family of cocyclic Hadamard matrices, *Ohio State Univ. Math. Res. Inst. Publ.* 10, de Gruyter, Berlin, 2002, 187–205.

[85] W. de Launey, personal communication, March 2005.

[86] W. de Launey, Generalised Hadamard matrices, Part V.5, *The CRC Handbook of Combinatorial Designs*, 2nd ed., C. J. Colbourn and J. H. Dinitz, eds., CRC Press, Boca Raton, 2006.

[87] W. de Launey and D. L. Flannery, *Cocyclic Development of Combinatorial Designs*, manuscript in preparation, 2006.

[88] W. de Launey, D. L. Flannery and K. J. Horadam, Cocyclic Hadamard matrices and difference sets, *Discr. Appl. Math.* 102 (2000) 47–61.

[89] W. de Launey and D. M. Gordon, A comment on the Hadamard conjecture, *J. Combin. Theory A* 95 (2001) 180–184.

[90] W. de Launey and K. J. Horadam, A weak difference set construction for higher dimensional designs, *Des. Codes Cryptogr.* 3 (1993) 75-87.

[91] W. de Launey and M. J. Smith, Cocyclic orthogonal designs and the asymptotic existence of cocyclic Hadamard matrices and maximal size relative difference sets with forbidden subgroup of size 2, *J. Combin. Theory A* 93 (2001) 37–92.

[92] W. de Launey and R. M. Stafford, On cocyclic weighing matrices and the regular group actions of certain Paley matrices, *Discr. Appl. Math.* 102 (2000) 63–101.

[93] W. de Launey and R. M. Stafford, On the automorphisms of Paley's Type II Hadamard matrix, *Discr. Math.* (2007), to appear.

[94] P. Dembowski and T. G. Ostrom, Planes of order n with collineation groups of order n^2, *Math. Z.* 103 (1968) 239–258.

[95] C. Ding and J. Yuan, A family of skew Hadamard difference sets, *J. Combin. Theory A* 113 (2006) 1526–1535.

[96] J. H. Dinitz and D. R. Stinson, eds., *Contemporary Design Theory: A Collection of Surveys*, Wiley, New York, 1992.

[97] D. Z. Djoković, Williamson matrices of orders 4.29 and 4.31, *J. Combin. Theory A* 59 (1992) 309 – 311.

[98] D. Z. Djoković, Williamson matrices of orders $4n$ for n = 33, 35, 39, *Discr. Math.* 115 (1993) 267–271.

[99] D. Z. Djoković, Two Hadamard matrices of order 956 of Goethals-Seidel type, *Combinatorica* 14 (1994) 375–377.

[100] H. Dobbertin, Construction of bent functions and balanced Boolean functions with high nonlinearity, *Fast Software Encryption: Second International Workshop*, B. Preneel, ed., LNCS 1008, Springer, Berlin, 1995, 61–74.

[101] H. Dobbertin, Almost perfect nonlinear power functions on $GF(2^n)$: the Welch case, *IEEE Trans. Inform. Theory* 45 (1999) 1271–1275.

[102] D. A. Drake, Partial λ-geometries and generalised Hadamard matrices, *Canad. J. Math.* 31 (1979) 617–627.

[103] R. Dutta, R. Barua and P. Sarkar, Pairing-based cryptographic protocols: A survey, Cryptology ePrint Archive, Report 2004/064, http://eprint.iacr.org/, last revised 24 June 2004.

[104] I. Duursma, T. Helleseth, C. Rong and K. Yang, Split weight enumerators for the Preparata codes with applications to designs, *Des. Codes Cryptogr.* 18 (1999) 103–124.

[105] D. F. Elliott and K. R. Rao, *Fast Transforms: Algorithms, Analyses, Applications*, Academic Press, New York, 1982.

[106] J. E. H. Elliott and A. T. Butson, Relative difference sets, *Illinois J. Math.* 10 (1966) 517–531.

[107] G. Ellis and I. Kholodna, Computing second cohomology of finite groups with trivial coefficients, *Homology, Homotopy Appl.* 1 (1999) 163–168 (electronic).

[108] D. Elvira and Y. Hiramine, On non-abelian semiregular relative difference sets, *Proc. Finite Fields and Applications* $\mathbb{F}_q5$, D. Jungnickel and H. Niederreiter, eds., Springer, Berlin, 2001, 122-127.

[109] P. G. Farrell, Recent developments in array error-control codes, Ninth IMA Int. Conf. Crypto. and Coding, Cirencester UK, 16-18 Dec. 2003, typescript of plenary talk. Abstract, LNCS 2898, Springer, Berlin, 2003, 1–3.

[110] D. L. Flannery, Transgression and the calculation of cocyclic matrices, *Australas. J. Combin.* 11 (1995) 67–78.

[111] D. L. Flannery, Calculation of cocyclic matrices, *J. Pure Appl. Algebra* 112 (1996) 181–190.

[112] D. L. Flannery, Cocyclic Hadamard matrices and Hadamard groups are equivalent, *J. Algebra* 192 (1997) 749–779.

[113] D. L. Flannery, personal communication, October 2004.

[114] D. L. Flannery and E. A. O'Brien, Computing 2-cocycles for central extensions and relative difference sets, *Comm. Algebra* 28 (2000) 1939–1955.

[115] R. Forré, The strict avalanche criterion: spectral properties of Boolean functions and an extended definition, *CRYPTO '88*, LNCS 403, Springer, Berlin, 1990, 450–468.

[116] M. D. Frau, *Cocyclic development of designs and applications*, Ph.D. Thesis (English version), University of Sevilla, Seville, Spain, 2003. Corrigenda: personal communication, March 2006.

[117] E. M. Gabidulin and V. V. Shorin, Unimodular perfect sequences of length p^s, *IEEE Trans. Inform. Theory* 51 (2005) 1163–1166.

[118] J. C. Galati, *A group extensions approach to relative difference sets*, Ph.D. Thesis, RMIT University, Melbourne, Australia, 2003.

[119] J. C. Galati, Application of Gaschütz' Theorem to relative difference sets in non-abelian groups, *J. Combin. Designs* 11 (2003) 307–311.

[120] J. C. Galati, A group extensions approach to relative difference sets, *J. Combin. Designs* 12 (2004) 279–298.

[121] J. C. Galati, On the non-existence of semiregular relative difference sets in groups with all Sylow subgroups cyclic, *Des. Codes Cryptogr.* 36 (2005) 29–31.

[122] M. J. Ganley, On a paper of Dembowski and Ostrom, *Arch. Math.* 26 (1976) 93–98.

[123] A. V. Geramita and J. Seberry, *Orthogonal Designs: Quadratic Forms and Hadamard Matrices*, Marcel Dekker, New York, 1979.

[124] J.-M. Goethals and J. J. Seidel, Orthogonal matrices with zero diagonal, *Canad. J. Math.* 19 (1967) 1001–1010.

[125] J.-M. Goethals and J. J. Seidel, A skew Hadamard matrix of order 36, *J. Austral. Math. Soc.* 11 (1970) 343–344.

[126] S. W. Golomb, *Shift Register Sequences*, Holden-Day, San Francisco, 1967. Reprinted by Aegean Park Press, 1982.

[127] S. W. Golomb, Construction of signals with favorable correlation properties, *Difference Sets, Sequences and Their Correlation Properties*, A. Pott et al., eds., NATO Science Series C, 542, Kluwer, Dordrecht, 1999.

[128] H. Gopalkrishna Gadiyar, K. M. Sangeeta Maini and R. Padma, Cryptography, connections, cocycles and crystals: a p-adic exploration of the Discrete Logarithm Problem, *INDOCRYPT 2004*, A. Canteaut and K. Viswanathan, eds., LNCS 3348, Springer, Berlin, 2004, 305–314.

[129] X. Gourdon, The 10^{13} first zeros of the Riemann Zeta function, and zeros computation at very large height, preprint October 2004. Available from http://numbers.computation.free.fr/Constants/constants.html. See also *New Scientist*, 27 November 2004, 11.

[130] J. Grabmeier and L. A. Lambe, Computing resolutions over finite p-groups, *Proc. ALCOMA-99*, A. Betten et al., eds., Springer, Berlin, 2001, 157–195.

[131] T. A. Gulliver, New quaternary linear codes of dimension 5, *Proc. 1995 ISITA*, IEEE (1995) 493.

[132] M. K. Gupta, *On some linear codes over $\mathbb{Z}_{2^s}$*, Ph.D. Thesis, IIT, Kanpur, India, 1999.

[133] J. Hadamard, Résolution d'une question relative aux déterminants, *Bull. Sciences Math.* (2), 17 (1893) 240–246.

[134] M. Hall Jr., *The Theory of Groups*, 2nd ed., AMS Chelsea, Providence RI, 1976.

[135] R. M. Hammaker et al., Hadamard Transform Raman spectrometry, Chapter 5, *Modern Techniques in Raman Spectroscopy*, J. J. Laserna, ed., Wiley, Chichester, 1996.

[136] J. Hammer and J. Seberry, Higher dimensional orthogonal designs and Hadamard matrices II, *Congr. Numer.* 27 (1979) 23–29.

[137] J. Hammer and J. Seberry, Higher dimensional orthogonal designs and Hadamard matrices, *Congr. Numer.* 31 (1981) 95–108.

[138] J. Hammer and J. Seberry, Higher dimensional orthogonal designs and applications, *IEEE Trans. Inform. Theory* 27 (1981) 772–779.

[139] A. R. Hammons, P. V. Kumar, A. R. Calderbank, N. J. A. Sloane and P. Solé, The $\mathbb{Z}_4$-linearity of Kerdock, Preparata, Goethals, and related codes, *IEEE Trans. Inform. Theory* 40 (1994) 301–319.

[140] M. Harada and V. D. Tonchev, Singly-even self-dual codes and Hadamard matrices, *AAECC-11*, G. Cohen, M. Giusti and T. Mora, eds., LNCS 948, Springer, Berlin, 1995, 279–284.

[141] H. F. Harmuth, *Transmission of Information by Orthogonal Functions*, 2nd ed., Springer, Berlin, 1972.

[142] M. Harwit and N. J. A. Sloane, *Hadamard Transform Optics*, Academic Press, New York, 1979.

[143] P. Hawkes and L. O'Connor, XOR and non-XOR differential probabilities, *EUROCRYPT-99*, LNCS 1592, Springer, Berlin, 1999, 272–285.

[144] A. S. Hedayat, N. J. A. Sloane and J. Stufken, *Orthogonal Arrays, Theory and Applications*, Springer, New York, 1999.

[145] A. S. Hedayat and W. D. Wallis, Hadamard matrices and their applications, *Ann. Statist.* 6 (1978) 1184–1238.

[146] T. Helleseth, C. Rong and D. Sandberg, New families of almost perfect nonlinear power mappings, *IEEE Trans. Inform. Theory* 45 (1999) 475–485.

[147] T. Helleseth and D. Sandberg, Some power mappings with low differential uniformity, *Applic. Algebra Eng. Commun. Comp.* 8 (1997) 363–370.

[148] T. Helleseth and P. V. Kumar, Sequences with low correlation, Chapter 21, *Handbook of Coding Theory*, V. S. Pless and W. C. Huffman, eds., North-Holland, Amsterdam, 1998.

[149] D. F. Holt, The calculation of the Schur multiplier of a permutation group, *Computational Group Theory (Durham 1982)*, Academic Press, London, 1984, 307–318.

[150] W. H. Holzmann and H. Kharaghani, A computer search for complex Golay sequences, *Australas. J. Combin.* 10 (1994) 251–258.

[151] A. F. Horadam, *A Guide to Undergraduate Projective Geometry*, Pergamon, Sydney, 1970.

[152] K. J. Horadam, Progress in cocyclic matrices, *Congr. Numer.* 118 (1996) 161–171.

[153] K. J. Horadam, Cocyclic Hadamard codes, *Proc. 1998 ISIT*, IEEE (1998) 246.

[154] K. J. Horadam, Sequences from cocycles, *AAECC-13*, M. Fossorier, H. Imai, S. Lin and A. Poli, eds., LNCS 1719, Springer, Berlin, 1999, 121–130.

[155] K. J. Horadam, An introduction to cocyclic generalised Hadamard matrices, *Discr. Appl. Math.* 102 (2000) 115–131.

[156] K. J. Horadam, Equivalence classes of central semiregular relative difference sets, *J. Combin. Des.* 8 (2000) 330–346.

[157] K. J. Horadam, Differentially 2-uniform cocycles — the binary case, *AAECC-15*, M. Fossorier, T. Hoeholdt and A. Poli, eds., LNCS 2643, Springer, Berlin, 2003, 150–157.

[158] K. J. Horadam, Differential uniformity for arrays, *Cryptography and Coding, Proc. 9th IMA International Conference*, LNCS 2898, Springer, Berlin, 2003, 115–124.

[159] K. J. Horadam, The shift action on 2-cocycles, *J. Pure Appl. Algebra* 188 (2004) 127–143.

[160] K. J. Horadam, A generalised Hadamard Transform, *Proc. 2005 ISIT*, IEEE (2005) 1006–1008.

[161] K. J. Horadam, A theory of highly nonlinear functions, *AAECC-16*, M. Fossorier et al., eds., LNCS 3857, Springer, Berlin, 2006, 87–100.

[162] K. J. Horadam and W. de Launey, Cocyclic development of designs, *J. Alg. Combin.* 2 (1993) 267–290, Erratum 3 (1994) 129.

[163] K. J. Horadam and W. de Launey, Generation of cocyclic Hadamard matrices, Chap. 20, *Computational Algebra and Number Theory*, W. Bosma and A. van der Poorten, eds., Kluwer Academic, Dordrecht, 1995, 279–290.

[164] K. J. Horadam and C. Lin, Construction of proper higher dimensional Hadamard matrices from perfect binary arrays, *J. Combin. Math. Combin. Comp.* 28 (1998) 237–248.

[165] K. J. Horadam and A. A. I. Perera, Codes from cocycles, *AAECC-12*, T. Mora and H. Mattson, eds., LNCS 1255, Springer, Berlin, 1997, 151–163.

[166] K. J. Horadam and A. Rao, Fourier Transforms from a weighted trace map, *Proc. 2006 ISIT*, IEEE (2006) 1080-1084.

[167] K. J. Horadam and P. Udaya, Cocyclic Hadamard codes, *IEEE Trans. Inform. Theory* 46 (2000) 1545–1550.

[168] K. J. Horadam and P. Udaya, A new construction of central relative $(p^a, p^a, p^a, 1)$-difference sets, *Des. Codes Cryptogr.* 27 (2002) 281–295.

[169] K. J. Horadam and P. Udaya, A new class of ternary cocyclic Hadamard codes, *Applic. Algebra Eng. Commun. Comp.* 14 (2003) 65–73.

[170] J. Horton, C. Koukouvinos and J. Seberry, A search for Hadamard matrices constructed from Williamson matrices, *Bull. ICA* 35 (2002) 75–88.

[171] J. F. Huang, C. C. Yang and S. P. Tseng, Complementary Walsh-Hadamard coded optical CDMA coder/decoders structured over arrayed-waveguide grating routers, *Opt. Commun.* 229 (2004) 241–248.

[172] D. R. Hughes and F. C. Piper, *Projective Planes*, GTM 6, Springer, New York, 1973.

[173] G. Hughes, Characteristic functions of relative difference sets, correlated sequences and Hadamard matrices, *AAECC-13*, M. Fossorier et al., eds., LNCS 1719, Springer, Berlin, 1999, 346–354.

[174] G. Hughes, Non-splitting abelian $(4t, 2, 4t, 2t)$ relative difference sets and Hadamard cocycles, *European J. Combin.* 21 (2000) 323–331.

[175] G. Hughes, The equivalence of certain auto-correlated quaternary and binary arrays, *Australas. J. Combin.* 22 (2000) 37–40.

[176] G. Hughes, Constacyclic codes, cocycles and a $u + v|u - v$ construction, *IEEE Trans. Inform. Theory* 46 (2000) 674–680.

[177] G. Hughes, Structure theorems for group ring codes with an application to self-dual codes, *Des. Codes Cryptogr.* 24 (2001) 5–14.

[178] N. Ito, Note on Hadamard matrices of type.Q, *Studia Sci. Math. Hungar.* 16 (1981) 389–393.

[179] N. Ito, Note on Hadamard groups of quadratic residue type, *Hokkaido Math. J.* 22 (1993) 373–378.

[180] N. Ito, On Hadamard groups, *J. Algebra* 168 (1994) 981–987.

[181] N. Ito, On Hadamard groups II, *J. Algebra* 169 (1994) 936–942.

[182] N. Ito, Some results on Hadamard groups, *Groups-Korea '94*, de Gruyter, Berlin and New York, 1995.

[183] N. Ito, Remarks on Hadamard groups, *Kyushu J. Math.* 50 (1996) 83–91.

[184] N. Ito, On Hadamard groups III, *Kyushu J. Math.* 51 (1997) 1–11.

[185] G. James and M. Liebeck, *Representations and Characters of Groups*, CUP, Cambridge, 1993.

[186] J. Jedwab, Generalised perfect binary arrays and Menon difference sets, *Des. Codes Cryptogr.* 2 (1992) 19–68.

[187] J. Jedwab, C. Mitchell, F. Piper and P. Wild, Perfect binary arrays and difference sets, *Discr. Math.* 125 (1994) 241–254.

[188] D. Jungnickel, On difference matrices, resolvable transverse designs and generalised Hadamard matrices, *Math. Z.* 167 (1979) 49–60.

[189] D. Jungnickel, On automorphism groups of divisible designs, *Can. J. Math.* 34 (1982) 257–297.

[190] D. Jungnickel, Difference sets, *Contemporary Design Theory: A Collection of Surveys*, J. H. Dinitz and D. R. Stinson, eds., Wiley, New York, 1992.

[191] D. Jungnickel and A. Pott, Difference sets: Abelian, *The CRC Handbook of Combinatorial Designs*, C. J. Colbourn and J. H. Dinitz, eds., CRC Press, Boca Raton, 1996.

[192] D. Jungnickel and A. Pott, Difference sets: an introduction, *Difference Sets, Sequences and Their Correlation Properties*, A. Pott et al., eds., NATO Science Series C, 542, Kluwer, Dordrecht, 1999, 259–295.

[193] D. Jungnickel and A. Pott, Perfect and almost perfect sequences, *Discr. Appl. Math.* 95 (1999) 331–359.

[194] D. Jungnickel and B. Schmidt, Difference sets: an update, *Geometry, Combinatorial Designs and Related Structures*, CUP, Cambridge, 1997, 89–112.

[195] D. Jungnickel and B. Schmidt, Difference sets: a second update, *Rend. Circ. Mat. Palermo* (2) Suppl. 53 (1998) 89–118.

[196] W. M. Kantor, Automorphism groups of Hadamard matrices, *J. Combin. Theory A* 6 (1969) 279–281.

[197] W. M. Kantor, Commutative semifields and symplectic spreads, *J. Algebra* 270 (2003) 96–114.

[198] G. Karpilovsky, *Projective Representations of Finite Groups*, Marcel Dekker, New York, 1985.

[199] T. Kasami, A Gilbert-Varshamov bound for quasi-cyclic codes of rate 1/2, *IEEE Trans. Inform. Theory* 20 (1974) 679.

[200] A. Kerber, *Applied Finite Group Actions*, 2nd ed., Springer, Berlin, 1999.

[201] H. Kharaghani and J. Seberry, The excess of complex Hadamard matrices, *Graphs Combin.* 9 (1993) 47–56.

[202] H. Kharaghani and B. Tayfeh-Rezaie, A Hadamard matrix of order 428, *J. Combin. Des.* 13 (2005) 435–440.

[203] H. Kimura and T. Niwasaki, Some properties of Hadamard matrices coming from dihedral groups, *Graphs Combin.* 18 (2002) 319–327.

[204] T. Kiran and B. Sundar Rajan, Consta-abelian codes over Galois rings, *IEEE Trans. Inform. Theory* 50 (2004) 367–380.

[205] T. Kiran and B. Sundar Rajan, Vandermonde-cocyclic codes and a suitable DFT, *Proc. 2004 ISIT*, IEEE (2004) 257.

[206] D. E. Knuth, Finite semifields and projective planes, *J. Algebra* 2 (1965) 182–217.

[207] L. E. Kopilovich, On perfect binary arrays, *Electron. Lett.* 24 (1988) 566–567.

[208] L. E. Kopilovich, Applications of difference sets to the aperture design in multielement systems in radio science and astronomy, *Difference Sets, Sequences and their Correlation Properties*, A. Pott et al., eds., NATO Science Series C, 542, Kluwer, Dordrecht, 1999, 297–330.

[209] L. E. Kopilovich and L. G. Sodin, Synthesis of coded masks for gamma-ray and X-ray telescopes, *Mon. Not. R. Astron. Soc.* 266 (1994) 357–359.

[210] I. S. Kotsireas, C. Koukouvinos and J. Seberry, Hadamard ideals and Hadamard matrices with two circulant cores, *European J. Combin.* 27 (2006) 658–668.

[211] C. Koukouvinos, personal communication, November 2005.

[212] C. Koukouvinos, website http://www.math.ntua.gr/people/ckoukouv/.

[213] P. V. Kumar, On the existence of square dot-matrix patterns having a specific three-valued periodic-correlation function, *IEEE Trans. Inform. Theory* 34 (1988) 271–277.

[214] E. Kupče and R. Freeman, Fast multi-dimensional NMR of proteins, *J. Biomolecular NMR* 25 (2003) 349–354.

[215] C. Lam, S. Lam and V. D. Tonchev, Bounds on the number of affine, symmetric and Hadamard designs and matrices, *J. Combin. Theory A* 92 (2000) 186–196.

[216] C. Lam, S. Lam and V. D. Tonchev, Bounds on the number of Hadamard designs of even order, *J. Combin. Des.* 9 (2001) 363–378.

[217] A. LeBel, *Shift actions on 2-cocycles*, Ph.D. Thesis, RMIT University, Melbourne, Australia, 2005.

[218] A. LeBel and K. J. Horadam, Direct sums of balanced functions, perfect nonlinear functions and orthogonal cocycles, preprint 2006.

[219] M. H. Lee, The complex reverse jacket transform, *Proc. 22nd Int. Symp. on Inf. Theory and its Applications (SITA 99)* Yuzawa, Niigata, Japan, Nov 30–Dec 3 1999, 423–426.

[220] M. H. Lee, A new reverse jacket transform and its fast algorithm, *IEEE Trans. Circuits Syst. II* 47(1) (2000) 39–47.

[221] M. H. Lee, B. Sunder Rajan and J. Y. Park, A generalized reverse jacket transform, *IEEE Trans. Circuits Syst. II* 48(7) (2001) 684–690.

[222] K. H. Leung, S. L. Ma and V. Tan, Planar functions from $\mathbb{Z}_n$ to $\mathbb{Z}_n$, *J. Algebra* 224 (2000) 427–436.

[223] R. Lidl and H. Niederreiter, *Finite Fields*, Vol. 20, Encyclopedia of Mathematics and its Applications, 2nd ed., CUP, Cambridge, 1997.

[224] C. Lin and W. D. Wallis, Barker sequences and circulant Hadamard matrices, *J. Combin. Inform. Sys. Sci.* 18 (1993) 19-25.

[225] C. Lin, W. D. Wallis and L. Zhu, Generalised 4-profiles of Hadamard matrices, *J. Combin. Inform. Sys. Sci.* 18 (1993) 397–400.

[226] S. Litsyn, An updated table of the best binary codes known, Chapter 5, *Handbook of Coding Theory*, V. S. Pless and W. C. Huffman, eds., North-Holland, Amsterdam, 1998.

[227] W-H. Liu, Y-Q. Chen and K. J. Horadam, Relative difference sets fixed by inversion. II. Character theoretical approach, *J. Combin. Theory A* 111 (2005) 175–189.

[228] O. A. Logachev, A. A. Salnikov and V. V. Yashchenko, Bent functions on a finite abelian group, *Discrete Math. Appl.* 7 (1997) 547–564.

[229] K. Ma, Equivalence classes of n-dimensional proper Hadamard matrices, *Australas. J. Combin.* 25 (2002) 3–17.

[230] K. Ma, *Properties of higher dimensional proper Hadamard matrices*, M. App. Sc. Thesis, RMIT University, Melbourne, Australia, 2003.

[231] S.-L. Ma, Planar functions, relative difference sets, and character theory, *J. Algebra* 185 (1996) 342–356.

[232] S.-L. Ma and A. Pott, Relative difference sets, planar functions and generalised Hadamard matrices, *J. Algebra* 175 (1995) 505–525.

[233] I. D. MacDonald, Some p-groups of Frobenius and extra-special type, *Israel J. Math.* 40 (1981) 350–364.

[234] C. Mackenzie and J. Seberry, Maximal q-ary codes and Plotkin's bound, *Ars Combin.* 26B (1988) 37–50.

[235] G. W. Mackey, *Induced Representations of Groups and Quantum Mechanics*, W. A. Benjamin and Editore Boringhieri, New York, 1968.

[236] S. Mac Lane, *Homology*, Springer, Berlin, 1975.

[237] F. J. MacWilliams and N. J. A. Sloane, *The Theory of Error-Correcting Codes*, North-Holland, Amsterdam, ninth impression, 1996.

[238] MAGMA computational algebra software system website, http://magma.maths.usyd.edu.au/magma/ .

[239] D. K. Maslen and D. N. Rockmore, Generalized FFTs - a survey of some recent results, *Groups and computation, II (New Brunswick, NJ, 1995)*, DIMACS Ser. Discrete Math. Theoret. Comput. Sci. 28, Amer. Math. Soc., Providence, RI, 1997, 183–237.

[240] S. Matsufuji and N. Suehiro, Factorisation of bent function type complex Hadamard matrices, *Proc. 1996 ISSSTA*, IEEE (1996) 950–954.

[241] M. Matsui, Linear cryptanalysis method for DES cipher, *EUROCRYPT-93*, LNCS 765, Springer, Berlin, 1994, 386–397.

[242] B. R. McDonald, *Finite Rings with Identity*, Marcel Dekker, New York, 1974.

[243] R. L. McFarland, Difference sets in abelian groups of order $4p^2$, *Mitt. Math. Sem. Giessen* 192 (1989) 1–70.

[244] A. J. Menezes, P. C. van Oorschot and S. A. Vanstone, *Handbook of Applied Cryptography*, CRC Press, Boca Raton, 1997.

[245] M. Miyamoto, A construction for Hadamard matrices, *J. Combin. Theory A* 57 (1991) 86–108.

[246] D. C. Montgomery, *Design and Analysis of Experiments*, 4th ed., Wiley, New York, 1997.

[247] P. Morandi, *Field and Galois Theory*, GTM 167, Springer, New York, 1996.

[248] W. H. Mow, A new unified construction of perfect root-of-unity sequences, *Proc. 1996 ISSSTA*, IEEE (1996) 955–959.

[249] A. A. Nechaev, Kerdock code in a cyclic form, *Diskretnaya Mat. (USSR)* 1 (1989) 123–139 (in Russian). English translation: *Discrete Math. Appl.* 1 (1991) 365–384.

[250] J-S. No and H-Y.Song, Expanding generalized Hadamard matrices over G^m by substituting several generalised Hadamard matrices over G, *J. Comm. and Networks* 3 (4) (2001) 361–364.

[251] K. Nyberg, Perfect nonlinear S-boxes, *EUROCRYPT-91*, LNCS 547, Springer, New York, 1991, 378–385.

[252] K. Nyberg, Differentially uniform mappings for cryptography, *EUROCRYPT-93*, LNCS 765, Springer, New York, 1994, 55–64.

[253] A. V. Oppenheim and R. W. Schafer, *Digital Signal Processing*, Prentice-Hall, Englewood Cliffs, 1975.

[254] W. Orrick, Switching operations for Hadamard matrices, http://www.arxiv.org/abs/math.CO/0507515.

[255] W. Orrick website, http://mypage.iu.edu/ worrick/.

[256] R. E. A. C. Paley, On orthogonal matrices, *J. Math. Phys.* 12 (1933) 311–320.

[257] I.B.S. Passi, *Group rings and their augmentation ideals*, LNM 715, Springer, Berlin, 1979.

[258] D.S. Passman, *The Algebraic Structure of Group Rings*, Wiley-Interscience, New York, 1977.

[259] A. A. I. Perera, *Orthogonal cocycles*, Ph.D. Thesis, RMIT University, Melbourne, Australia, 1999.

[260] A. A. I. Perera and K. J. Horadam, Cocyclic generalised Hadamard matrices and central relative difference sets, *Des., Codes Cryptogr.* 15 (1998) 187–200.

[261] M. Petrescu, *Existence of continuous families of complex Hadamard matrices of certain prime dimensions*, Ph.D. Thesis, UCLA, USA, 1997.

[262] N. Pinnawala and A. Rao, Cocyclic simplex codes of type α over $\mathbb{Z}_4$ and $\mathbb{Z}_{2^s}$, *IEEE Trans. Inform. Theory* 50 (2004) 2165–2169.

[263] V. S. Pless and W. C. Huffman, eds., *Handbook of Coding Theory*, North-Holland, Amsterdam, 1998.

[264] L. Poinsot and S. Harari, Generalized Boolean bent functions, *INDOCRYPT 2004*, A. Canteaut and K. Viswanathan, eds., LNCS 3348, Springer, Berlin, 2004, 107–119.

[265] A. Pott, On the structure of abelian groups admitting divisible difference sets, *J. Combin. Theory A* 65 (1994) 202–213.

[266] A. Pott, *Finite Geometry and Character Theory*, LNM 1601, Springer, Berlin, 1995.

[267] A. Pott, A survey on relative difference sets, *Groups, Difference Sets and the Monster*, de Gruyter, New York, 1996, 195–232.

[268] A. Pott, Nonlinear functions in abelian groups and relative difference sets, *Discr. Appl. Math.* 138 (2004) 177–193.

[269] A. Pott, P. V. Kumar, T. Helleseth and D. Jungnickel, eds., *Difference Sets, Sequences and their Correlation Properties*, NATO Science Series C, 542, Kluwer, Dordrecht, 1999.

[270] B. Preneel et al., Propagation characteristics of Boolean functions, *EURO-CRYPT '90*, LNCS 473, Springer, Berlin, 1991, 161–173.

[271] C. Qu, J. Seberry and J. Pieprzyk, Homogeneous bent functions, *Discr. Appl. Math.* 102 (2000) 133–139.

[272] S. Rahardja and B. J. Falkowski, Classifications and graph-based representations of switching functions using a novel complex spectral technique, *Int. J. Electron.* 86 (1997) 731-742.

[273] S. Rahardja and B. J. Falkowski, Family of unified complex Hadamard Transforms, *IEEE Trans. Circuits Syst. II* 46 (1999) 1094–1100.

[274] A. Rao, Shift-equivalence and cocyclic self-dual codes, *J. Combin. Math. Combin. Comput.* 54 (2005) 175–185.

[275] N. A. Riza and M. A. Arain, Code-multiplexed optical scanner, *Applied Optics* 8 (2003) 1493–1502.

[276] D. J. S. Robinson, *A Course in the Theory of Groups*, 2nd ed., Springer, New York, 1996.

[277] R. M. Roth, Maximum-rank array codes and their application to crisscross error correction, *IEEE Trans. Inform. Theory* 37 (1991) 328–336.

[278] O. S. Rothaus, On "bent" functions, *J. Combin. Theory A* 20 (1976) 300–305.

[279] R. E. Sabin, On determining all codes in semi-simple group rings, *AAECC-10*, LNCS 673, Springer, Berlin, 1993, 279–290.

[280] B. Schmidt, Williamson matrices and a conjecture of Ito's, *Des. Codes Cryptogr.* 17 (1999) 61–68.

[281] B. Schmidt, Cyclotomic integers and finite geometry, *J. Amer. Math. Soc.* 12 (1999) 929–952.

[282] O. Schreier, Über Erweiterungen von Gruppen I, *Monatsh. Math. Phys.* 34 (1926) 165–180.

[283] O. Schreier, Über Erweiterungen von Gruppen II, *Abh. Math. Sem. Hamburg Univ.* 4 (1926) 321–346.

[284] I. Schur, Über die Darstellung der endlichen Gruppen durch gebrochene lineare Substiutionen, *J. Reine Angew. Math.* 127 (1904) 20–50.

[285] I. Schur, Untersuchungen über die Darstellung der endlichen Gruppen durch gebrochene lineare Substiutionen, *J. Reine Angew. Math.* 132 (1907) 85–137.

[286] I. Schur, Über die Darstellung der symmetrischen und der alternierenden Gruppe durch gebrochene lineare Substiutionen, *J. Reine Angew. Math.* 139 (1911) 155–250.

[287] J. Seberry, website http://www.uow.edu.au/ jennie/.

[288] J. Seberry and M. Yamada, Hadamard matrices, sequences, and block designs, Chapter 11, *Contemporary Design Theory: A Collection of Surveys*, J. H. Dinitz and D. R. Stinson, eds., Wiley, New York, 1992.

[289] Y. Shaked and A. Wool, Cracking the Bluetooth PIN, *Proc. MobiSys '05*, ACM Press, New York, 2005, 39–50.

[290] C. E. Shannon, A mathematical theory of communication, *Bell System Tech. J.* 27 (1948) 379–423, 623–656.

[291] V. Shashidhar and B. Sundar Rajan, Consta-dihedral codes and their transform domain characterization, *Proc. 2004 ISIT*, IEEE, 2004, 256.

[292] P. J. Shlichta, Three- and four-dimensional Hadamard matrices, *Bull. Amer. Phys. Soc.* 16(8) (1971) 825-826.

[293] P. J. Shlichta, Higher dimensional Hadamard matrices, *IEEE Trans. Inform. Theory* 25 (1979) 566-572.

[294] V. M. Sidel'nikov, On the mutual correlation of sequences, *Soviet Math. Dokl.* 12 (1971) 197–201.

[295] M. K. Siu, The combinatorics of binary arrays, *J. Stat. Plann. Inf.* 62 (1997) 103–113.

[296] D. Slepian, Group codes for the Gaussian channel, *Bell Syst. Tech. J.* 47 (1968) 575–602.

[297] N. Sloane, website http://www.research.att.com/ njas/hadamard/.

[298] E. D. J. Smith, R. J. Blaikie and D. P. Taylor, Performance enhancement of spectral-amplitude coding optical CDMA using pulse-position modulation, *IEEE Trans. Communications* 46 (9) (1998) 1176–1185.

[299] K. W. Smith, Nonabelian Hadamard difference sets, *J. Combin. Theory A* 70 (1995) 144–156.

[300] P. Solé, A quaternary cyclic code and a family of quadriphase sequences with low correlation properties, *Coding Theory and Applications*, LNCS 388, Springer, Berlin, 1989, 193–201.

[301] H. Y. Song and S. W. Golomb, On the existence of cyclic Hadamard difference sets, *IEEE Trans. Inform. Theory* 40 (1994) 1266–1268.

[302] T. St Denis, Fast Pseudo-Hadamard Transforms, Cryptology ePrint Archive, Report 2004/010, http://eprint.iacr.org/, last revised 2 February 2004.

[303] J. J. Sylvester, Thoughts on inverse orthogonal matrices, simultaneous sign successions and tesselated pavements in two or more colours, with applications to Newton's rule, ornamental tile work and the theory of numbers, *Phil. Mag.* 34 (1867) 461–475.

[304] W. Tadej and K. Zyczkowski, A concise guide to complex Hadamard matrices, *Open Sys. & Information Dyn.* 13 (2006) 133–177.

[305] V. Tarokh, H. Jafarkhani and A. R. Calderbank, Space-time block codes from orthogonal designs, *IEEE Trans. Inform. Theory* 45 (1999) 1456–1467.

[306] V. D. Tonchev, Hadamard matrices of order 36 with automorphisms of order 17, *Nagoya Math. J.* 104 (1986) 163–174.

[307] V. D. Tonchev, Self-orthogonal designs and extremal doubly even codes, *J. Combin. Theory A* 52 (1989) 197–205.

[308] R. J. Turyn, Complex Hadamard matrices, *Combinatorial Structures and Their Applications*, Gordon and Breach, New York, 1970, 435–437.

[309] R. J. Turyn, An infinite class of Williamson matrices, *J. Combin. Theory A* 12 (1972) 319–321.

[310] P. Udaya and K. J. Horadam, Cocyclic Hadamard codes from semifields, Proc. 2000 ISIT, IEEE, 2000, 31.

[311] S. A. Vanstone and P. C. van Oorschot, *An Introduction to Error Correcting Codes with Applications*, Kluwer, Boston, 1989.

[312] S. Verdu, *Multiuser Detection*, CUP, Cambridge, 1998.

[313] J. S. Wallis, On the existence of Hadamard matrices, *J. Combin. Theory A* 21 (1976) 188–195.

[314] W. D. Wallis, *Combinatorial Designs*, Marcel Dekker, New York, 1988.

[315] W. D. Wallis, A. P. Street and J. S. Wallis, *Combinatorics: Room Squares, Sum-Free Sets, Hadamard Matrices*, LNM 292, Springer, Berlin, 1972.

[316] J. L. Walsh, A closed set of normal orthogonal functions, *Amer. J. Math.* 55 (1923) 5–24.

[317] Z.-X. Wan, *Quaternary Codes*, World Scientific, Singapore, 1997.

[318] J. Wang and T. S. Ng, eds., *Advances in 3G Enhanced Technologies for Wireless Communications*, Artech House, Norwood, 2002.

[319] A. F. Webster and S. E. Tavares, On the design of S-boxes, *CRYPTO 85*, LNCS 218, Springer, Berlin, 1986, 523–534.

[320] Z. Wen and Y. Tao, Orthogonal codes and cross-talk in phase-code multiplexed volume holographic data storage, *Opt. Commun.* 148 (1998) 11–17.

[321] S. B. Wicker, *Error Control Systems for Digital Communication and Storage*, Prentice-Hall, Upper Saddle River, 1995.

[322] W. Willems, A note on self-dual group codes, *IEEE Trans. Inform. Theory* 48 (2002) 3107–3109.

[323] J. Williamson, Hadamard's determinant theorem and the sum of four squares, *Duke Math J.* 11 (1944) 65–81.

[324] R. M. Wilson and Q. Xiang, Constructions of Hadamard difference sets, *J. Combin. Theory A* 77 (1997) 148–160.

[325] A. Winterhof, On the non-existence of generalised Hadamard matrices, *J. Statist. Plann. Inference* 84 (2000) 337–342.

[326] J. Wolfmann, Bent functions and coding theory, *Difference Sets, Sequences and Their Correlation Properties*, A. Pott et al., eds., NATO Science Series C, 542, Kluwer, Dordrecht, 1999, 393–418.

[327] M.-Y. Xia, Some infinite classes of special Williamson matrices and difference sets, *J. Combin. Theory A* 61 (1992) 230–242.

[328] M. Yamada, Hadamard matrices of generalised quaternion type, *Discr. Math.* 87 (1991) 187-196.

[329] K. Yamamoto, On a generalised Williamson equation, *Colloq. Math. Soc. Janos Bolyai* 37 (1981) 839-850.

[330] X. Y. Yang and S. Jutamulia, Three-dimensional photorefractive memory based on phase-code and rotation multiplexing, *Proc. IEEE* 87 (11) (1999) 1941–1955.

[331] Y. X. Yang, The proofs of some conjectures on higher dimensional Hadamard matrices, *Kexue Tongbao* (English translation) 31 (1986) 1662-1667.

[332] Y. X. Yang, Existence of one-dimensional perfect binary arrays, *Electron. Lett.* 23 (1987) 1277-1278.

[333] Y. X. Yang, On the H-Boolean functions, *J. Beijing Uni. Posts and Telecomm.* 11 (1988) 1–9.

[334] Y. X. Yang, *Theory and Applications of Higher-Dimensional Hadamard Matrices*, Kluwer, Dordrecht, 2001.

[335] G. J. Yu, C. S. Lu and H. Y. Liao, A message-based cocktail watermarking system, *Pattern Recognition* 36 (2003) 957-968.

Index